# COAL DESULFURIZATION PRIOR TO COMBUSTION

# COAL DESULFURIZATION PRIOR TO COMBUSTION

Edited by Robert C. Eliot

NOYES DATA CORPORATION
Park Ridge, New Jersey, U.S.A.
1978

# FOREWORD

The detailed descriptive information in this book is based on federally funded studies, conferences, other publications, and U.S. patents relating to the desulfurization of coal prior to combustion.

Coal is a major source of energy in the USA and most probably will continue to be such for many years. However, one of the problems with coal as a source of energy is its sulfur content, present mostly as iron sulfide (pyritic sulfur) and in coal-borne organic systems, some containing the thiophene ring. Upon combustion of the coal as mined, harmful sulfur-containing gases and ashes are emitted into the atmosphere with deleterious effects upon animal and plant life.

In order to make coal burning environmentally safe and acceptable, some method of desulfurization must be applied. Stack gas scrubbing is most often used, but is economical only for large industrial and power plants. Fluidized bed processes are also available for desulfurization during combustion. True, there are minable beds of low sulfur coal west of the Mississippi, concentrated in such states as Montana and Wyoming, far removed from the electric power demand centers of the east, thus adding to the many difficulties the U.S. is currently experiencing with mining, transporting, and using coal.

Coal cleaning, and especially desulfurization prior to combustion, appears to be an attractive alternative. Many pollution controls and much monitoring can be omitted when desulfurized coal is available to plants carrying out relatively small operations.

This book is an accurate state-of-the-art report and can serve as a practical manual to those more intimately connected with the present-day problems of burning coal.

Because the information in this book is taken from many sources, it is possible that certain portions of this book may disagree or conflict with other parts of the book. This is especially true of monetary values and opinions of future potential. We chose to include these different points of view, however, in order to make the book more valuable to the reader. Cost figures provided are those given in the report cited, the date of which is always given. When the dates of the cost figures themselves are given, we have included them.

Advanced composition and production methods developed by Noyes Data are employed to bring these durably bound books to you in a minimum of time. Special techniques are used to close the gap between "manuscript" and "completed book." Industrial technology is progressing so rapidly that time-honored, conventional typesetting, binding and shipping methods are no longer suitable. We have bypassed the delays in the conventional book publishing cycle and provide the user with an effective and convenient means of reviewing up-to-date information in depth.

The Table of Contents is organized in such a way as to serve as a subject index and provides easy access to the information contained in this book. The bibliography at the end of the volume lists the highly important government reports and their source of purchase.

Some of the illustrations in this book
may be less clear than could be desired;
however, they are reproduced from the
best material available to us.

# CONTENTS AND SUBJECT INDEX

# COAL CLEANING
# AN INTRODUCTION AND OVERVIEW

The material in this section is taken from the *Coal Preparation Environmental Engineering Manual,* written by David C. Nunenkamp of J.J. Davis Associates (PB-262 716). Further information on this manual will be found in the bibliography on page 301. This material will serve as an introduction to the subject of coal desulfurization, although it applies to the total beneficiation of coal, of which desulfurization is only a part.

## BACKGROUND

As a direct result of the energy crisis, the United States requirements for coal in 1985 may be as much as 1.7 billion tons per year. With the annual production of coal in the early 1970s running between 575 to 600 million tons per year, this estimate indicates that the U.S. production of coal must triple in about 15 years. More conservative estimates, some of which were made before the energy crisis and oil embargo of 1973-74, indicated that the United States requirements would be about one billion tons per year in the early 1980s. The published goals for Project Independence (our country's plan to achieve energy self-sufficiency by 1985) include a requirement for 1.2 billion tons of coal to be produced annually by 1985.

The mere setting of this goal to double or triple coal production over a 10-year period is not sufficient. A concerted effort by the entire country, including consumers, producers, and governmental agencies, must be made in order to obtain these goals. The projected demands which may be placed upon the coal industry come at a time when coal production and, specifically, productivity have encountered many setbacks. Coal production in recent years has been considerably below the projected 1985 demands. In fact, the total tonnage of mechanically cleaned coal in this country was actually decreasing until 1972, e.g., from 335 million tons in 1969 to 271 million tons in 1971. In 1972, the real tonnage of mechanically cleaned coal increased for the first time since 1967 to a total of 289 million tons.

While U.S. production of coal fluctuated between 400 and 600 million tons annually since 1950, the productivity (production per man shift) enjoyed a nearly uninterrupted rate of increase. This increase in productivity in all types of mines held true until the enactment of the Coal Mine Health and Safety Act of 1969 which appears to have reduced the productivity of underground coal mining. Strip mining has continued to enjoy increases in productivity, however. Figure 1.1 shows the productivity of U.S. coal mines from 1910 to 1974.

**FIGURE 1.1: U.S. SOFT COAL PRODUCTIVITY BY MINE TYPE**

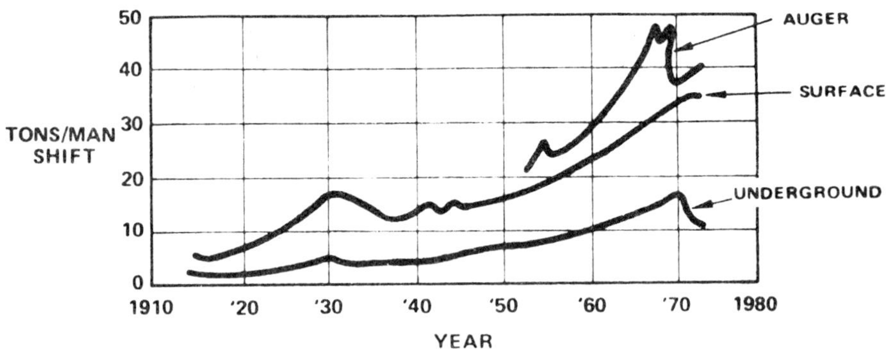


Source:  PB-262 716

Assuming that the industry is unable to make dramatic improvements in productivity in existing mines and that 600 million additional tons of coal annually are required by 1985, then 70% of the projected 1.2 billion annual tonnage must come from mines not now in existence. Specifically, enough new mines must be opened to produce an additional 600 million tons of coal annually over the next decade, in addition to mines needed to replace those that are being closed as they are worked out.

According to Dr. John Fallon, then Director of the Federal Energy Administration, April 7, 1975, in a speech given to the Institute of Electrical and Electronics Engineers, it will be necessary to take the following actions to achieve the 1985 production levels:

   Develop 140 new 2 million ton per year eastern underground mines,

   Develop 30 new 2 million ton per year eastern surface mines,

   Develop 100 new 5 million ton per year western surface mines,

   Recruit and train 80 thousand new eastern coal miners,

   Recruit and train 45 thousand new western coal miners.

This plan of action is ambitious, to say the least. Disregarding the long-term problems confronting the coal industry, the short-run obstacles alone are considerable. To open a new coal mine takes many years lead time; normally eighteen months are required to develop a new surface mine, and 5 to 9 years are required to develop a new underground mine. To achieve an increase in coal production of 600 million tons per year by 1985 will require that, on the average, one new underground mine (2 million tons/yr) and one new surface mine (5 million tons/yr) be brought into production every month for the next ten years. In contrast, only 13 mines with capacity greater than 2 million tons per year were brought into production during the decade of the 1960s.

It is certainly feasible for the industry to open the new mines and produce the extraction equipment required. Assuming it can also solve the manpower requirements, the next step toward increased coal production is coal beneficiation equipment and the facilities in which the coal is cleaned. It will be necessary to design and construct as many coal preparation plants as new coal mines. The old philosophy that one need only extract the coal from the ground and allow the consumer (primarily electric utilities) to worry about the processing and consumption of the coal is being altered rapidly.

With the current emphasis on coal utilization and with the mounting concerns over the waste disposal practices of the coal mining industry, it is imperative that individuals involved with the coal production industry, and specifically those involved with the monitoring of this industry, have a basic understanding of the processes and techniques of the physical cleaning of coal, the known potential pollutants, and the current practices for control of these pollutants.

## GEOLOGICAL FACTORS WHICH AFFECT THE NATURE OF COAL

Many geological factors influence thickness, continuity, quality and mining conditions of coal. Some geological features occurred during peat accumulation or shortly thereafter, others occurred millions of years later. The recognition of the nature of these features is important in the mining operation and ultimately affects the physical cleaning of the coal. Several of the more common features that affect coal cleaning are described below.

*Shale Partings:* Streams periodically flood the peat swamps where the vegetable material accumulates, depositing mud and silt layers that become bands of slate and siltstone after the vegetable material is coalified. In general, the closer the peat beds were to the flooding stream, the thicker the deposits left and the more total was the disruption to the bed.

*Washouts:* After the plant material has been accumulated and buried by various sediments, it may be removed by the erosive actions of streams. This activity is called a washout. Washouts may occur shortly after deposition of the peat or after coalification is complete.

*Faults:* Faults are fractures in the rock sequence along which the strata on each side of the fracture appear to have moved in different directions. The movement may be measured from inches to miles and in any direction from horizontal to vertical.

*Clay Veins:* Irregular, vertical to inclined tabular masses of clastic material (clay, sand or silt) that interrupt the coal seam are called clastic dikes or "clay veins." These clay veins may be from a fraction of an inch to several feet thick and may extend for some distance into the strata overlying the coal. They frequently contribute to roof instability as the coal is mined. The clay veins tend to be numerous in some areas and commonly intersect each other. They add to the waste material that must be removed from the salable coal as well as creating safety hazards and drainage problems.

*Concretions:* The coal as well as the associated rocks commonly contains aggregations of minerals in spherical, disclike or irregular forms. They may be microscopic or several feet across, although the most commonly observed size is several inches wide. Mine and roof shales commonly contain concretions made up of calcite ($CaCO_3$), dolomite [$CaMg(CO_3)_2$], siderite ($FeCO_3$), and pyrite ($FeS_2$). The presence of large concretions in mine roof material may have a considerable effect upon roof stability creating safety hazards and adding to the waste material. In the coal headed for a preparation plant, pyritic concretions are common, ranging from less than an inch to several feet, and are usually referred to as sulfur balls.

*Igneous Intrusions:* In some areas, the coal and associated strata may have been intruded by once-molten igneous rocks forcibly injected into the sedimentary sequence from below. The igneous rock is commonly seen as a dike which is a nearly vertical tabular mass cutting across the bedding of the sediments. Depending on the size of the igneous mass and its temperature, the coal is thermally affected, being either advanced in rank or coked immediately adjacent to the igneous body. The igneous rocks that occur within a coal seam are much harder than the coal which may cause mining problems and contribute to preparation problems.


## PROPERTIES OF COAL

The material we call coal is classified by a series of chemical analyses and physical tests which define the coal in its various stages of metamorphism. Coal increasingly metamorphoses (responds to pressure and heat) from lignite and subbituminous ranks through the high-volatile, medium-volatile, low-volatile bituminous coal ranks to anthracite and meta-anthracite. Coalification is a gradual process and the classification of coal by ranks is just an identification of the various stages of that process and is based upon such properties as the percentage of fixed carbon, the percentage of volatile matter, calorific value and the agglomerating character. However, the classification by ranks does little to describe the overall complexities of the chemical and physical composition of different coals.

Coal is a very complex material and its chemical composition varies widely. The principle differences between coals can be traced to the different plant assemblages in the original forest, and to the history of the coal bed since it was formed.

The original peat bogs and coastal swamps were occasionally subjected to flooding by streams from adjacent hills. As this happened additional clay and silt were deposited in the swamp. These additional deposits became mixed with the plant debris and are responsible for the ash content of the coal: the muddier the original bog, the greater the ash content of the coal. As the peat became buried, other changes occurred. The deeper it was buried, the greater the compression and heat experienced by the bed. The greater the compression and heat, the more the volatile constituents were removed: the more volatiles removed, the greater the carbon content of the coal.

In order to classify coal, we must be able to recognize the different classes. This recognition is accomplished on the basis of identification of unique characteristics. The characteristics which permit the distinction between two specimens of coal are called properties. The physical properties are concerned with the characteristics of coal in its natural state, or prior to its end use as a fuel. For example, the hardness of coal determines the maintenance cost on coal handling equipment; the specific gravity of coal determines the coal preparation techniques used in a cleaning plant as well as the capacity of coal bins, boats and size of cargo and other coal storage facilities. The physical properties are, of course, dependent upon the chemical constituents that make up coal.

The chief physical properties important to coal preparation are specific gravity, size stability and uniformity, friability, resistance to weathering, grindability and presence of impurities. The chemical constituents that are important to coal preparation relate primarily to the impurities in the coal, i.e., those that are not carbon such as moisture, ash, pyrite, sulfur, etc.

**Specific Gravity**

The density of coal is its weight per unit of volume. The specific gravity of coal is its density referred to the density of water at 4°C. Various values ranging from 1.23 to 1.72 are recorded in literature for the specific gravity of "pure" coal. The variations are due to differences in rank, differences in moisture and ash content, and differences in methods used to determine specific gravity. The specific gravity of clean coal increases with rank and ranges from lignite to anthracite. Coal of a given rank has a higher apparent specific gravity when wet than when dry, and similarly, a change in specific gravity is exhibited with the change in ash content: higher ash content gives higher specific gravity.

The most important use of this physical characteristic is the part that it plays in the cleaning of coal by wet cleaning methods. The basic principle on which these operate is that the specific gravity of coals differs from their associated impurities and that there is a relationship between the velocity with which the particles fall in water and their relative densities.

Shale, clay and sandstone, if pure, have a specific gravity of about 2.6. Carbonaceous shale ranges in specific gravity from 2.0 to 2.6 depending upon the quantity of carbonaceous material present. Other impurities such as gypsum, kaolin and calcite have specific gravities of 2.3, 2.6 and 2.7, respectively, while the specific gravity of pyrite is about 5.0. Since the specific gravities of all these impurities are considerably greater than the specific gravity of coal, these impurities will fall to the bottom of a container filled with water more rapidly than coal. If the water is given a pulsating motion by compressed air, for example, causing the water to move up and down, the impurities will be kept at the bottom and the coal at the top where it can be recovered.

**Size Stability and Uniformity**

Size stability and uniformity of a given coal are critical to the coal cleaning operation because the cost of cleaning the coal increases dramatically as the percentage of fine size coal in the preparation plant increases. The size stability of coal may be expressed as a function of friability and/or weathering.

*Friability:* The strength of coal is displayed, among other ways, in its ability to withstand degradation of size upon handling. The tendency towards breakage during handling, termed "friability," depends to some extent on the toughness, elasticity and fracture characteristics as well as upon strength. The greater the friability of a given coal, the greater the chance for size degradation, e.g., very friable coal will produce a larger percentage of fines when the coal is fed to a crusher.

Friability normally increases with coal rank (with the exception of anthracites) reaching a maximum in coals of the low-volatile group. Coals of somewhat lower rank than low volatile are usually relatively nonfriable and, hence, resist degradation in size with its accompanying increase in the amount of surface exposed to oxidation. With coals of subbituminous rank, degradation by slacking or weathering supplements that due to breakage or handling. Anthracites are compared in friability to the subbituminous coals; both are harder than bituminous coals and decidedly more resistant to breakage than the very friable low-volatile coals. Lignites were found to be the least friable of all coals.

*Weathering:* Weathering is the tendency of coals to disintegrate or slack on exposure to weather, particularly when alternately wetted and dried or subjected to hot sunshine. Lower ranked coals like lignite slack very readily; subbituminous coals slack to some extent but less readily than lignite; and bituminous coals are affected only slightly by weathering. The size degradation caused by slacking is expressed as a percentage and termed slack index. Slack indexes of 5% or less characterize bituminous coals whereas the slack indexes for lignite approach 100%.

## Grindability

Grindability of coal, or the ease with which it may be pulverized, is a composite physical property embracing other specific properties such as hardness, strength, tenacity and fracture. A general relationship exists between the grindability of a specific coal and its rank. Coals that are the easiest to grind are found in the medium-volatile and low-volatile groups. These coals are decidedly easier to grind than coal of the high-volatile bituminous, subbituminous and anthracite ranks. The most common index of grindability is the Hardgrove grindability index. The capacity and power input for pulverizing and repair costs of pulverizers vary with the grindability index. The higher the index the easier the coal is to grind.

## Impurities in Coal

Coal is not a uniform substance, but rather a mixture of combustible metamorphosed plant remains that vary in both physical and chemical composition. The diversity of the original plant materials and the degree of metamorphism or coalification that have affected these materials are the two major reasons for the variety of physical components in coal. This widely varying composition greatly affects the preparation characteristics of the coal.

*Moisture:* The percentage of moisture present in a given coal bed, commonly called "bed moisture," is more or less constant throughout a given mine and is a general characteristic of the rank of the coal. Bed moisture may range from a low of 1, 2 or 3% in bituminous coal to a high of 45% in lignite. The actual moisture content of a given coal as it enters a preparation plant or a steam generator is dependent upon a number of factors in addition to its bed moisture. The mining methods used to extract the coal, the storage techniques of both the raw and the clean coal products, the method of cleaning and drying of the coal and the method of transporting the coal to user may all affect the moisture content of a coal.

The moisture in the coal, whether inherent or surface, can be considered as an impurity from the viewpoint of utilization. It is, of course, a dilutant in that it reduces available energy yield of the coal in proportion to the amount of moisture present and even in excess of this amount for some uses, especially for coal's largest single customer—electric power generation. Not only does moisture replace potential energy in proportion to the amount present, but it further robs Btu output because the moisture must be heated to stack temperatures in the boiler furnace before it is expelled.

*Minerals:* The mineral impurities occurring in coal may be classified broadly into those that form ash and those that contribute sulfur. From the standpoint of coal cleaning, both the ash-forming and the sulfur-containing impurities may be subdivided into two classes—impurities that are structurally a part of the coal and hence not separable by physical means, and inorganic impurities that can be eliminated to a greater or lesser extent by crushing and ordinary cleaning methods. The relative rate at which the mineral and the organic materials accumulated in the swamp determines the physical character and ash content of the product that resulted. If organic matter predominated, the product formed was coal containing some inherited impurities. If silt predominated, a carbonaceous shale was formed. Products intermittent between these two are classified as bone or boney coal depending upon the amount of silt incorporated in their structure.

Coal ash varies greatly in its chemical composition. It is a mixture of silica ($SiO_2$) and alumina ($Al_2O_3$) which came from sand, clay, slate and shale; iron oxide ($Fe_2O_3$) from pyrite and marcasite; magnesia ($MgO$) and lime ($CaO$) from limestone and gypsum; the alkalis, sodium oxide and potassium oxide ($Na_2O$ and $K_2O$); phosphorus pentoxide ($P_2O_5$); and miscellaneous amounts of trace elements. Table 1.1 shows the important minor and trace elements found in most coals. Much more detailed listings may be found in the referenced literature. The residue from these minerals after the coal has been burned is called ash. The average ash content of the entire thickness of a coal bed is at least 2 or 3%, even for very pure coals, and 10% and more for coals found in most commercial mines. Coal material that is too high in ash for ordinary use may be called bone coal, bituminous shale or black slate.

Some ash-forming impurities are so finely divided and so intimately mixed with pure coal substances that they may be considered a structural part of the coal. Impurities of this type cannot be separated from the coal by physical preparation. The chief value of determining them quantitatively is that they fix a minimum ash content of the cleanest portion of the raw coal—the so-called true, fixed, normal or inherent ash content. In the washing processes for eliminating impurities, the value of inherent ash may be approached as a limiting

## TABLE 1.1: MINOR AND TRACE ELEMENTS IN COAL

| Minor Elements (about 1% or more, on ash) | Trace Elements (about 0.1% or less, on ash) | |
|---|---|---|
| **Pollutant:** | **Named as Hazardous:** | |
| Sulfur | Beryllium | Cadmium |
| Nitrogen | Fluorine | Mercury |
| | Arsenic | Lead |
| **Ash-Forming:** | Selenium | |
| Sodium | | |
| Potassium | **Others Analyzed:** | |
| Iron | **Coal Basis** | **Ash Basis** |
| Calcium | | |
| Magnesium | Boron | Lithium |
| Silica | Vanadium | Scandium |
| Alumina | Chromium | Manganese |
| Titanium | Cobalt | Strontium |
| | Nickel | Zirconium |
| | Copper | Barium |
| | Zinc | Ytterbium |
| | Gallium | Bismuth |
| | Germanium | |
| | Tin | |
| | Yttrium | |
| | Lanthanum | |
| | Uranium | |

minimum to designate the portion of the ash content of coal that is structurally part of the coal itself and, therefore, cannot be separated by mechanical means. Other impurities are interbedded with coal and may be in thin layers or in thick rocklike deposits. Clay is the most common substance in banded impurities consisting mainly of one or more of the three common clay minerals—kaolinite, illite and montmorillinite.

*Clay and Shale:* One of the principal contaminants of raw coal is clay or shale from the roof and floor or from interbedded partings. Clay presents major problems to the coal preparation plant. Approximately 95% of the coal cleaned in this country is cleaned using some type of wet processing. The majority of these wet process techniques use the difference in density between coal and its associated impurities as the basis for separating the coal from the impurities.

The pronounced tendency of clays to disintegrate in water and to form plastic masses has definite implications in terms of the design and operation of preparation plants, i.e., they show up as an additional capital cost in plant design and as an operational cost on a daily basis. The direct operational difficulties (cost) associated with the particle disintegration and the resulting dispersion of colloidal matter appear in the form:

    ....of contamination of and increased viscosity of dense-medium
        suspensions,
    ....difficulties in dewatering and drying of the fine coal sizes,
    ....difficulties in the filtration of froth-flotation products,
    ....handling difficulties in the disposal of fine refuse.

In addition to the problems listed above and with specific reference to the low-ranked lignite and subbituminous coals, other operational difficulties arise when the lattice structure of the particular clays associated with these coals renders them susceptible to swelling. These clays may swell to such a degree that their apparent specific gravity is altered significantly. This alteration brings the specific gravity of the clay down to 1.60, very close to that of the coal itself. As the specific gravity of the clays approaches that of the coal being washed, several things may happen. First, the clay becomes extremely difficult to separate from the coal. Secondly, the apparent density of the wash bath is altered significantly allowing slate to be discharged with the coal at the top of the washer.

The problems generated by clay and shales in a washing plant appear to be related to the rank of the coal. In anthracite coal, the shale is so well indurated and compacted that it is called slate and it shows very little tendency toward particle disintegration. On the other hand, clay and shale in low-rank coals, such as subbituminous, exhibit a maximum amount of particle disintegration and an amplification of the difficulties discussed.

*Sulfur:* Of the minerals found in coal, sulfur is the most important single element impeding the utilization of coal as a clean fuel. Many U.S. steam coals contain high percentages of sulfur which must be reduced as air pollution regulations become increasingly more stringent. The reduction of sulfur in coal is a difficult problem which has long been under study. Sulfur in coal is reported in detailed chemical analysis as sulfate sulfur, pyritic sulfur and organic sulfur. The sulfur content of coals varies from 0.1 to 10.0% by weight.

Sulfate sulfur, or that part of the total sulfur that can be extracted by treatment with hydrochloric acid, is usually of only minor importance (less than 0.1 weight percent). The sulfate sulfur occurs in combination with either calcium or iron and is usually water soluble, originating from in situ pyrite oxidation. The amount of sulfate sulfur in a coal increases rapidly with weathering as the oxidation of iron sulfides gives rise to ferrous and ferric sulfates.

The term pyritic (sulfide) sulfur is used to refer to either of the two dimorphous forms of ferrous disulfide ($FeS_2$)—pyrite or marcasite. The two minerals have the same chemical composition, but have different crystalline forms. Pyrite is isometric (cubic) and marcasite is orthorhombic. The Victorian brown coals of Australia are an exception in that marcasite is virtually the only sulfide material reported.

Microscopic pyrites occur predominantly in coal in four forms:

    1.    Veins—generally thin and filmlike along the vertical joints
           (cleat), but may be up to several inches wide and contain
           large pyrite crystals with well-developed crystal faces.

2.  Lenses—extremely variable in shape and size but generally flattened and elongated in cross sections, ranging in size from a fraction of an inch thick to several inches in diameter.

3.  Nodules or balls—roughly spherical in shape and from inches to several feet in diameter.  These sulfur balls are usually not pure pyrite but include one or more of the following:  calcite, siderite, clay minerals and organic matter.

4.  Pyritized plant tissue—often included with the carbonate minerals in a "coal ball," which is a portion of coal in which the plant material has undergone replacement by inorganic material rather than coalification.

Sulfide sulfur occurs as individual particles (0.1 micron to 25 cm in diameter) disseminated throughout all coal deposits.  Pyrite is a dense mineral (4.5 g/cc) compared with bituminous coal (1.30 g/cc), but like coal is quite water-insoluble unless oxidized.

The organic sulfur is a part of, and chemically bonded to, the coal; it cannot be removed unless the chemical bonds holding it are broken.  The amount of organic sulfur present, therefore, defines the theoretical lowest limit at which a coal can be cleaned by physical methods.  Where organic sulfur is associated with certain constituents of coal, gravimetric reductions may be possible; however, organic sulfur is generally considered to be uniformly distributed throughout the coal and not amenable to reductions by conventional mechanical cleaning.

Only the sulfide and sulfate sulfur forms in coal may be removed by mechanical cleaning.  The extent of that removal, which is possible (10 to 90%), is primarily a function of particle size of the pyrite and the nature of its dissemination.  Very small and highly disseminated pyrite particles are nearly impossible to separate from coal.  The pyrite may be of microscopic size and so intimately mixed with the coal that it cannot be liberated, or it may be predominantly coarse and readily released from the coal when crushed.  For a given situation, the removable sulfur is the total sulfur less the sum of the organic sulfur and that portion of the finely disseminated pyrite which cannot be removed.

## COAL RESERVES

Coal is found on every continent of the world, including Antarctica, although most of the coal deposits are found in the Northern Hemisphere.  According to the "Survey of Energy Resources," World Energy Conferences, coal has been mined in 70 countries of the world; however 80% or more of all identified coal reserves occur in the United States, the Soviet Union and China.

Due to the many different methods used to estimate coal reserves, and because available information on coal varies widely, comparisons of the reserves between or among countries is very difficult.  The United States Bureau of Mines and the United States Geological Survey data indicate that the United States has at least one-fifth to one-sixth of all the coal in the world.  Approximately one-eighth of the land area of the United States is underlain by coalbearing strata.  These strata occur in at least 37 states.

Nearly all the bituminous and anthracite coal is found in the eastern half of the country. Although the full range of coal ranks is found in the western half of the United States, most western coal reserves are subbituminous coal or lignite. In most of the coalbearing areas, more than one coal seam is present (from a few seams to 117 that have been identified in West Virginia). The individual seams range in thickness from a fraction of an inch to more than 100 feet. Most of the bituminous coal seams are 20 feet thick or less, and most mining has been in seams from 3 to 10 feet thick.

According to the 1974 Keystone Coal Industrial Manual, "The identified and hypothetical reserves of coal in the United States amount to some 3,224 billion tons. However, based on current technology, economics and environmental regulations, only some 150 billion tons could reasonably be extracted."

There are three main classes of reserves. They are: measured, indicated and inferred. They may be described as follows.

1. Measured (proven) reserves lie within ½ mile of a point of observation and are considered to be within 20% of true tonnage.

2. Indicated (probable) reserves are based on points of observation approximately one mile apart, but not more than 1½ miles, covering a band 1½ miles wide surrounding the area of proven reserves.

3. Inferred reserves, in general, lie more than 2 miles from points of observation. Sometimes this category is broken into strongly inferred reserves, which are estimated by projections beyond the 4 mile limit. The Bureau of Mines frequently reports known reserves that represent the sum of measured and indicated reserves.

In computing the volume of reserves in each of the thickness categories for each bed, the total thickness of coal is used, exclusive of partings greater than ⅜ " thick. Beds or parts of beds made up of alternating layers of thin coal and partings are omitted if the total partings exceed one-half the total thickness, or if the ash content exceeds 33%. Frequently, the distribution of reserves is also categorized according to thickness of overburden: 0 to 1,000 feet, 1,000 to 3,000 feet and 3,000 to 6,000 feet.

The breakdown of total U.S. coal resources according to Keystone is as follows.

|  | **Billion Tons** |
|---|---|
| Mapped and explored (identified) | |
| 0 - 3,000 ft overburden | 1,581 |
| Unmapped and unexplored (indicated and probable) | |
| 0 - 3,000 ft overburden | 1,306 |
| 3,000 - 6,000 ft overburden | 337 |
|  | 3,224 |

However, the economically exploitable coal, which is defined as "material having a thickness of more than 28 inches and less than 1,000 feet overburden. . ." and from identified reserves, is stated to be less than 260 billion tons. Of this figure, the United States Bureau of Mines says we will recover 50% of the underground reserves (105 billion tons) and 90+% (45 billion tons) of the surface reserves for a total of 150 billion tons.

The coal fields of the Appalachian Region, which stretches northeastward from Alabama through Tennessee, Virginia, West Virginia, Ohio and Pennsylvania, contain the largest deposit of high-rank bituminous coal in the world, and most of the anthracite coal in the United States.

One of the characteristics of the Appalachian Region coals which enhance their value is their ability to form coke or agglomerate when heated in the absence of, or with a limited supply of, air. All of the coals are not used for coke-making, however, because some contain more sulfur than is desirable for metallurgical-grade coke. There is more information on the quality of these coals than for those found in any other region in the country. This is due to the many analyses of the coals made by Federal and State agencies in connection with the use of these coals, not only for coke-making, but for light, power and heat in the industrial, commercial and residential sectors of the economy.

West Virginia ranks second to Illinois in total bituminous coal reserves, but first in reserves of bituminous coal among the states in the Appalachian Region. Approximately 46% of West Virginia's reserves are low-sulfur coals (here defined as 1.6% sulfur or less), making a total of 91% of the reserve having relatively little sulfur.

West Virginia coals vary so greatly that it is convenient to separate them as northern and southern coals. In the North, the Pittsburgh bed produces medium-sulfur coals, and the upper Freeport and Sewell beds produce low-sulfur coals that are excellent for steam generation. In the South, the Lower Kittanning, No. 2 Gas, Peerless, Cedar Grove and Sewell beds produce some of the finest steam quality coal mined in the United States. As the sulfur content of these coals is generally low, only the ash content needs to be reduced.

In Pennsylvania, large quantities of bituminous coal are produced for electric utilities. Most of this coal comes from the Upper and Lower Freeport, Upper and Lower Kittanning and Pittsburgh coal beds. These are generally medium-sulfur coals (85% of the reserve contains 3% or less sulfur and 35% has a sulfur content of no more than 2%). The Central Pennsylvania beds, including both medium- and low-volatile coals, generally contain less sulfur than those in the western part of the state and are upgraded primarily to reduce the ash content before they are used for steam generation.

In Ohio the principal coal beds mined are extensions of Pennsylvania's Pittsburgh, Middle and Lower Kittanning, Upper and Lower Freeport and Sewickley (Meigs Creek) bed; these coal beds usually contain medium-ash and high-sulfur. They are used primarily for steam generation.

Maryland's coals are similar to those of the eastern portion of the bituminous fields of Pennsylvania, but these usually have low-sulfur content. In eastern Kentucky and Virginia, the coals are of low-sulfur content. In Tennessee and Alabama, the sulfur content of the coal ranges from low to high.

Of the bituminous deposits, about two-thirds are located in the states east of the Mississippi River. The coal fields or deposits in Illinois, Indiana and western

Kentucky contain 29% of the estimated remaining bituminous coal reserve, but Illinois alone has the largest bituminous reserve of all states. Coals in these states are generally higher in sulfur, especially organic sulfur, with almost 80% of the reserved containing more than 3% sulfur. There are, however, several small deposits of low-sulfur coals in southern Illinois and Indiana where sulfur content averages 1.5% or less.

The interior western region contains large deposits of medium- to high-volatile bituminous, which have not been extensively mined because they are too far from the eastern centers of population and industry. These deposits extend across Iowa, Missouri, eastern Nebraska, Kansas and into Oklahoma, with a related bed in Texas. A smaller area of low-volatile bituminous and anthracite extends over into Arkansas.

The small lignite beds in Texas and Arkansas extend over into Alabama and are properly in the Gulf Province. They are of only fair quality and few analyses for them are available. They have been included with the interior western region in the USBM studies for convenience. Coals in the northern Great Plains province comprise enormous deposits of lignite and subbituminous, which have scarcely been touched. Lignite is characterized by a high content of water and ash, and an ash content of alkaline earths which is significantly higher than other coals.

The western region is defined here, as in the USBM studies of coals by regions, to include the deposits in the Rocky Mountain states and a few isolated deposits in the Pacific Northwest. A southwest subregion at the Four Corners area of Arizona, New Mexico, Utah and Colorado has been established for washability data collection. The coals of the western United States are geologically younger than the eastern coals, and 70% are subbituminous or lignitic in ranks. Although the lower rank western coals are generally of low-sulfur content and often contain only medium amounts of ash, they also are of lower calorific value and are mostly used for steam generation where they can be mined easily and utilized close to their source. However, in some recent applications, these coals are being shipped to eastern steam generators.

On a broad regional level, only the bituminous coals of South Appalachia and some of the lignites of the West will directly, or with the best coal cleaning technology, meet the most strict sulfur emission levels, although there are other seams with substantial reserves which can comply. The coals of North Appalachia, as a group, can be prepared to meet some regional state implementation plans. Overall, the cleaning of northeastern coals combusted for power generation would result in 34% sulfur reduction (nearly 3 million tons of sulfur annually) utilizing current cleaning practice; this level would be increased to 46% (over 4 million tons annually) by the application of the best known preparation technology.


## OBJECTIVES OF COAL PREPARATION

### Background

Coal often exists in its natural state with many impurities, i.e., sulfur, clay, rock, shale and other inorganic materials generally called ash. During the past decade increasing emphasis has been placed on removing the impurities, especially those which result in sulfur oxide emissions upon combustion of the coal. Historically, in the United States coal preparation has been utilized only for specific coals destined for carbonization. The reasons are varied: primarily to reduce their sulfur content, to provide a specific uniform product, to enhance salability, and to improve the economic advantages for coal marketing by developing a superior product. The technological and economic growth of the last 25 years, the resulting degradation of our nation's environment and the introduction of emission standards for air pollution control (sulfur oxides) have changed this picture considerably in recent years.

Years ago, in the hand-loading days of our coal industry, the quality of coal produced was generally satisfactory (regardless of use) because only the cleanest seams were mined and the majority of impurities inherent in mining operations were not loaded out with the coal. However, productivity per man was very low. Mechanization improved productivity, but impurities increased to the extent that some form of cleaning became necessary at many mines, even those in the cleanest seams.

The transformation from hand loading to mechanical mining was quite rapid during the mid-1930s. Tipples and earlier type cleaning devices became inadequate almost overnight. The quality of coal was jeopardized again with the adoption of full-seam mining throughout the industry. Cleaning units were installed on coarse coal sizes to eliminate the manpower required for hand picking the coal as it came from the mine. In addition, due to the marked increase in finer sizes in the run-of-the-mine coal called "ROM," cleaning units were installed to pick up the slack in the coal output.

Today with the thinner, dirtier seams being mined, the impurities in the raw coal may be not only from the seam itself, but also in extraneous material taken in mining of the roof or floor. With increased mechanization, a higher proportion of top and bottom material is taken in mining, which increases the tonnage of reject to be handled. Also, the effects on mining practice of the coal mine Health and Safety Act of 1969 have contributed significantly to the increase in impurities in the ROM coal. For example, the water sprays on continuous miners used to ally the dust at the face seem to add significantly to the moisture content of the raw coal while excessive rock dusting adds other incombustibles to the ROM coal.

**Current Practice**

Coal is providing an increasing share of energy consumed by stationary sources (utility, industrial, commercial and residential). Demand for electrical energy, the shortage of available oil and gas and stagnation of nuclear power development, have made critical the issue as to whether energy can be made available, in its desired forms, to meet future demands without sacrificing the environment.

Today raw coal is cleaned to remove as much noncarbonaceous material as is economically feasible in order to produce a uniform high-quality feedstock for any desired use. Some of the reasons for coal preparation are:

    ....removal of substantial quantities of sulfur from coal,
    ....concentration of carbon in the clean coal,
    ....removal of ash,

....reduction in concentration of trace elements,

....uniform quality of product including ash, moisture and
　　　Btu content.

Coals have highly variable characteristics by seam and by geographical location. Coals are prepared by size reduction and sorting, based upon particle size and density, to create uniform products of high calorific content and reduced mineral levels, especially sulfur. However, only the pyritic sulfur fraction of the total sulfur content is amenable to separation by physical processing. This limitation of sulfur reduction to the natural organic sulfur level of a particular coal means that the level of coal quality improvements attainable is variable, being constrained by processing objectives, cost, processing technology and coal characteristics.

The specific ways of preparing coal are of course determined by its end use. Most of the coal produced in this country is consumed either by carbonization—for the production of metallurgical and chemical coals, or by combustion—for electric power generation, process heating and steam for manufacturing and mining industries or space heating. Although many of the same methods are used in evaluating coals for different uses, the problems, bodies of knowledge and approaches associated with carbonization and combustion in each area are sufficiently dissimilar that coal evaluation in each area merits separate discussion.

## Metallurgical Coke

Another fuel form, metallurgical coke, is almost universally used in blast furnaces, both in ferrous and nonferrous smelting. Coke is the hard, condensed residue resulting from the slow combustion of bituminous coal in the absence of air. This process distills and drives off the volatiles and leaves a high-carbon product, i.e., coke.

During decomposition, the coal mass fuses and swells and becomes plastic. The volatile substances driven off during the coking process range from simple gases such as $CO$, $CO_2$, $H_2O$, $H_2$, $N_2$, $CH_4$, $H_2S$, $S_2$ and $NH_3$ to various complex hydrocarbons and other organic compounds, some containing nitrogen and sulfur. Gradually the mass solidifies as the process reaches completion.

The by-product coke oven, as shown in Figure 1.2, is the primary tool for processing coke in the United States. The oven is externally heated and allows for the recovery of the coal gases, coal tar, and other valuable by-products.

Not all coals are suitable for coking purposes and those that are selected must be carefully prepared before carbonization to produce a high quality coke. The main purpose in cleaning coals is to reduce moisture, ash and sulfur content; however, coal is also prepared to obtain a uniform product. This is important because coal often varies in quality in different areas of a mine. By preparing the coal, a blending of the various qualities can be achieved to assure a uniform coke with minimum ash and sulfur content.

When used for metallurgical purposes, the presence of sulfur compounds in the fuel represents a genuine problem. For example, in high or vertical furnace processes a lowering of the sulfur content in coke by 1% saves from 18 to 20%

## FIGURE 1.2: BY-PRODUCT COKE OVEN

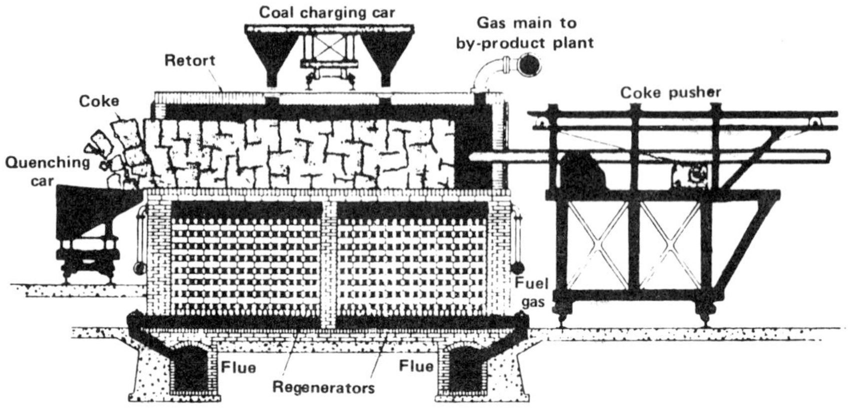


Source: PB-262 716

of the fuel, considerably increasing the efficiency of the metallurgical aggregates
and contributing to an improvement in the quality of the metal. Also, sulfur
in coal used for metallurgy is apt to contaminate the metal. This holds equally
true for several other elements which comprise the ash content of coal such as
phosphorus and arsenic.

### Steam Coal

About two-thirds of the electric energy in the United States is generated by coal-
fired plants. Many of these plants use high-sulfur coal, although increasingly
more stringent Federal, State and local air pollution regulations have intensified
the demand for clean fuels and superior control devices.

The major problem of coal-burning power plants is reducing the air pollutants
in stack gases. In most of these plants, a chief pollutant is sulfur dioxide from
the combustion of organic sulfur compounds present in the coal. Stack gas
cleaning systems are expensive to install and operate, and in some cases would
not be needed if most of the pollutants were removed from the coal prior to
combustion.

The sulfur dioxide standards now applicable to the power industry include Federal regulations which primarily relate to new facilities and those imposed by the
individual State Implementation Plans (SIPs). These regulations apply to steam
generating facilities which were started or modified after August 17, 1971, within 180 days of the time they came on-line. They apply to all facilities having
more than 250 million Btu/hour input (about 10 tons of coal). Besides the
maximum 2 hour average value of 1.2 pounds $SO_2$ per million Btu fired, corresponding values for particulate matter are 1.0 pound and no greater than 20%
opacity, and for nitrogen oxides, 0.7 pound per million Btu fired.

Estimates made in accordance with Project Independence call for the demand of coal to expand to between 1.2 and 1.7 billion tons per year by 1985. About 94 billion tons of naturally occurring low-sulfur coal can be foreseen as a supply that meets air quality regulations. The remaining portion will have to be regulated by using control devices or by coal preparation.

Available methods for controlling sulfur oxide emissions from stationary combustion sources fall into the following major categories:

    ....the physical removal of pyritic sulfur by physical coal cleaning
        prior to combustion
    ....the scrubbing of sulfur oxides from the combustion flue gas
    ....the conversion of coal to a clean fuel by such processes as
        gasification, liquefaction and chemical extraction.

Of these methods, physical removal of pyritic sulfur is the least expensive and the most highly developed method. The degree of sulfur reduction possible depends upon the characteristics of the raw coal and its amenability to sulfur release upon crushing. These characteristics are unique to specific coals and vary from coal to coal. Until such time as new coal conversion technology becomes available and economical, most sulfur oxide emission control will be affected by physical coal cleaning, flue gas scrubbing or a combination of both.

Additionally, the use of coal with high ash content in coal-fired plants results in a greater loss of efficiency, yields a greater amount of ash, and leads to greater losses in the flue gases. Also, the loss of sensible heat and combustible matter in the ash is greater and the cost of drying is correspondingly increased.

With the exception of coal used by some stokers or wet bottom furnaces, coal used in utility power plants is normally pulverized. The costs of grinding and the wear and tear of the pulverizers are disproportionately increased if the coal has a high ash content because the shale is harder to grind than the coal. Furthermore, the mineral matter in the dust entering the combustion chamber must be heated to the flame temperature without contributing anything to the heating and the incombustible dust must be discharged from the furnace. The flue gases generally carry large quantities of incombustible dust which is either discharged through the stack or accumulates on the stack walls.

Other than poor design or operation, the quality of the coal greatly affects the efficiency of the combustor. In addition to the operational costs and problems, the increased transportation costs (transporting moisture and other impurities) and the increased disposal cost of the ash add considerable emphasis to the merits of clean coal.

## Summary

Coal is used in sintering, pelletizing, zinc retort smelting, blast furnace smelting and other metallurgical processes. For these processes, special coals prepared to rigid specifications are used to get the desired metallurgical results at lowest cost. By far the largest application is in the form of coke for the iron blast furnace.

Coal is also becoming the primary fuel for steam generation for electric utilities. The mechanical coal cleaning process will allow certain coals to be combusted without additional sulfur emission controls and in those situations where such controls are still necessary, prior coal cleaning helps reduce the emission control costs.

For whatever purpose coal or coke is used, it is to the advantage of the consumer that the fuel should contain the minimum amount of ash. Incombustible material in the fuel reduces its gross calorific value, increases the weight that must be handled and transported, gives rise to difficulties of combustion and involves further expense in its disposal. Also, ash in coal increases the production of smoke and results in the discharge of fine dust from chimney stacks, especially from the stacks of pulverized fuel boilers.

It is clear that with an increasing electric load generated by coal, the emission of $SO_2$ into the atmosphere must be kept at an acceptable level. There have, however, been difficulties in perfecting $SO_2$ clean-up systems and processes. Most estimates indicate that these processes will not reach widespread commercial usefulness before the mid-1980s because of chemical and mechanical problems. This fact, coupled with the need to meet stringent air quality standards passed by the Federal Government, provides the rationale for preparing raw coal to remove as much pyritic sulfur as possible before firing.

Clean coal's greatest applicability is to: (1) installations which are not able to use flue gas desulfurization, such as industrial boilers of small size; and (2) existing combustors which require clean coal to meet State Implementation Plans (SIPs).


## THE PREPARATION PROCESS

### Overview

The coals of the United States have highly variable characteristics depending on seam and geographic location. Since coals vary so widely, coal cleaning processes are typically engineered for each coal source and designed with respect to the use to be made of the coal. There is a considerable process uniformity among plants, but each plant is usually individually designed.

Coals are prepared by size reduction and subsequent particle sorting based upon particle size and density. The level of coal quality improvements attainable is variable, being constrained by processing objectives, cost, processing technology and coal characteristics.

For years, preparation plants were designed to produce multiple sizes of coal for various customers, such as lump, egg, stove, stoker and nut sizes. Today, however, plants are designed to produce only one product of definitive characteristics for one specific customer. The preparation plant is designed to remove the noncombustibles from the coal at the minimum practical operating cost and at the optimum practical yield. However, the run-of-the-mine (ROM) coal is prepared only to the extent that is necessary to make the produce salable.

The range of coal cleaning processes now being practiced in the United States may be generalized into four individual levels of preparation. These levels may be defined in the manner described on the following page.

*Level 1*—no preparation, direct utilization of the run-of-the-mine product.

*Level 2*—removal of gross noncombustible impurities, plus control of particle size and promotion of uniformity (typically 95% material yield and 99% thermal recovery). Little change in sulfur content.

*Level 3*—single-stage cleaning allowing little component liberation. Particle sizes less than ⅜ inch usually are not prepared. 80% material yield and 95% thermal recovery. Limited ash and sulfur content.

*Level 4*—multi-stage cleaning with controlled pyrite liberation. Usually incorporated dewatering and thermal drying. 70% material yield and 90% thermal yield. Maximum ash-sulfur rejection and calorific content of product.

Preparation practice for most coals used by electric utilities lies between levels 2 and 3. The preparation practices for metallurgical coals are typically level 4. The extent to which a specific coal can be cleaned is dependent upon the characteristics of the coal and the sophistication of the preparation process. The limitations are often both economic and technical.

The technical limitations of the preparation process relate primarily to the very small component particles existing in coal. Many of these particles are residual structures of vegetation and minerals, generally irregular in shape. The pyrite particles in many coals are less than 1 micron (0.0004") in their longest dimension. Particles smaller than 50 microns cannot be practically separated from each other, and separating them is usually inefficient. Larger particles, or those less homogeneous in composition, respond more readily to separation.

To be separable, impurity-containing particles must have masses greater than the pure coal particles. The difficulty in separating small size particles (less than 50 microns) results from their slower response to the acceleration of gravity than larger particles; they literally float within the coal. Moreover, since most of the separation is done in water systems, a further complication exists in working with small particles in that removal of the water from them is significantly more difficult and more costly than removing water from the larger-sized particles due to the smaller porosity of the smaller particles, or of the combination of particles.

Because of the technical difficulty in separating small particles, the separation costs increase as the particle size decreases. The processes which will remove more pyrite from the coal necessarily utilize smaller particle sizes and are considerably more costly. Accordingly, coal cleaned primarily for ash removal is cleaned with as large a particle size as is practical. It is for this reason that coal processing plants which were not designed for sulfur removal often do not function well as pyrite removers.

The economic limitations of coal preparation are varied and numerous. Cleaning of coarse coal is relatively simple and less costly than cleaning of the finer sizes. The fine coal portion in the raw coal feed has materially increased as mechanization of mining process has increased, thus adding considerably to cleaning plant costs. Wet cleaning units for fine coal are not themselves expensive; it is the equipment necessary to dewater and dry the product that adds significantly to the cost. Clarifying the process water and thermal drying substantially increase plant capital investment. Yet many modern cleaning plants must contain this equipment in order to obtain the desired ash, sulfur and moisture in the product and still recover the greatest amount of salable coal.

The disposal of waste refuse developed during the coal cleaning process (CCP) represents an additional cost which must be attributed to the preparation plant. Coal preparation processes are consumers of energy—they both utilize it in the processing and lose some of it in the rejected refuse. Most energy consumed during coal processing is utilized for one of the following:

    ....to move the coal components through the cleaning system
    ....to create new surface area by breaking or crushing
    ....to activate equipment to manipulate the particle separation
    ....to remove water from the coal
    ....to operate environmental protection systems.

In general, processing energy requirements increase with the beneficiation level and decrease as particle size increases.

Among the factors which may determine the final delivered cost of coal to an electric generating station are:

    ....the cost of run-of-the-mine coal at the mine portal
    ....the cost of cleaning
    ....the cost of handling and disposal of preparation plant refuse
    ....the level of clean coal yield and thermal recovery
    ....the cost of coal storage at the mine, preparation plant and
        generation station
    ....the cost of coal loading at the mine or preparation plant and
        unloading at the generating station
    ....the transportation costs.

Other economic impacts which must be compared between use of run-of-the-mine coal and clean coal are:

    ....the pulverization costs (power consumed and plant maintenance)
    ....the disposal cost of ash developed during coal combustion.

The economic implications of coal preparation are presented in Table 1.2 comparing sample costs for the same coal burned "as mined" and cleaned. The clean coal, with a 0.6% lower cost, on a weight basis is 5.2% higher cost in terms of ¢/MM Btu or mils/kWh generated. This cost comparison model neglects several factors which are difficult to quantify, but would undoubtedly enhance the value of prepared coal. Among the factors favoring clean coal are:

    ....greater reliability of power plant performance
    ....reduced coal handling costs and storage costs
    ....greater heat-release capabilities—boiler capacity design
    ....reduced slag-fouling maintenance in boiler and
        heat transfer systems
    ....reduced quantities of fly ash for collection.

## TABLE 1.2: COMPARATIVE COAL COSTS FOR UTILITY CONSUMPTION UTILIZING CLEANED COAL AND RUN-OF-MINE FROM THE SAME MINE

**Basis: 1 Ton Cleaned Coal**

| | Prepared Coal* | Run-of-Mine Coal |
|---|---|---|
| Value at shipping point | | |
| $ expression | 14.46** | 13.31*** |
| ¢/MM Btu | 52.20 | 45.30 |
| mils/kWh | 5.35 | 4.64 |
| Value at utility (includes transportation)† | | |
| $ expression | 18.25 | 17.86 |
| ¢/MM Btu | 65.90 | 60.70 |
| mils/kWh | 6.76 | 6.23 |
| Value as fired (includes coal grinding costs)†† | | |
| $ expression | 18.38 | 18.14 |
| ¢/MM Btu | 66.40 | 61.70 |
| mils/kWh | 6.80 | 6.33 |
| Total fuel costs at utility (includes ash disposal)††† | | |
| $ expression | 18.62 | 18.73 |
| ¢/MM Btu | 67.20 | 63.70 |
| mils/kWh | 6.89 | 6.53 |

**Basis for Comparative Cost Calculations**

| | |
|---|---|
| Coal Data, % | |
| Clean coal yield | 83.20 |
| Thermal loss in cleaning | 5.85 |
| Heat content, Btu/lb | |
| Run-of-mine | 12,240 |
| Clean coal | 13,850 |
| Percent increase | 13.20 |
| Ash content, wt % | |
| Run-of-mine | 16.40 |
| Cleaned coal | 7.90 |
| Percent decrease | 51.80 |

*Assumed cleaning cost $1.50/T of clean coal. A constant moisture content of run-of-mine and cleaned coal is assumed.

**Based on Central Pennsylvania low-volatile bituminous coal. Assessed at $14.46/T based on average U.S. selling price for utility coal, May 1974. This price was equivalent to 65.8¢/MM Btu, and represents an average calorific content of 11,000 Btu/lb. Source: Federal Power Commission Data. *Coal News,* No. 4226, 1/14/74. National Coal Association, Washington, D.C. It is further assumed for this example that the figure of $14.46/T includes $1.80/T contribution to the UMWA Royalty Fund.

***Value of run-of-mine coal $9.28/T. 1.20 tons required to prepare 1.00 ton of clean coal. Upon direct sale of the run-of-mine product, the $1.80/T UMWA Royalty would be added.

†Assumed shipping cost $3.79/T (for 1973). Source: Coal Traffic Annual, 1974 edition, p 27. National Coal Association, Washington, D.C. The cost advantages of storage and handling 20% less coal in cleaned form at the power station have not been included.

††The grinding of coal for pulverized firing to 70% minus 200 mesh requires energy consumption which varies with coal hardness. Hardness is usually expressed as Hardgrove Grindability Index. A 55 HGI coal uses 7.9 kWh/T, while a 100 HGI coal uses 4.4 kWh/T. For these calculations power was charged at 3¢/kWh. The value for the softer coal was utilized for clean coal while the harder coal value was used for run-of-mine coal. Source: Private communication, Mr. Richard Borio, Combustion Engineering, Inc., Windsor, Connecticut (February 1975).

†††Calculations based upon $3.00/T for ash disposal at the utility.

The state of the art of coal cleaning has been reviewed in a
report by A.W. Deurbrouck and P.S. Jacobsen of the Pittsburgh
Energy Research Center (NTIS CONF-741025-5), upon which
the material in this section is based.  Further information on
this report will be found in the bibliography on page 301.
Only the material from the report which most concerns sulfur
removal is used here.


## SULFUR AND ITS REMOVAL

### Introduction

Raw coal is cleaned to remove as much noncarbonaceous material as is economically feasible so as to produce a uniform high-quality feedstock for any desired use.  Five of the more significant benefits derived from coal preparation
are:

1.  Removal of substantial quantities of sulfur from coal—the only
    commercially proven reliable method.  Most of the sulfur that
    has been removed from coal has been removed by the process of
    coal cleaning.

2.  Concentration of carbon in the clean coal—important because
    the carbon content of the feedstock will determine the capacity
    of a coal utilization unit.  Therefore, the capital costs can be
    minimized for any capacity of unit built.

3.  Removal of ash—an impurity which is abrasive and which causes
    disposal problems, increased pulverization costs, and high trans-
    portation cost per unit of energy, plus air and water pollution
    problems.

4.  Reduction in concentration of trace elements—reducing corro-
    sion and erosion effects, air and water pollution, and potential
    of catalyst poisoning.

5.  Uniform quality of product including ash, moisture, and Btu
    content—a big plus for controlling almost any chemical re-
    action or combustion-type process.

In addition, there are miscellaneous benefits including easier handling and production of a visually more attractive fuel.

Total tonnage of coal cleaned annually has increased for the first time since
1967(1).  In 1972 almost 293 million tons of cleaned coal were produced, an
increase of approximately 22 million tons over 1971.  As shown in Figure 1.3,
the production of clean (or prepared) coal has actually been decreasing as an
energy-hungry nation has been inclined to forego the benefits derived from using
a clean coal in a desire to reclaim all the Btus in the raw coal.  Interestingly,
new coal preparation facilities recently going on line or in the planning stage
are being designed to provide Btu recoveries well in excess of 90%.

Today's coal-cleaning plant is a modern, environmentally acceptable installation
capable of providing coals to meet exacting specifications with minimal loss of

## FIGURE 1.3:  TOTAL AND PREPARED COAL PRODUCTION

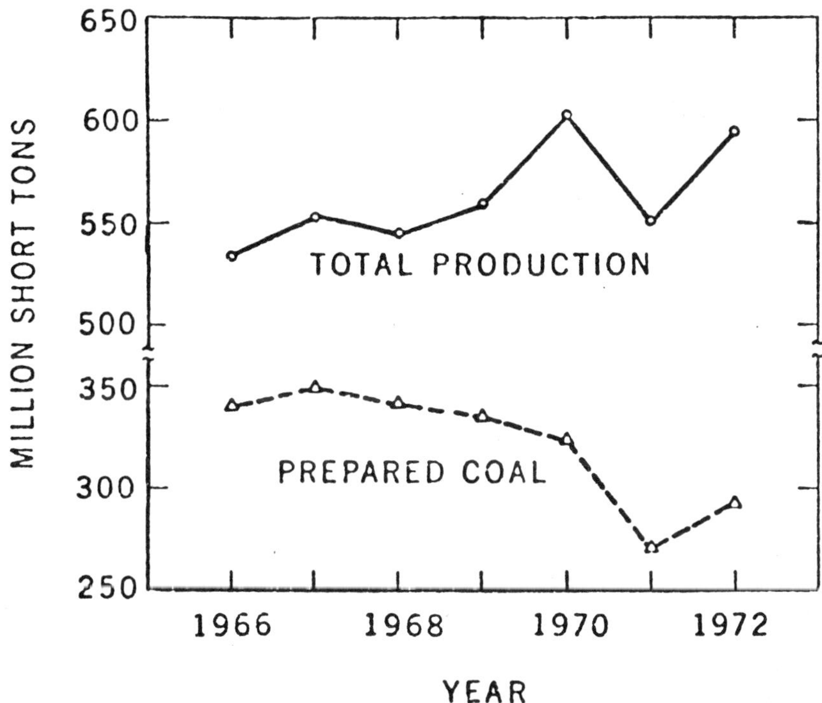


Source:  CONF-741025-5

usable Btus to the refuse product.  Coal-cleaning jigs and dense-medium cyclones and vessels clean approximately 75% of all the clean coal produced (1)
(Table 1.3).

## TABLE 1.3:  MECHANICAL CLEANING OF COAL BY EQUIPMENT TYPE, PERCENTAGE OF CLEAN COAL PRODUCED (1)

| Washer Type | Year | | | |
| --- | --- | --- | --- | --- |
| | 1942 | 1952 | 1962 | 1972 |
| Jigs | 47.0 | 42.8 | 50.2 | 43.6 |
| Dense-medium processes | 8.8 | 13.8 | 25.2 | 31.4 |
| Concentrating tables | 2.2 | 1.6 | 11.7 | 13.7 |
| Flotation | --- | --- | 1.5 | 4.4 |
| Pneumatic | 14.2 | 8.2 | 6.9 | 4.0 |
| Classifiers | 7.4 | 8.5 | 2.1 | 1.0 |
| Launders | 13.1 | 5.2 | 2.2 | 1.9 |

Since 1942, the use of dense-medium process and concentrating tables has increased at a very rapid rate at the expense of classifiers, launders, and pneumatic

methods. Froth flotation processing of coal in the United States is relatively new but has gained rapid acceptance, first for in-plant water clarification, and more recently as one of the standard processes for cleaning fine coal.

### Resources of Low-Sulfur Coal

The coal resources in the states east of the Mississippi River, containing 1.0% sulfur or less, are estimated at 95.3 billion tons (2)(Figure 1.4). Although this is only 20% of the total remaining resource, it is still a significant amount. Available washability data indicate that much of the medium-sulfur content coal can be upgraded to 1.0% sulfur or less using state-of-the-art coal-cleaning techniques.

**FIGURE 1.4: ESTIMATED REMAINING COAL RESOURCES OF ALL RANKS, BY SULFUR CONTENT, IN STATES EAST OF THE MISSISSIPPI RIVER, JANUARY 1, 1965**

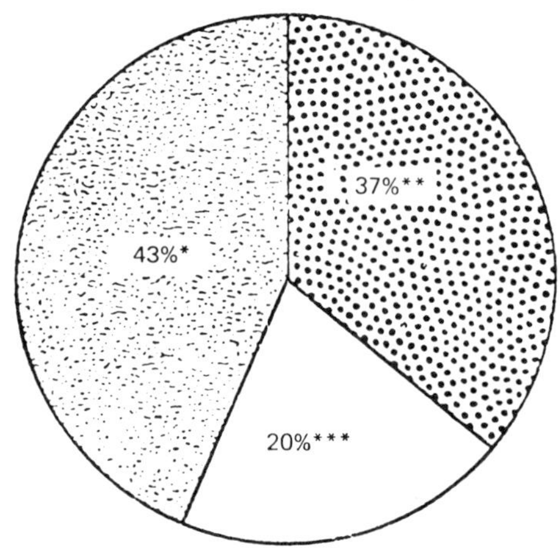


*High-sulfur coals (over 3.0%)—206.5 billion tons
**Medium-sulfur coals (1.1 to 3.0%)—177.3 billion tons
***Low-sulfur coals (1.0% or less)—95.3 billion tons

Source: CONF-741025-5

### Sulfur Forms

Sulfur in coal occurs in three forms: organic, sulfate, and pyritic. Organic sulfur, which is an integral part of the coal matrix and which generally cannot be removed by direct physical separation, comprises from 30 to 70% of the total sulfur of most coals. Generally, the organic sulfur/total sulfur ratio is highest for low-sulfur coals and decreases as the total sulfur content increases. The sulfate content is normally an oxidation product which is water soluble and therefore readily removed during coal cleaning. Sulfate sulfur contents are usually less than 0.05%. Pyritic sulfur is the mineral pyrite, which in coal occurs as discrete and sometimes microscopic particles. It is a heavy mineral with a specific gravity of about 5.0, whereas coal has a maximum specific gravity of only 1.8. The pyrite content of a coal generally can be reduced significantly by size reduction and subsequent specific gravity separation.

Pyrite can be released from coal by stage crushing. Figure 1.5 shows the results of such stage crushing on a sample of Lower Freeport bed coal from Ohio and a sample of Pittsburgh bed coal from West Virginia.

## FIGURE 1.5: PYRITIC SULFUR REDUCTION POTENTIAL FOR TWO APPALACHIAN REGION COALS AS A RESULT OF STAGE CRUSHING

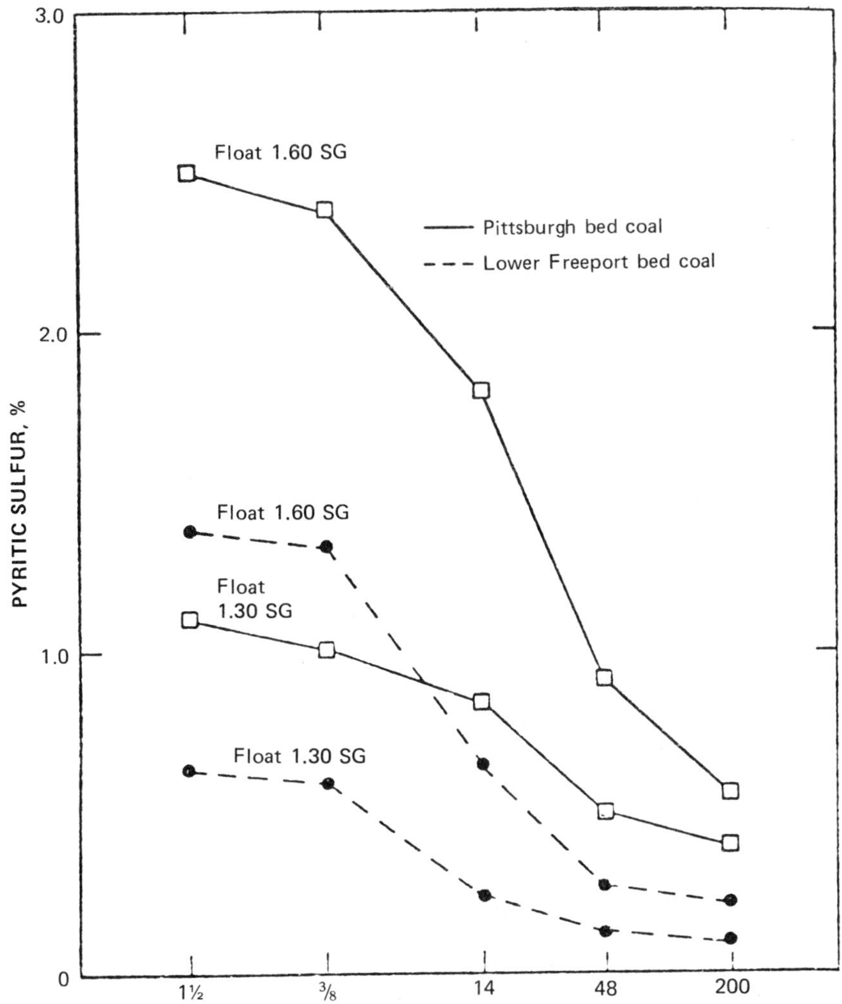


Source:  CONF-741025-5

For the Lower Freeport bed coal sample, crushing from 3/8" to 14-mesh provided the maximum pyritic sulfur reduction for the float 1.30 SG material, but crushing from 14- to 48-mesh provided the maximum percentage of pyritic sulfur reduction for the float 1.60 SG material. For the Pittsburgh bed coal sample, maximum sulfur reduction of about 0.8 percentage point was obtained by crushing from 14- to 48-mesh regardless of the specific gravity product.

Interestingly, further crushing of the Pittsburgh bed coal sample from 48- to 200-mesh showed an additional pyritic sulfur reduction of about 0.5 percentage point for the float 1.60 SG material. Regardless of how difficult and expensive the processing of coal crushed to 48-mesh or even 200-mesh top size is, the potential benefits of obtaining low-sulfur products may well justify the additional problems encountered.

## Controlling Sulfur Oxide Emissions

Available methods for controlling sulfur oxide emissions from stationary combustion sources have been previously discussed. They are:

1. The physical removal of pyritic sulfur by coal cleaning prior to combustion.
2. The removal of sulfur oxides from the combustion flue gas.
3. Conversion of coal to a clean fuel by such processes as gasification, liquefaction, and chemical extraction.

Of these methods, physical removal of pyritic sulfur is the least expensive and most highly developed method. The degree of sulfur reduction possible depends on the characteristics of the raw coal and its amenability to sulfur release upon crushing; these characteristics are unique to the coals and vary from coal to coal.

A preliminary study was undertaken by the National Air Pollution Control Administration (NAPCA), the predecessor of the Environmental Protection Agency (EPA), in 1965 to determine the effect of pyrite reduction through coal cleaning on sulfur oxide emissions from stationary power stations (3). The result of this investigation indicated that this effect could not be quantified owing to lack of information. The detailed information which would have to be obtained to establish this relationship included:

1. Investigation of the sulfur forms and their distribution in all major high-sulfur coal beds in the United States.
2. The applicability of available commercial coal preparation methods for pyrite removal, and the adaptation of these techniques to maximize sulfur reduction.
3. Development of processes that would economically recover by-products from coal-cleaning rejects, thus reducing cleaning costs and air and water pollution.

## Sulfur Forms and Coal Washability

Later in 1965, NAPCA funded a study to determine washability of United States coals (4). The samples to be evaluated were representative of the raw coal occurring at a particular location. The washability analyses were used to evaluate the effect of size reduction and specific gravity separation on pyritic sulfur liberation and subsequent removal. Representative portions of the samples were stage-crushed to three top sizes: coarse (1½"); intermediate (⅜"); and fine (14-mesh). Washability studies were performed on the various sized fractions with organic liquids of standardized specific gravities to determine the effects of top size and specific gravity parameters on pyrite liberation and separation.

To date, a total of 455 samples from six coal-producing regions have been evaluated. [Editor's Note: See Chapter 3.] All of the samples were collected from working mines producing steam coals; where possible the largest mines from a specific bed were sampled. Samples collected to date represent mines which provide more than 50% of the annual steam coal production.

*Northern Appalachian Region Coals:* Two hundred twenty-seven samples collected from Maryland (34), Ohio (58), Pennsylvania (103), and northern West Virginia (32) have been evaluated.

The curves in Figure 1.6 show the pounds of $SO_2$/million (M) Btu emission for the fired raw coal (curve **a**), the coal at a 50% Btu recovery when crushed to 14-mesh top size (curve **b**), and the coal at a 90% Btu recovery when crushed to 1½" top size (curve **c**).

**FIGURE 1.6: EMISSION OF NORTHERN APPALACHIAN REGION COALS**

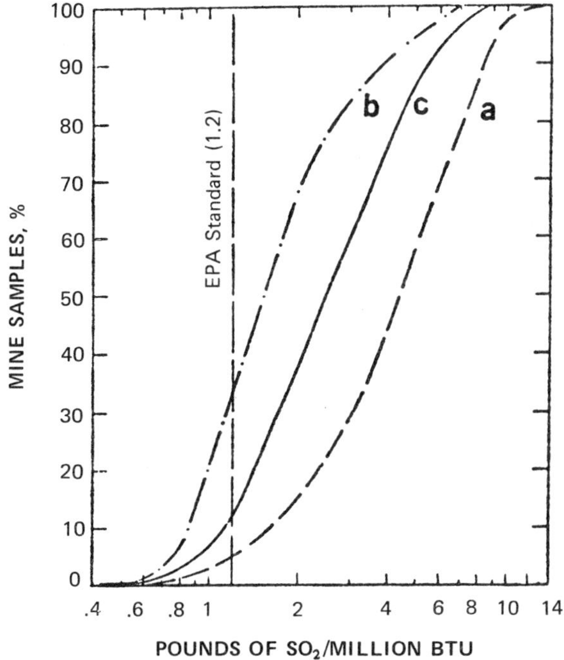


Source:  CONF-741025-5

The data show that less than 5% of all the raw coal samples would comply with the current EPA emission standard of 1.2 pounds of $SO_2$/M Btu with no preparation.  Only 35% of the samples would comply at a Btu recovery of 50% when crushed to 14-mesh top size.  However, if the emission standard were raised to 2.0 pounds of $SO_2$/M Btu (which is equivalent to 1.3% total sulfur for a coal having a heating value of 13,000 Btu/lb), then 15% of the raw coal samples would meet this standard with no preparation.  Approximately 70% of the samples would meet this standard at a 50% Btu recovery when crushed to 14-mesh top size.

If a stack $SO_2$ removal system capable of achieving only 60% efficiency were used at the power plant, then 30% of the raw coal samples producing emissions of 3.0 lb of $SO_2$/M Btu could be burned and still meet the current EPA standard. 60% of the samples would meet this standard at a Btu recovery of 90% when crushed to 1½" top size.

*Midwest Region Coals:*  95 samples collected from three states, Illinois (40), Indiana (20), and western Kentucky (35), have been evaluated.  The curves in Figure 1.7 show the pounds of $SO_2$/M Btu emissions for the fired raw coal (curve **a**), the coal at a 50% Btu recovery when crushed to 14-mesh top size (curve **b**), and the coal at a 90% Btu recovery when crushed to 1½" top size (curve **c**).

FIGURE 1.7:  EMISSION OF MIDWEST REGION COALS

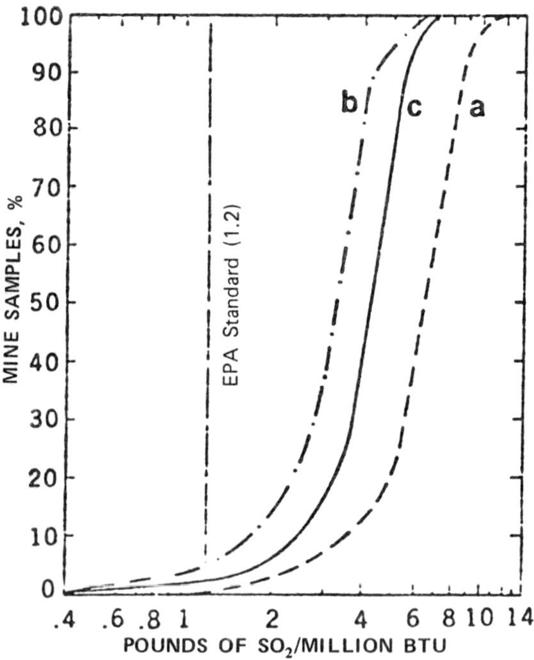


Source:  CONF-741025-5

The data show that less than 1% of all the raw coal samples would comply with the current EPA emission standard of 1.2 lb of $SO_2$/M with no preparation, and less than 5% of the samples would comply at a Btu recovery of 50% when crushed to 14-mesh top size. Even if the emission standard were raised to 2.0 pounds of $SO_2$/M Btu, only 3% of the raw coal samples would meet the standard with no preparation. Only 14% would meet the standard at a Btu recovery of 50% when crushed to 14-mesh top size.

The coals of the Midwest Region contain higher percentages of organic sulfur than do the Northern Appalachian Region coals; consequently, it will be more difficult to produce a clean coal that will meet the current EPA emission standard of 1.2 lb of $SO_2$/M Btu of coal fired. If a stack $SO_2$ removal system capable of achieving 70% efficiency were used at the power plants, then 10% of the raw coal samples producing emissions of 4.0 lb of $SO_2$/M Btu could be burned and still meet the current EPA criteria; 40% could be burned at a Btu recovery of 90% when crushed to 1½".

*Other Region Coals:* Enough coal samples from the other four regions have not been collected to date to permit an extensive evaluation.

**Sulfur Reduction Potential of Coal Cleaning Devices**

Studies also were funded by NAPCA to determine the sulfur reduction potential of conventional and unconventional coal cleaning devices including wet concentrating tables, concentrating spirals, hydrocyclones, electrokinetic separators, and froth flotation (5). Results of these studies have shown a great potential for significant pyritic sulfur reduction when feed top sizes are reduced adequately.

*Wet Concentrating Table:* Tests conducted by Bituminous Coal Research, Inc. showed that the wet concentrating table effectively removed free pyrite from coal (6). Two-stage cleaning tests, in which a coarse sample was tabled and the clean fraction pulverized and retabled, indicated that pyrite removal was enhanced relative to single-stage cleaning. The wet concentrating table is an easy-to-operate, inexpensive, and reliable coal cleaner.

*Humphreys Coal Cleaning Spiral:* The Humphreys spiral is a machine capable of removing appreciable amounts of pyritic sulfur from fine size coal. Although not generally considered as a coal-cleaning device, the spiral has many attributes which deserve further investigation.

In a recent study conducted by the Bureau of Mines, the spiral provided excellent pyritic sulfur reduction down to 200 mesh on a 14-mesh top size feed (7). The maximum unit capacity for these tests was approximately 2.0 tph.

*Electrokinetics:* The electrokinetic process for the physical cleaning of coal uses the principles of electrophoresis for separating pyrite from coal. A laboratory-scale investigation indicated that pyrite and other impurities such as silica could be separated from the coal (8). The technique, unfortunately, was deemed uneconomical on the basis of power-input requirements for scaling up to the pilot stage as a commercial-scale coal-cleaning procedure.

*Froth Flotation:* A unique two-stage froth flotation process was developed and patented by the Bureau to remove pyritic sulfur from fine-size coal (9)(10). The process consists of a first-stage standard coal flotation step in which high-ash refuse and the coarse or shale associated pyritic sulfur are removed as tailings. The first-stage coal-froth concentrate is then repulped in fresh water, pH is maintained below 7, and coal depressant, a pyrite collector, and a frother are added to float any pyritic material carried over in the first-stage froth; the second-stage underflow is left as the final clean coal product.

Laboratory and pilot plant flotation tests with coals from various coalbeds throughout the Appalachian Region showed that pyritic sulfur reductions of up to 80% could be achieved using this technique.

The coal depressant used is a carbohydrate colloid (a starchy substance) that is recommended for depression of carbonaceous gangue in the flotation of minerals. Examples of these hydrophilic carbohydrate substances include cornstarch and quebracho extract.

The pyrite collector most frequently used is potassium amyl xanthate, one of the more powerful xanthates available, and one that is particularly useful in operations where a strong nonselective sulfide collector is needed.

For most of the two-stage flotation work, methyl isobutyl carbinol (MIBC) was used as the frother, but any of the other commonly used aliphatic alcohols or water-soluble polyglycol-type frothers would probably be satisfactory. Any pH adjustment necessary was made with either hydrochloric or sulfuric acid. Sulfuric acid is preferred for large-scale industrial application because it is less expensive.

The pilot plant work was done in a specially designed two-stage flotation facility. Two banks of flotation cells are used. Both the first- and second-stage flotation machines shown consist of four 0.6 cubic-foot-capacity Wemco Fagergren flotation cells.

Table 1.4 shows first- and second-stage coal-pyrite flotation test results from the pilot plant study treating a Lower Freeport bed coal from Cambria County, Pa. The analyses show a pyritic sulfur reduction from 2.19 to 0.87% in the first stage, with a further reduction to as low as 0.42% in the second stage. Second-stage recoveries for all tests shown were over 90%.

**TABLE 1.4: TWO-STAGE PILOT PLANT FLOTATION RESULTS FROM TREATING LOWER FREEPORT BED COAL**

| Product | Weight Percent | Ash | Pyritic Sulfur | Total Sulfur |
|---|---|---|---|---|
| | | | . . . . . . . . .Analysis, %. . . . . . . . . | |
| | | . . . . . . . . . . . . . .First Stage. . . . . . . . . . . . . . . | | |
| 2 tests (average): 0.1 lb MIBC/ton | | | | |
| Clean coal | 62.3 | 7.8 | 0.87 | 1.40 |
| Reject | 37.7 | 71.4 | 4.36 | 4.50 |
| Feed | 100.0 | 31.8 | 2.19 | 2.57 |

(continued)

**TABLE 1.4:  (continued)**

| Product | Weight Percent | Ash | Pyritic Sulfur | Total Sulfur |
|---|---|---|---|---|
| | | . . . . . . . . .Analysis, %. . . . . . . . | | |
| | . . . . . . . . . . . . . Second Stage . . . . . . . . . . . . . | | | |
| Test 1:  0.6 lb Aero Depressant 633/ton | | | | |
| 0.7 lb potassium amyl xanthate/ton | | | | |
| 0.1 lb MIBC/ton | | | | |
| Clean coal | 95.1 | 7.5 | 0.55 | 1.12 |
| Reject | 4.9 | 13.3 | 6.85 | 7.41 |
| Feed | 100.0 | 7.8 | 0.86 | 1.43 |
| Test 2:  0.6 lb Aero Depressant 633/ton | | | | |
| 0.7 lb potassium amyl xanthate/ton | | | | |
| 0.1 lb MIBC/ton | | | | |
| Clean coal | 93.3 | 7.5 | 0.49 | 1.02 |
| Reject | 6.7 | 12.4 | 6.28 | 6.42 |
| Feed | 100.0 | 7.8 | 0.88 | 1.38 |
| Test 3:  0.6 lb Aero Depressant 633/ton | | | | |
| 0.8 lb potassium amyl xanthate/ton | | | | |
| 0.1 lb MIBC/ton | | | | |
| Clean coal | 90.5 | 7.4 | 0.42 | 0.97 |
| Reject | 9.5 | 10.7 | 5.07 | 5.53 |
| Feed | 100.0 | 7.7 | 0.86 | 1.40 |

On-site testing of this coal-pyrite flotation process is currently being piloted at
a preparation plant in Pennsylvania.  The laboratory flotation machine and associated equipment have been set up at the plant for daily testing of slurry feed
to the plant's froth flotation machines.  The results to date are promising, and
a production-scale, continuous evaluation of the process is planned for the near
future.

## REFERENCES


(1)  U.S. Bureau of Mines, *Mineral Yearbooks, 1942, 1952, 1962, and 1972.* Chapter on
Coal—Bituminous and Lignite.

(2)  J.A. Decarlo, E.T. Sheridan, and Z.E. Murphy, *Sulfur Content of United States Coals.*
BuMines IC 8312, 1966, 44 pp.

(3)  Paul Weir Company, Inc., *An Economic Feasibility Study of Coal Desulfurization,* vol.
1 and 2.  (A Study for the Division of Air Pollution, Public Health Service, U.S. Department of Health, Education and Welfare, Contract No. PH 86-65-29, Oct. 1965.)

(4)  A.W. Deubrouck, *Sulfur Reduction Potential of the Coals of the United States,* BuMines
RI 7633, 1972, 289 pp.

(5)  L. Hoffman, J.B. Truett, and S.J. Aresco, *An Interpretative Compilation of EPA Studies
Related to Coal Quality and Cleanability,* The Mitre Corp.  Prepared for Office of Research and Development, U.S. Environmental Protection Agency, Washington, D.C.
20460, May, 1974.  EPA-650/2-74-030, 266 pp.

(6)  Bituminous Coal Research, Inc., *An Evaluation of Coal Cleaning Process and Techniques for Removing Pyritic Sulfur from Fine Coal.*  Sept. 1969, 268 pp.

(7)  J.E. Zeilinger and A.W. Deurbrouck, *Physical Desulfurization of Fine-Size Coals by
Spiraling.*  BuMines RI to be published.

(8)   K.J. Miller and A.F. Baker, *Evaluation of a Novel Electrophoretic Separation Method to Remove Pyritic Sulfur from Coal.*  BuMines RI 7960, 1974, 14 pp.

(9)   K.J. Miller (assigned to U.S. Government); Flotation of Pyrite from Coal; U.S. Patent 3,807,557, April 30, 1974.

(10)  K.J. Miller and A.F. Baker, *Flotation of Pyrite from Coal.*  BuMines TPR 51, 1972, 7 pp.

# LOW TEMPERATURE PROCESSES FOR COAL DESULFURIZATION

The material for this chapter has been based upon a report by
E. Mendizabal of the University of California Lawrence Berkeley
Laboratory (LBL-5216).  Further information on this report will
be found in the bibliography on page 301.


## INTRODUCTION

### Occurrence and Forms of Sulfur

Sulfur is present in coal in three different forms: as metallic sulfides, as metallic
sulfates and as organic sulfur.  Pyritic sulfur in coal is present as the minerals
pyrite and marcasite which have the same composition, $FeS_2$, but differ in crystalline structure.  Because it is difficult to distinguish between them, they are
usually designated simply as pyrites or iron pyrites.

Pyrites are distributed in coal in many ways.  They can occur in lenses and bands,
balls or nodules, joints or cleats, and as finely dispersed particles.  The size and
distribution affect the amount that can be removed by conventional coal preparation methods.  Coarse crushing may release much of the pyrites in the lenses,
bands, cleats and joints, for subsequent removal by mechanical cleaning.  In order
to free the fine particles from the coal, it is necessary to grind the coal very fine
prior to cleaning.  However, this procedure has an adverse effect on process efficiency and severely restricts the methods of cleaning that can be applied.

The amount of sulfate sulfur in freshly mined coals in general is less than 0.05%,
but it gradually increases due to oxidation of the pyrites.  In fact, its presence
in more than small amounts indicates that the coal has been weathered.  Because
sulfate sulfur in commercial coals is so low, it is not an important factor in coal
utilization.

Organic sulfur is distributed through the coal matrix as an integral part of the
coal molecular structure and accordingly it can not be removed by conventional
cleaning or preparation processes, but requires chemical treatment.

In general, organic sulfur is the predominant form of low-sulfur coals; however, as the total sulfur increases, both the organic and pyritic forms increase. There is no fixed direct relationship between the amounts of each form. Although there is no exact knowledge of the forms in which organic sulfur is present in the coal matrix, some groups have been suggested as follows:

(1)  Bahtnagar and Duti reported mercaptans and disulfides in the extracts. However, the bulk of the sulfur compounds remained in the residue.

(2)  Roy obtained extracts from an Indian coal using benzene and ethylenediamine, in which he was able to identify mercaptans, sulfide, disulfides, and thiophenes. He also treated the Indian coal with $KMnO_4$ and found that 70% of the sulfur remained in the oxidized residue, so he concluded that 70% of the sulfur is in ring structure.

(3)  Kavcic similarly concluded that 70% of the organic sulfur present in a high sulfur Yugoslavian coal is a part of the heterocyclic aromatic ring structure.

(4)  Gusev demonstrated the presence of both thioether and bis-thioether groups in Russian coals.

(5)  Postowsky demonstrated the presence of organically fixed thioether sulfur in coals.

(6)  Lissner and Nemes, studying the distribution of sulfur in various fractions obtained when coal is carbonized in an atmosphere of nitrogen and steam or with steam alone at temperatures from 200° to 560°C, found that the organic sulfur occurs in four forms. Between 100° to 300°C sulfur is evolved from phenolic sulfur groups and probably from $\equiv C-SH$ or $\equiv CSSC\equiv$ groups containing nitrogen. Superheated steam hydrolyzes sulfidic sulfur at 380°C, and retards the formation of new organic sulfur compounds with the evolution of inorganic sulfur.

Figure 2.1 displays a representative coal molecule of Hill and Lyon which includes sulfur-containing groups such as:

$$-C-C-, \quad R-S-R' \quad \text{and} \quad R-SH$$

In another coal structure proposed by Wiser shown in Figure 2.2, the above groups are also present, but in addition Wiser includes:

$$R-S-S-R'$$

where R and R' are in general cyclic compounds.

In general, the fraction of the sulfur present as mercaptans, sulfides, or as a part of a heterocyclic ring structure is unknown. This is unfortunate, because the form of the sulfur in coal has a great bearing on the chemical treatment necessary to remove it. When coal is coked, Tiessen found that the coke product contained approximately 45% of the original organic sulfur. If coal is hydrodesulfurized at moderate temperatures and without catalysis (solvent refined coal process), around 40% of the organic sulfur remains in the coal. However, it is known that sulfur appearing in ring structures is more difficult to hydrogenate or to remove by decomposition than other forms of organic sulfur in coal.  Thus, the above observations suggest that the major portion of the sulfur appearing in the coke and in the coal product originates from ring structure, and that the easily removed portions are other forms of sulfur.

It follows that approximately 40 to 60% of the organic sulfur in American coals is present in ring structure.

## FIGURE 2.1:  MODEL FOR HIGH-VOLATILE COAL

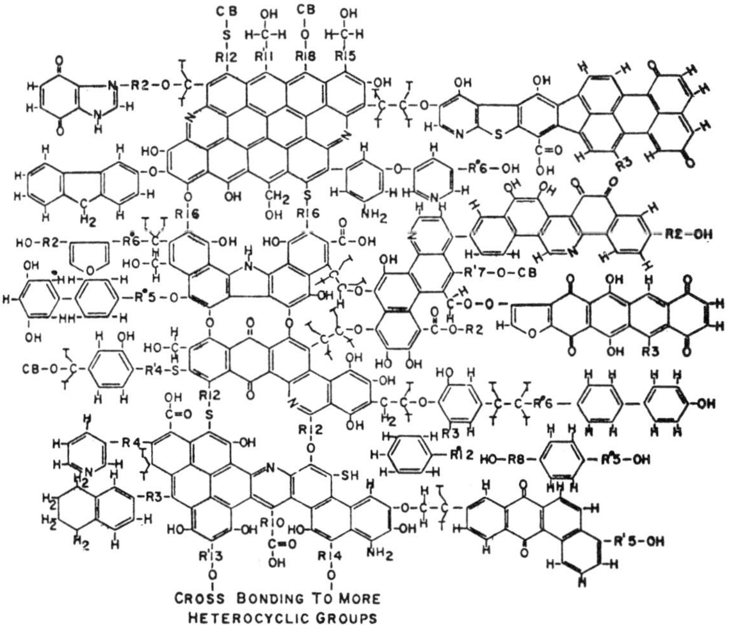

R°N = Alicyclic rings of N carbons.
RN = Alkyl side chain of N carbons.
R'N = Unsaturated alkyl side chain of N carbons.
CB = Cross bonding by O or S to new heterocyclic groups with side chains.
T = Tetrahedral 3 dimensional C—C bonds, C—O bonds and C—S bonds.

Source:  LBL-5216

FIGURE 2.2: A MODEL FOR BITUMINOUS COAL STRUCTURE*

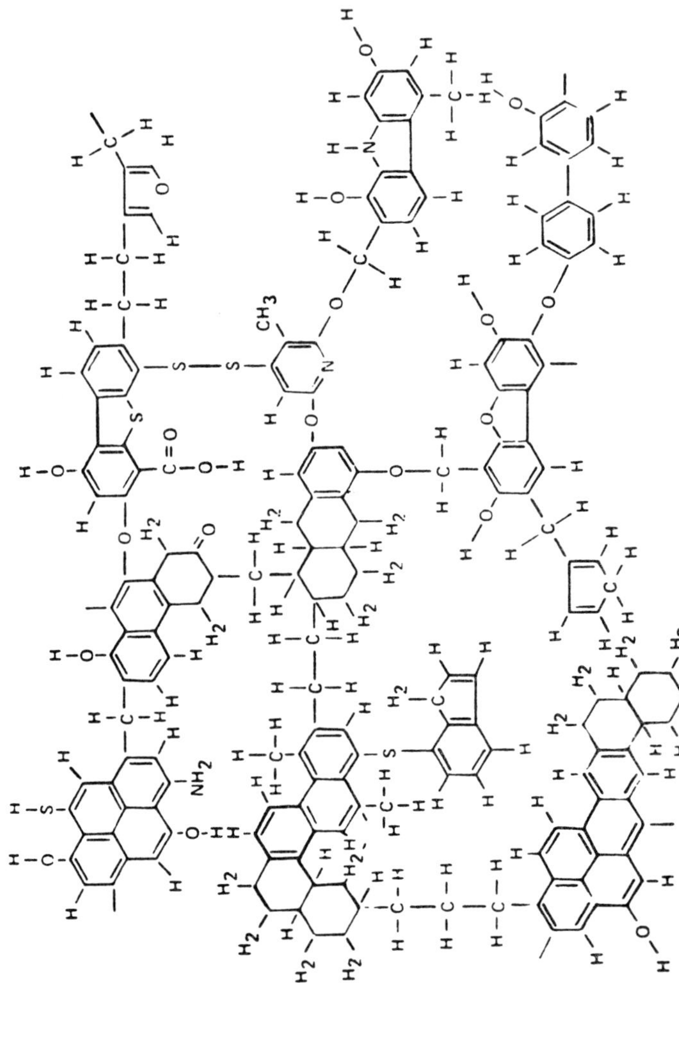

*According to W.H. Wiser, *Coal Catalysis*

Source: LBL-5216

**Chemistry of Organic Sulfur**

Little quantitative information is available on either the amount or kind of sulfur compounds present in coal.  Consequently, the chemistry of model sulfur-containing compounds of the type believed to exist in the coal structure was studied.

*Hydrogenation:*  Equilibrium constants for the hydrogenation of various type compounds to yield saturated hydrocarbons and hydrogen sulfide are summarized by Emmett and are plotted in Figure 2.3 for the following reactions:

1.   Ethanethiol

$$CH_3CH_2SH + H_2 \longrightarrow CH_3CH_3 + H_2S$$

2.   2-Propanethiol

$$(CH_3)_2CHSH + H_2 \longrightarrow CH_3CH_2CH_3 + H_2S$$

3.   3-Thiapentane

$$CH_3CH_2SCH_2CH_3 + 2H_2 \longrightarrow 2CH_3CH_3 + H_2S$$

4.   Thiacyclohexane

$$CH_2 \underset{CH_2-CH_2}{\overset{CH_2-CH_2}{\diagup\diagdown}} S + 2H_2 \longrightarrow n\text{-}C_5H_{12} + H_2S$$

5.   3,4-Dithiahexane

$$CH_3CH_2SSCH_2CH_3 + 3H_2 \longrightarrow 2CH_3CH_3 + 2H_2S$$

6.   Thiophene

$$\underset{S}{\underset{CH\;\;\;CH}{\overset{CH-CH}{\|\quad\|}}} + 4H_2 \longrightarrow n\text{-}C_4H_{10} + H_2S$$

The graph shows that hydrogenation of sulfur compounds is thermodynamically favorable, especially for disulfides, in the range of 700°K (427°C) and lower temperatures.

Some information on the relative hydrodesulfurization rates of various compounds is available from work done on petroleum fractions.

Byrns, Bradley, and Lee observed that thiophenes were the most difficult sulfur compounds to decompose during the hydrodesulfurization of feed stocks in the gasoline boiling range.

Casagrande, Meerbott, Sartor, and Trainer found that hydrodesulfurization of a pressure distillate tops (0.34% S) under pressure over a tungsten-nickel sulfide catalyst, gave a good removal of sulfides, mercaptans, and disulfides.  The residual sulfur compounds included thiophenic rings.  Figure 2.4 shows their results.

## FIGURE 2.3: EQUILIBRIUM CONSTANTS FOR THE HYDROGENATION OF SULFUR COMPOUNDS TO FORM $H_2S$

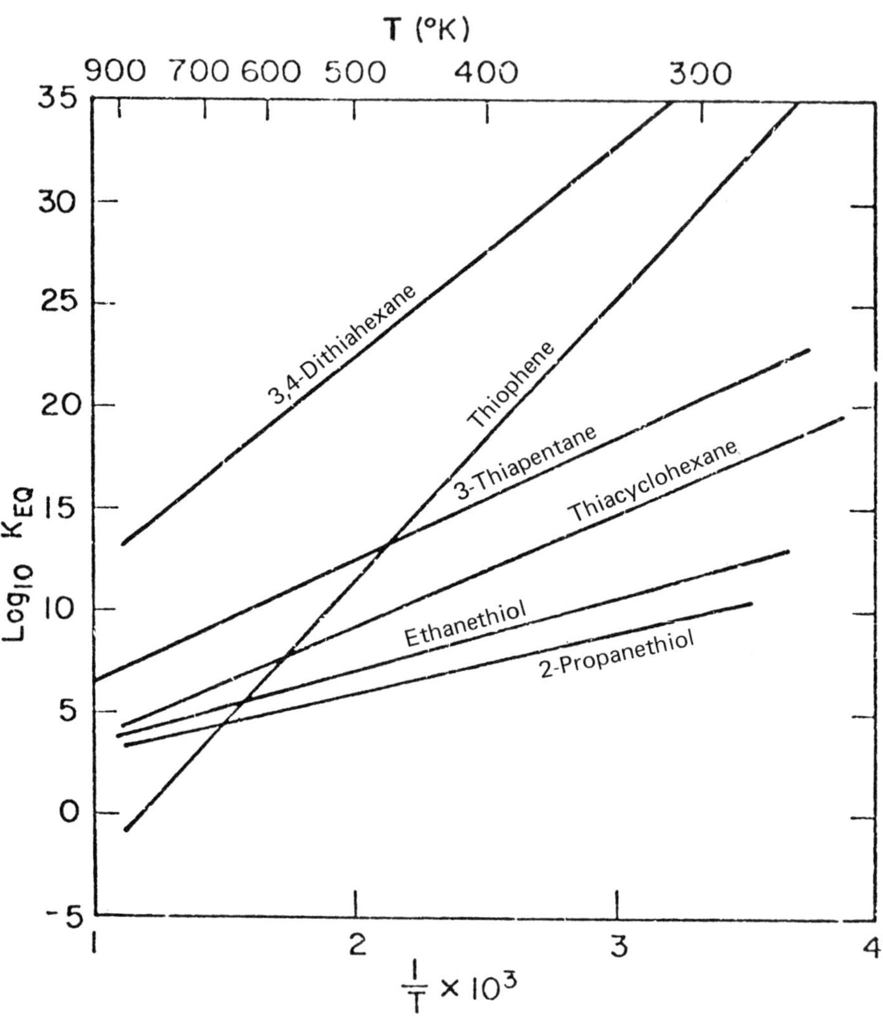


Source: LBL-5216

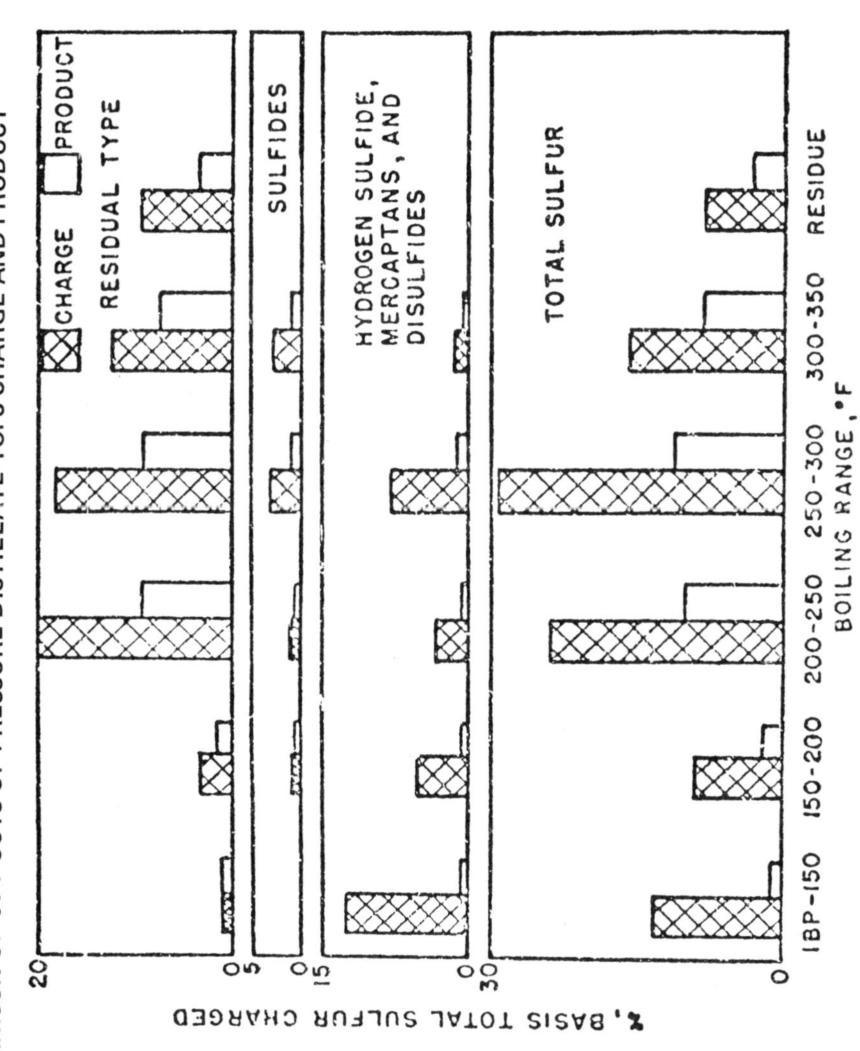


Source:  LBL-5216

*Decomposition Reactions:* The decomposition of the model compounds to yield unsaturated hydrocarbons and $H_2S$ can be represented by the following stoichiometric equations:

1.  Ethanethiol

$$CH_3CH_2SH \longrightarrow CH_2{=}CH_2 + H_2S$$

2.  2-Propanethiol

$$(CH_3)_2CHSH \longrightarrow CH_3CH{=}CH_2 + H_2S$$

3.  3-Thiapentane

$$CH_3CH_2SCH_2CH_3 \longrightarrow 2CH_2{=}CH_2 + H_2S$$

4.  Thiacyclohexane

$$(CH_2)_5S \longrightarrow CH_2{=}CHCH_2CH{=}CH_2 + H_2S$$

5.  3,4-Dithiahexane

$$CH_3CH_2SSCH_2CH_3 \longrightarrow CH_2{=}CH_2 + CH{\equiv}CH + 2H_2S$$

The equilibrium constants for the reactions shown are summarized by Emmett and plotted in Figure 2.5

**FIGURE 2.5:   EQUILIBRIUM CONSTANTS FOR THE DECOMPOSITION OF SULFUR COMPOUNDS TO FORM $H_2S$**

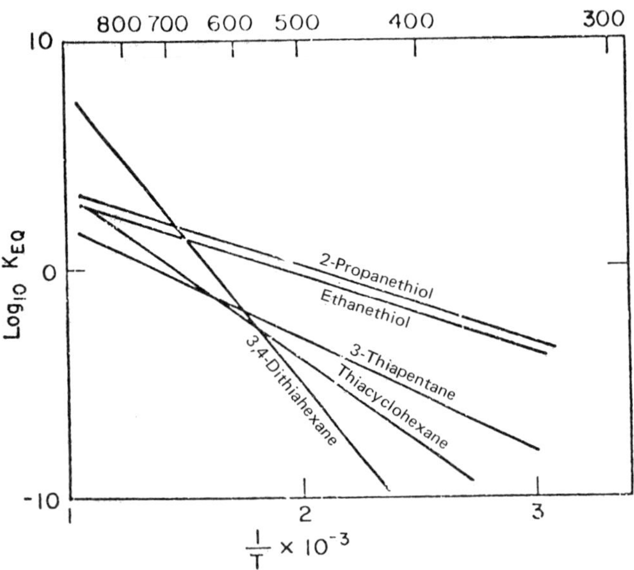


Source:   LBL-5216

From the graph it can be seen that the decomposition reactions are favored at high temperatures, but at lower temperatures reactions 1 and 2 are still possible. Some of the works that confirm these data are:

(1)    Woolhouse mentions that $H_2S$ was produced by the decomposition of organic sulfur upon carbonization of a coal.

(2)    Jolly studied the effect of rate of heating in the decomposition of the organic sulfur to give $H_2S$.

(3)    Powell concluded that one-quarter to one-third of the organic sulfur decomposes to give hydrogen sulfide at temperatures below 500°C. He also found that a small part of the sulfur was evolved as volatile organic sulfur compounds.

(4)    Scott concluded that the heterocyclic sulfur bond is very stable, and that if the sulfur is not a part of an aromatic system, the C–S bond scission is relatively easy at low-temperature carbonization levels.

*Oxidation:*  Mercaptans can be oxidized to sulfonic acids with various types of oxidizing agents, such as oxygen and hydrogen peroxide.

$$RSH + 1\tfrac{1}{2}O_2 \longrightarrow RSO_2OH$$

Alkylsulfides can react with $O_2$:

$$2RSSR + 2H_2O + 5O_2 \longrightarrow 4RSO_3H$$

Wallace and Sckriescheim studied the oxidation of thiols and disulfides with $O_2$ under basic conditions. They conclude that both aliphatic and aromatic compounds can be oxidized to their corresponding sulfonic acids.

*Sodium Hydroxide Treatment:*  Treatment with sodium hydroxide has been used for many years to remove mercaptans from gasoline cuts in the petroleum industry. The reaction that takes place is:

$$RSH + NaOH \rightleftharpoons RSNa + H_2O$$

The solubility of the mercaptans in sodium hydroxide decreases as the molecular weight increases. The removal of mercaptans reaches a maximum at 2 M concentration. Some solubilities of mercaptans in 1 N sodium hydroxide solution are:

| Mercaptan | Solubility, g/l |
|---|---|
| Methyl to butyl | very soluble |
| Amyl | 328.0 |
| Hexyl | 94.0 |
| Heptyl | 27.6 |
| Octyl | 8.0 |
| Nonyl | 2.3 |

When mercaptans are treated with steam in the presence of NaOH, they can decompose as follows:

$$RSH \longrightarrow H_2S + R'CH{=}CH_2$$

$$H_2S + 2NaOH \longrightarrow Na_2S + 2H_2O$$

The overall reaction is:

$$RSH + 2NaOH \longrightarrow Na_2S + 2H_2O + R'CH=CH_2$$

When mercaptans are treated with 3 N NaOH at 260°C, two reactions can take place:

$$RSNa + NaOH \longrightarrow ROH + Na_2S$$

or

$$2RSNa \longrightarrow R_2S + Na_2S$$

Sulfides can react in the following way:

$$RCH_2SCH_2R' + 2NaOH \longrightarrow R=CH_2 + R'=CH_2 + Na_2S + H_2O$$

Disulfides can decompose according to this stoichiometry:

$$2RSSR + 4OH^- \longrightarrow 3RS^- + RSO_2^- + 2H_2O$$

The $RS^-$ can decompose further to yield the $Na_2S$ and the unsaturated hydrocarbon.

### Chemistry of Inorganic Sulfur

*Hydrogenation:* Pyritic sulfur, $FeS_2$, can be treated with hydrogen to yield FeS and hydrogen sulfide according to the equation:

$$FeS_2 + H_2 \longrightarrow FeS + H_2S$$

at temperatures greater than 230°C.

The reaction:

$$FeS + H_2 \longrightarrow Fe + H_2S$$

is not very favorable because the equilibrium values:

$$K = \frac{P_{(H_2S)}}{P_{(H_2)}}$$

are 0.0008 and 0.003 at 723° and 994°C, respectively. It follows that the hydrogen sulfide partial pressure must be very small in order to make the latter reaction approach completion.

*Oxidation:* $FeS_2$ can be easily oxidized with O to $FeSO_4$, even at ordinary temperatures:

$$FeS_2 + 3O_2 \longrightarrow FeSO_4 + SO_2$$

$FeS_2$ in the presence of water and oxygen can react as follows:

$$FeS_2 + H_2O + 3\tfrac{1}{2}O_2 \longrightarrow FeSO_4 + H_2SO_4$$

$$2FeSO_4 + H_2SO_4 + \tfrac{1}{2}O_2 \longrightarrow Fe_2(SO_4)_3 + H_2O$$

Both sulfates are soluble in water. $FeS_2$ can react with ferric sulfate as follows:

$$FeS_2 + Fe_2(SO_4)_3 \longrightarrow 3FeSO_4 + 2S$$

$$7Fe_2(SO_4)_3 + 8H_2O + FeS_2 \longrightarrow 15FeSO_4 + 8H_2SO_4$$

With NaOH, the $FeS_2$ can react as follows:

$$8FeS_2 + 30NaOH \longrightarrow 4Fe_2O_3 + 14Na_2S + Na_2S_2O_3 + 15H_2O$$

## COAL DESULFURIZATION PROCESSES

[Editor's Note: Many of these processes are proprietary, and will be found discussed in greater detail in the Coal and Coke Desulfurizing Processes, Chapters 5, 6, and 7.]

Several approaches to achieve the desulfurization of coal have been proposed. Among these are hydrogenation, oxidation, and sodium hydroxide leaching. These will be considered separately.

### Hydrogenation

Hydrodesulfurization appears to be the most successful treatment for the removal of sulfur from coal. The principal processes using hydrogen are:

*Hydrocarbon Research, Inc. Process (H-Coal):* In this process, coal (-40 mesh) is slurried in a derived solvent oil, mixed with compressed hydrogen, and fed to an ebullated-bed reactor containing catalyst. The coal is catalytically hydrogenated as the dissolution occurs to form products which depend strongly on the reaction operating conditions and the particular type of coal used. Typical operating conditions are 850°F and 3,000 psig. Gases and lighter hydrocarbons are separated using a flash drum.

The slurry is separated, using hydroclones, into an overhead recycle solvent stream of low solid content and an underflow stream containing mineral and undissolved carbonaceous matter. The latter stream is further processed using a precoated rotary drum filter, and the filtrate is subsequently separated by distillation. A schematic drawing of the process is shown in Figure 2.6.

This process gives liquid and gaseous products, and reduces the sulfur content to as low as 0.1 to 0.2%. Mineral matter tends to accumulate on the catalyst, decreasing its activity so that catalyst regeneration is necessary.

*Synthoil Process:* In this process, a slurry of recycled solvent and coal is mixed, preheated, and fed to a fixed-bed reactor with turbulently flowing hydrogen. The reactor is packed with a commercial catalyst as used in hydrodesulfurization of petroleum derivatives (cobalt-molybdate on silica-activated alumina catalyst). Operating conditions in the reactor are usually 840°F and 2,000 to 4,000 psig. The effluent gases are separated from the extract oil in the high-pressure receivers, the hydrogen sulfide and ammonia are removed, and the hydrogen is recycled. The pressure in the extract oil is then reduced, followed by either centrifugation or filtration to remove mineral matter and undissolved organic matter. A flow sheet of the process is shown in Figure 2.7.

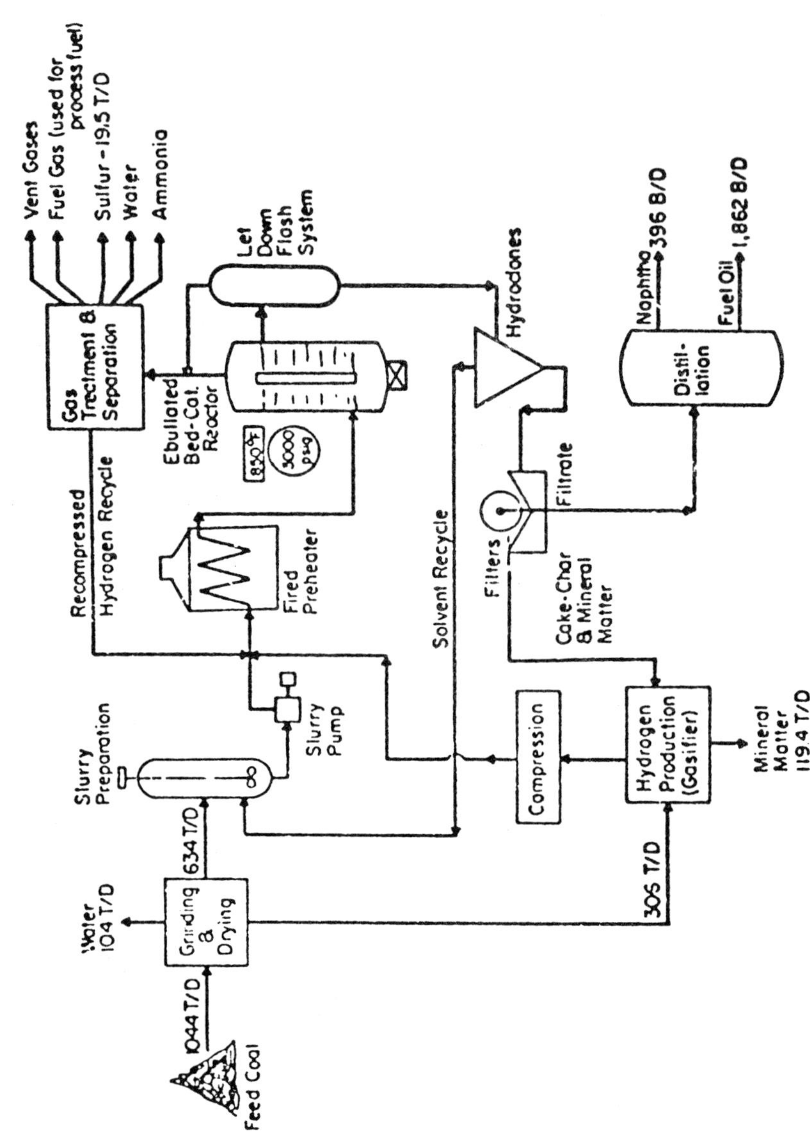


Source: LBL-5216

FIGURE 2.7: SCHEMATIC DRAWING OF THE US BUREAU OF MINES SYNTHOIL PROCESS UNIT

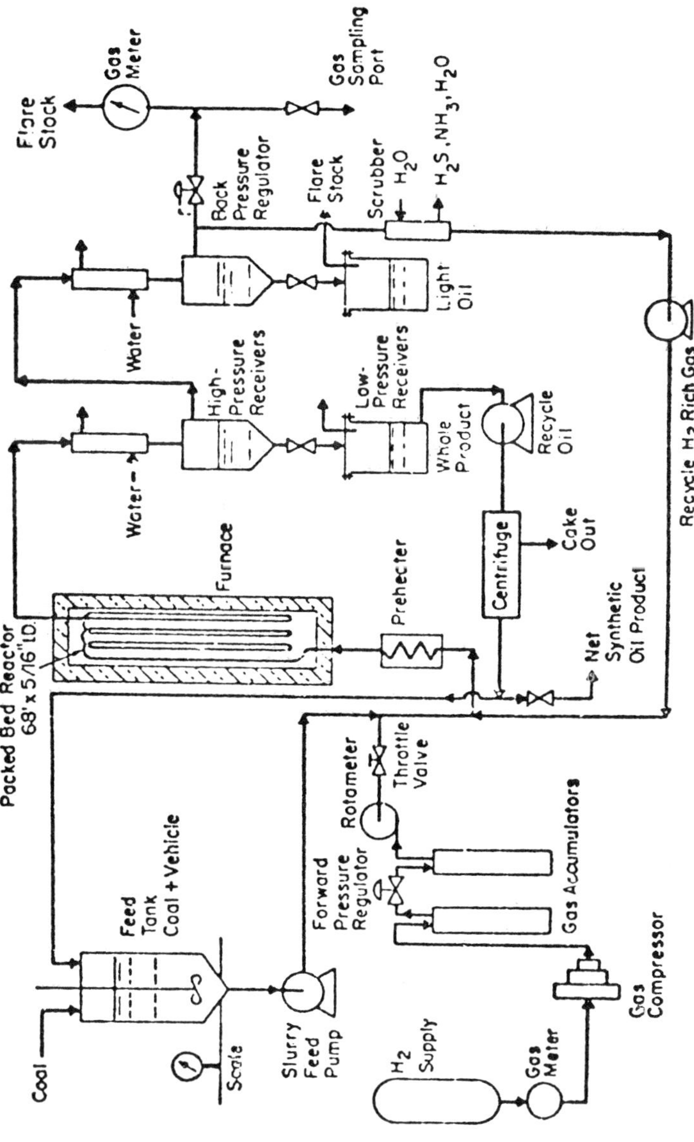


Source: LBL-5216

The process yields a low sulfur oil that flows freely at room temperature. The sulfur present in the oil product is less than 0.3% even for coals containing over 5% sulfur.

In this system, the hydrogenation is milder and there is less hydrogen consumption than the H-Coal process, but the amount of hydrogen recycle is large, leading to high compression costs. Further problems are that the bed plugs easily, and that the reactor must shut down to replace catalyst.

*Solvent Refined Coal (SRC):*  In this process, coal ground to -200 mesh is mixed with three parts of a recycle solvent having a boiling range of about 550° to 850°F. This slurry is mixed with hydrogen preheated and fed into the reactor or dissolver where typical operating conditions are 825° to 850°F and 1,000 to 2,000 psig. The reactor effluent is collected in a high-pressure receiver where the gas and liquid phases are separated. The liquid slurry is then filtered and the liquid is cooled until it solidifies. A flow sheet is shown in Figure 2.8.

**FIGURE 2.8:  SCHEMATIC FLOW SHEET OF THE SOUTHERN SERVICES, INC. PILOT PLANT**

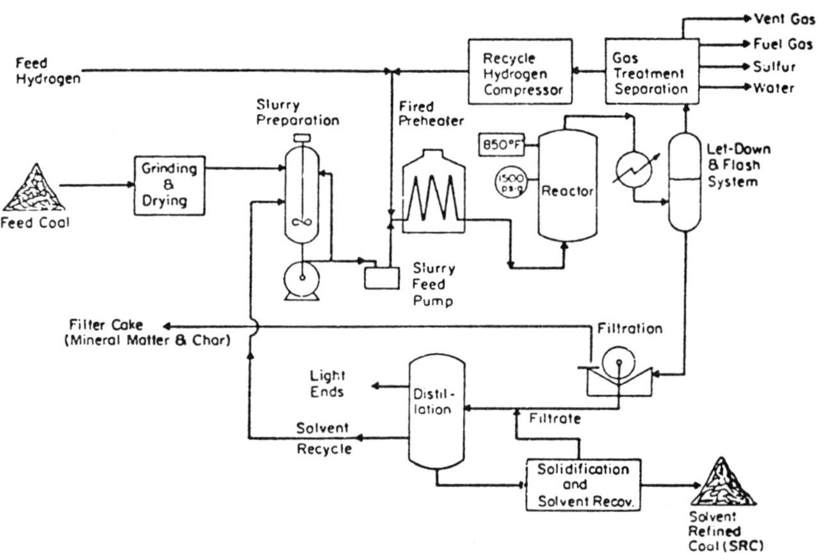


Source: LBL-5216

This process produces a low sulfur ash-free material, which has a very consistent heating value of 16,000 Btu/lb regardless of the coal feedstock processed. In this process, all the pyritic sulfur and up to 70% organic sulfur is removed. Most of the carbon in the coal is dissolved in the reactor, and hydrogen consumption is 1 to 2% of the weight of the coal processed. Advantages of this process are low hydrogen consumption, no catalyst, and operating costs lower than either H-Coal or the Synthoil process. The low cost is due principally to less severe operating conditions.

A major disadvantage of this process is that it gives a higher sulfur content in the coal product than the H-Coal and Synthoil process. Another problem area is that there appears to be no reliable and economical filter that can operate continuously at the conditions required. The filtration step is essential for the removal of the ash and the inorganic sulfur.

## Oxidation

Another method for removing sulfur is to oxidize and leach out the sulfur compounds.

*Dillon Process:*  Dillon slurried coal with water and air at 620 psia and 175°C for 10 minutes. Initially the coal contained 2.55% total sulfur, of which 0.94% was organic sulfur and, after treatment, it had only 0.48% total sulfur. All the pyritic and half of the organic sulfur was removed.

*Mukai Process:*  Mukai reported that practically 100% of the pyritic sulfur was removed from a bituminous coal by a room temperature treatment of the coal with a 3% solution of hydrogen peroxide, without affecting the properties of the coal's coking ability.

*Meyers Process:*  Meyers proposed a process in which coal is mixed with either ferric sulfate or ferric chloride solution and fed to a reactor where the pyritic sulfur is selectively oxidized to sulfate and free sulfur. The reaction temperature is approximately 100°C. The aqueous solution containing sulfate ion is separated by filtration from the coal and washed to remove the residual ferric salt. The elemental sulfur is removed by steam or by solvent extraction (toluene). The oxidizing agent can be regenerated and recycled. A flow sheet of the process is shown in Figure 2.9.

## FIGURE 2.9:  SCHEMATIC FLOW SHEET OF MEYERS PROCESS

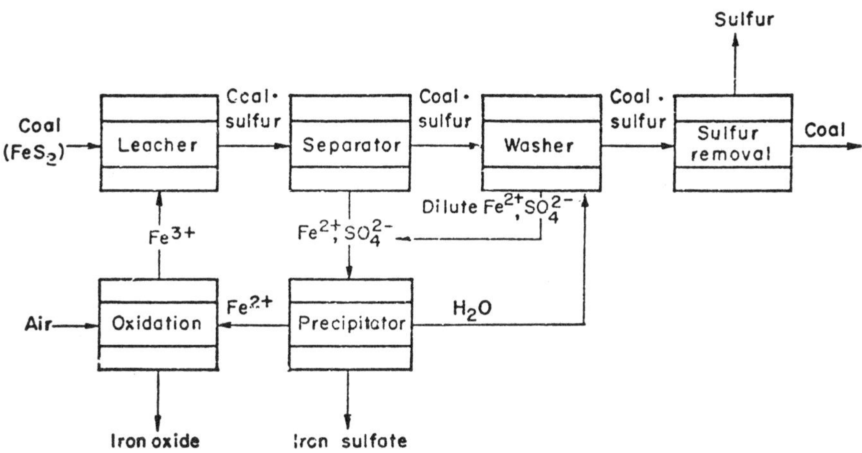


Source:  LBL-5216

Meyers claims that the process can remove practically 100% of the pyritic sulfur present in the coal without affecting the coal matrix; however, in order to achieve this removal, a multibatch leaching process is needed. Otherwise, a single batch on the order of 8 hr is needed to remove more than 80% of the pyritic sulfur, although higher temperatures can shorten the time.

*Ledgemont Process:*  Kennecott Copper Corporation proposed a different process for removal of the pyritic sulfur.  This process uses a coal slurry in a leaching reactor where, under the conditions of operation, the pyritic sulfur is oxidized only to soluble sulfates and elemental sulfur.  The slurry is then separated, and the coal fraction is washed with water.  The water collected from the washings and the leaching operations is neutralized with lime or limestone.

The insoluble iron compounds and gypsum are separated from the water and discarded, and the water is returned to the process.  Reaction conditions are 300 psig of $O_2$ and 130°C, the reaction time is 2 hr.  A flow sheet for the process is shown in Figure 2.10.  They claim that in some cases close to 100% of the pyritic sulfur is removed.

## FIGURE 2.10:  LEDGEMONT PROCESS FOR REMOVAL OF PYRITE SULFUR FROM COAL

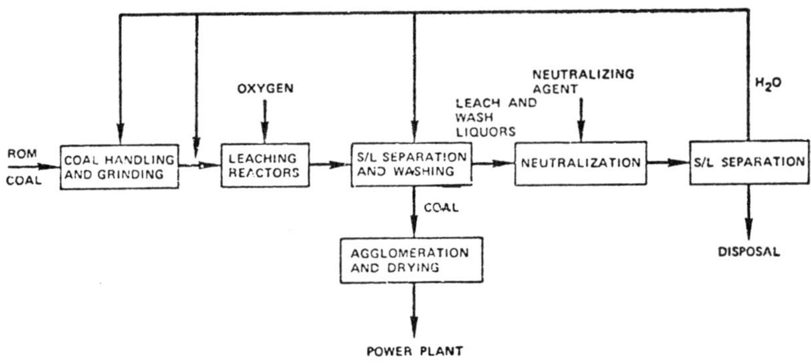


Source:  LBL-5216

## Sodium Hydroxide Treatment

Leo Kasehagen studied the action of aqueous alkali at different concentrations and temperatures on a bituminous coal.  He studied a range of temperatures from 250° to 400°C and concentrations ranging from 1 N to 100% NaOH in water.  His data are shown in Figure 2.11.

The data show that for the dilute concentrations an increase in temperature gives an increase in the amount of organic sulfur removed, but the use of higher concentrations seems to have adverse effects on the amount of organic sulfur that is removed.  Moreover, the higher-concentration behavior with respect to temperature is opposite to that of the dilute solutions.

## FIGURE 2.11:  EFFECT OF TEMPERATURE AND CONCENTRATION ON REMOVAL OF ORGANIC SULFUR

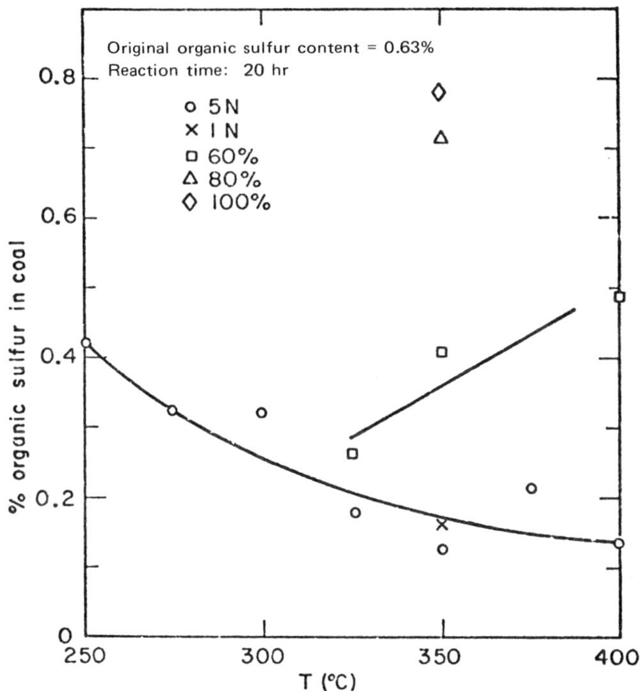


Source:  LBL-5216

*Battelle Process:*  Recently, Battelle proposed a hydrothermal coal-treatment process to remove sulfur from coal using sodium hydroxide.  In this process, coal is ground to permit 70% to pass through a 200 mesh screen.  The coal is mixed in a slurry tank with the leaching agent composed of 5 to 10% sodium hydroxide and $Ca(OH)_2$ in water, pumped into the reactor at 350 to 2,500 psig, and heated to 225° to 350°C.

These conditions were necessary to extract the sulfur and a portion of the ash from the coal.  The sulfur is mainly converted to sodium sulfide.  The desulfurized coal is separated from the coal slurry in a centrifugal filter and transported to a dryer.  A flow sheet of the process is shown in Figure 2.12.

The leaching agent used is regenerated with carbon dioxide:

$$Na_2S + CO_2 + H_2O \longrightarrow Na_2CO_3 + H_2S$$

The hydrogen sulfide is converted to elemental sulfur by the Clauss or Stretford process.  Regeneration of the sodium hydroxide is completed by reacting the sodium carbonate solution with lime, yielding sodium hydroxide and a precipitate of calcium carbonate:

$$Na_2CO_3 + CaO + H_2O \longrightarrow 2NaOH + CaCO_3$$

The calcium carbonate is calcined to lime and $CO_2$.

$$CaCO_3 \longrightarrow CaO + CO_2$$

The developers claim that the leaching removes almost all the pyritic sulfur and depending on the coal up to 70% of the organic sulfur, and that if the product from the alkaline desulfurization is washed with dilute acid the majority of the ash is extracted. The product from the hydrothermal treatment is a good stock for the commercial production of synthetic oils.

## FIGURE 2.12:  BATTELLE HYDROTHERMAL COAL PROCESS

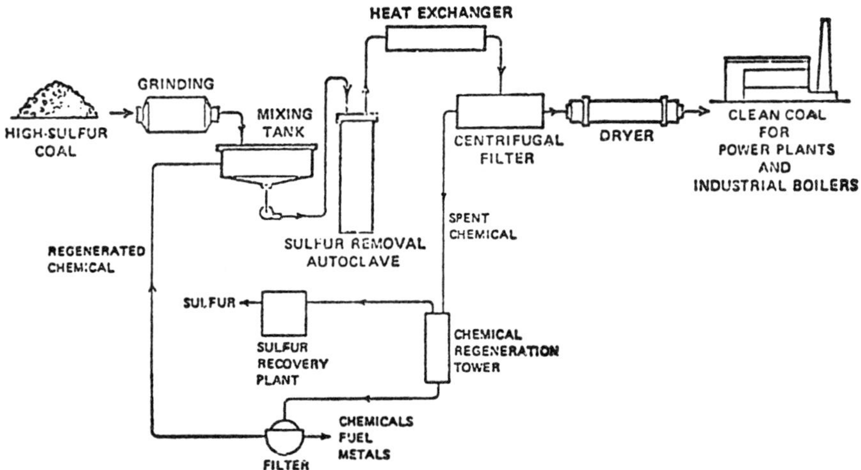


Source:  LBL-5216

## Other Proposals

Meyers tried to remove the organic sulfur from coal using different chemical leaching agents. He found that weak organic acids can remove part of the organic sulfur from coals. Among the weak organic acids investigated, it was found that the best is nitrobenzene, followed by cresol. Aqueous caustic removes small amounts of organic sulfur from some coals but not from others.

Hamersma and Meyers found that p-cresol was a more efficient organic sulfur extraction agent than nitrobenzene. Organic sulfur removal at 200°C ranged from 10 to 50%. One serious problem in this process is the high level of solvent retention in the processed coal.

## CONCLUSIONS

> [Editor's Note:   Mr. Mendizabal has studied the effectiveness
> of methods for removing sulfur from #6 Illinois Coal with
> aqueous solutions at low temperatures and pressures.   His
> conclusions are as follows.]

The sodium hydroxide process does not appear to be very promising for the removal of organic sulfur from Illinois #6 coal.  The highest organic sulfur removal was 36% (AFB), and was only reached at the highest temperature and pressure used.  The success of this process appears to depend largely on the form in which the organic sulfur is present in the coal.

Because approximately 80% of the pyritic sulfur was extracted at 200°C with the sodium hydroxide solution, this process appears to be effective for the removal of this form of sulfur.  At this temperature, the pressure is only 175 psig, a very positive feature.  Moreover, coal does not need to be finely ground to accomplish pyritic sulfur removal.  The time needed to achieve this removal at 200°C is estimated to be less than 30 minutes although this needs to be studied in more detail.

The large amount of coal dissolved in these experiments would have been much less than that obtained if the coal had not been weathered.  This process is not suitable to use in subbituminous coal because the lower rank coal has higher solubility in sodium hydroxide solutions.

Oxygen treatment could be promising if it were possible to increase the rate at which the pyritic sulfur oxidized with respect to the rate of oxidation of the coal matrix.  This process has the advantage of being carried out at low temperature and pressure.

Ferric sulfate treatment seems to be very promising for coals where only pyritic sulfur removal is necessary because almost 100% can be removed at low temperature and pressure.  None of the organic solvents at the conditions used seemed to be a good leaching agent for the sulfur.

In summary, when more than 40% of the organic sulfur must be removed, the high-temperature high-pressure hydrogen treatment appears to be necessary because these processes attack the very stable sulfur forms encountered in coal.  Therefore, it is recommended that hydrogenation processes be studied to find a catalyst that will work at low temperatures and pressures.

# SULFUR CONTENT AND
# SULFUR REDUCTION POTENTIAL
# OF U.S. COALS

The material in this section is taken from a report by
L. Hoffman, et al of the Mitre Corporation (PB-211 505).
Further information on this report will be found in the
bibliography on page 301. References for this section are
given on page 60.

## PRESENT AND POTENTIAL AVAILABILITY OF COAL

This study examines the present and potential availability of commercial steam
coal in terms of coal rank and sulfur content. Included are data on present and
possible coal production growth rates in terms of coal source region, coal rank,
sulfur content, present production levels, and considerations associated with new
mining activities.

A provisional estimate of United States coal resources is given in Table 3.1 below. The provided Btu values are based on weighted values of coal in place by
rank and differs from the national average of coal burned.

### TABLE 3.1: PROVISIONAL ESTIMATES OF COAL RESOURCES OF THE UNITED STATES (1)

| Category | Amount |
|---|---|
| Known recoverable reserves | $198 \times 10^9$ tons ($4.2 \times 10^{18}$ Btu) |
| Undiscovered recoverable reserves | $150 \times 10^9$ tons ($3.2 \times 10^{18}$ Btu) |
| Known marginal and submarginal resources | $1,380 \times 10^9$ tons ($29 \times 10^{18}$ Btu) |
| Undiscovered marginal and sub-marginal resources | $1,500 \times 10^9$ tons ($31 \times 10^{18}$ Btu) |
| Total | $3,228 \times 10^9$ tons ($67.4 \times 10^{18}$ Btu) |

The distribution of known recoverable reserves by major region or basin is given
in Table 3.2.

**TABLE 3.2: DISTRIBUTION OF KNOWN RECOVERABLE RESERVES (1)(2)**

| Region or Basin | Known Recoverable Reserves ($10^9$ tons) |
|---|---|
| Northern Appalachian | 15 |
| Southern Appalachian | 12 |
| Eastern Interior | 62 |
| Western Interior | 11 |
| Northern Rocky Mountains | 80 |
| Southern Rocky Mountains | 17 |
| West Coast | 1 |
| Total | 198 |

The estimated distribution of known recoverable reserves according to coal rank and sulfur content is as follows in Table 3.3.

**TABLE 3.3: ESTIMATED RANK AND SULFUR CONTENT* DISTRIBUTION OF KNOWN RESERVES (2)**

| | Part of Total | S 1% | S 1-2% | S 2% |
|---|---|---|---|---|
| Bituminous | 46% | 13.7% | 6.2% | 26.2% |
| Subbituminous | 24.7% | 24.6% | 0.1% | ----- |
| Lignite | 28.4% | 25.8% | 2.6% | ----- |
| Anthracite | 0.9% | 0.9% | ---- | ----- |

(* on a dry basis)

The total estimated recoverable reserves in million of tons (bituminous, subbituminous, and lignite) by sulfur contents up to a maximum 2% sulfur by weight are provided in Table 3.4.

**TABLE 3.4: LOW AND MEDIUM SULFUR RECOVERABLE RESERVES JANUARY 1965 (MILLION TONS) (2)**

| Sulfur Content % Wt. (dry) | 0.7 | 0.8-1.0 | 1.0-1.5 | 1.5-2.0 |
|---|---|---|---|---|
| Bituminous and Subbituminous | 28,650 | 25,445 | 7,530 | 6,805 |
| Lignite | 37,905 | 6,750 | 4,530 | 0 |

## Coal Reserves in the Rocky Mountain States

More than half of the potential coal reserves of the United States lie in a belt approximately 500 miles wide which straddles the Rocky Mountains and runs from Canada to Mexico. The quality of these coals varies both in carbon content and volatile matter.  The magnitude of these reserves is such that, with an assumed recovery rate of 50 percent and a national consumption rate of 600 million tons per year, they could satisfy the needs of the United States for hundreds of years.  Estimates of coal reserves of Rocky Mountain states (as of January 1965) by rank are given in Table 3.5.

### TABLE 3.5:  COAL RESERVES OF ROCKY MOUNTAIN STATES (JANUARY 1965) (MILLION TONS) (2)(3)

| | Bituminous | Sub-bituminous | Lignite | Anthracite | Total |
|---|---|---|---|---|---|
| Arizona | -- | 4,000 | -- | -- | 4,000 |
| Colorado | 62,416 | 18,300 | -- | 90 | 80,806 |
| Montana | 2,105 | 132,117 | 87,482 | -- | 221,704 |
| New Mexico | 10,686 | 50,735 | -- | 6 | 61,427 |
| Utah | 27,658 | 150 | -- | -- | 27,808 |
| Wyoming | 12,820 | 107,904 | -- | -- | 120,724 |
| North Dakota | -- | -- | 350,698 | -- | 350,698 |
| South Dakota | -- | -- | 2,031 | -- | 2,031 |
| TOTAL | 115,685 | 313,206 | 440,211 | 96 | 869,198 |

In order to estimate the western coal reserves in each sulfur category, it was necessary to build a data base of coal analyses.  Rocky Mountain region coal characteristic statistics obtained from the data base used in this report are given in Table 3.6.

### TABLE 3.6:  ROCKY MOUNTAIN REGION COAL CHARACTERISTICS

| | Average Value | Standard Deviation | Number of Analyses |
|---|---|---|---|
| I.  Bituminous coal | | | |
| Moisture content, wt % | 7.7 | 3.49 | 1,210 |
| Volatile matter, wt % dry | 40.3 | 4.24 | 949 |
| Fixed carbon, wt % dry | 51.4 | 3.78 | 949 |
| Ash, wt % dry | 8.0 | 4.33 | 1,210 |
| Total sulfur, wt % dry | 0.92 | 0.96 | 1,210 |
| Pyritic sulfur, wt % dry | 0.29 | 0.53 | 261 |
| Organic sulfur, wt % dry | 0.60 | 0.43 | 261 |
| Grindability index | 50 | 8.6 | 107 |
| Btu/lb, as received | 11,879 | 956.1 | 949 |
| II.  Subbituminous coal | | | |
| Moisture content, wt % | 19.6 | 5.79 | 527 |
| Volatile matter, wt % dry | 40.0 | 3.72 | 359 |
| Fixed carbon, wt % dry | 51.0 | 4.82 | 359 |
| Ash, wt % dry | 8.4 | 5.20 | 526 |
| Total sulfur, wt % dry | 0.80 | 0.71 | 527 |
| Pyritic sulfur, wt % dry | 0.22 | 0.30 | 168 |
| Organic sulfur, wt % dry | 0.53 | 0.53 | 168 |
| Grindability index | 51 | 9.2 | 30 |
| Btu/lb, as received | 9,235 | 939.2 | 359 |

(continued)

**TABLE 3.6:  (continued)**

|  | Average Value | Standard Deviation | Number of Analyses |
|---|---|---|---|
| III. Lignite |  |  |  |
| Moisture content, wt % | 36.8 | 4.72 | 221 |
| Volatile matter, wt % dry | 42.7 | 3.78 | 193 |
| Fixed carbon, wt % dry | 46.5 | 5.02 | 193 |
| Ash, wt % dry | 10.5 | 3.61 | 221 |
| Total sulfur, wt % dry | 0.96 | 0.58 | 221 |
| Pyritic sulfur, wt % dry | 0.15 | 0.26 | 28 |
| Organic sulfur, wt % dry | 0.58 | 0.35 | 28 |
| Grindability index | 48 | 10.2 | 43 |
| Btu/lb, as received | 6,763 | 522.9 | 193 |

Total sulfur content statistics by state from the data base are shown in Table 3.7.

**TABLE 3.7:  TOTAL SULFUR (DRY BASIS) STATISTICS FROM ANALYSES OF ROCKY MOUNTAIN REGION COAL SAMPLES**

| | Min. | Avg. | Max. | Std. Dev. | No. of Analyses |
|---|---|---|---|---|---|
| | | (% wt dry) | | | |
| **I.   BITUMINOUS COAL** | | | | | |
| Arizona | .40 | .79 | 1.30 | .31 | 13 |
| Colorado | .20 | .78 | 5.10 | .55 | 561 |
| Montana | .40 | 2.16 | 10.30 | 1.94 | 110 |
| New Mexico | .40 | .97 | 4.60 | .82 | 89 |
| Utah | .30 | .83 | 6.67 | 1.08 | 231 |
| Wyoming | .30 | .71 | 1.60 | .23 | 206 |
| | .20 | .92 | 10.30 | .96 | 1210 |
| **II.  SUBBITUMINOUS COAL** | | | | | |
| Arizona | .40 | 1.32 | 2.30 | .73 | 8 |
| Colorado | .24 | .49 | 2.42 | .26 | 171 |
| Montana | .30 | 1.19 | 6.10 | 1.02 | 136 |
| New Mexico | .46 | .78 | 1.54 | .35 | 14 |
| Utah | .51 | 1.90 | 6.74 | 1.74 | 12 |
| Wyoming | .18 | .70 | 2.70 | .32 | 185 |
| | .18 | .80 | 6.74 | .71 | 526 |
| **III. LIGNITE** | | | | | |
| Montana | .30 | .94 | 3.20 | .67 | 65 |
| North Dakota | .29 | .98 | 2.80 | .55 | 144 |
| | .29 | .96 | 3.20 | .58 | 209 |

Rocky Mountain strippable resources and reserves of coal and lignite as defined by the Bureau of Mines are shown in Table 3.8.

### TABLE 3.8: STRIPPABLE RESOURCES AND RESERVES (MILLIONS OF TONS) (4)

| Rank | Remaining Strippable Resources | Strippable Reserves | . . . . Strippable Reserves . . . . | | |
| --- | --- | --- | --- | --- | --- |
| | | | Low Sulfur | Medium Sulfur | High Sulfur |
| Bituminous coal | | | | | |
| Colorado | 870 | 500 | 476 | 24 | 0 |
| Utah | 252 | 150 | 6 | 136 | 8 |
| Subtotal | 1,122 | 650 | 482 | 160 | 8 |
| Subbituminous coal | | | | | |
| Arizona | 400 | 387 | 387 | 0 | 0 |
| Montana | 7,813 | 3,400 | 3,176 | 224 | 0 |
| New Mexico | 3,307 | 2,474 | 2,474 | 0 | 0 |
| Wyoming | 22,028 | 13,971 | 13,377 | 65 | 529 |
| Subtotal | 33,548 | 20,232 | 19,414 | 289 | 529 |
| Lignite | | | | | |
| Montana | 7,058 | 3,497 | 2,957 | 540 | 0 |
| North Dakota | 5,239 | 2,075 | 1,678 | 397 | 0 |
| South Dakota | 399 | 160 | 160 | 0 | 0 |
| Subtotal | 12,696 | 5,732 | 4,795 | 937 | 0 |

Notes: Bituminous coal resource and reserve not estimated for Montana, New Mexico and Wyoming.
Subbituminous coal resource and reserve not estimated for Colorado.
Some high sulfur content coal is not included in above.

Rocky Mountain strippable reserves of coal that are most attractive for servicing eastern markets are in the Montana-Wyoming area. These strippable coal beds, east of the Continental Divide, are located in the contiguous Fort Union region in Montana and the Powder River region in Wyoming. The less than 1% sulfur content (by dry weight) subbituminous strippable reserves (in million tons) in these two areas are "conservatively" estimated to be:

| | . . . . . . .Cumulative Values. . . . . . . | | |
| --- | --- | --- | --- |
| Sulfur content, % by weight | ≤0.5 | ≤0.7 | ≤1.0 |
| Strippable reserves, million tons | 4,000 | 10,700 | 14,700 |

The annual maximum production growth potential in million of tons per year based on a 20-year mining lifetime is:

| | | | |
| --- | --- | --- | --- |
| Sulfur content, % by weight | ≤0.5 | ≤0.7 | ≤1.0 |
| Production growth potential | 9.5 | 26 | 36 |

The annual production of Rocky Mountain coal is continually increasing. New mine capacities (in million tons per year) by state and year for the period of 1970 through 1974 are shown in Table 3.9.

## TABLE 3.9:  CAPACITY OF NEW MINES IN ROCKY MOUNTAIN REGION FOR PERIOD OF 1970 THROUGH 1974 (IN MILLION TONS PER YEAR)(3)

| State | 1970 | 1971 | 1972 | 1973 | 1974 | Total |
|---|---|---|---|---|---|---|
| Arizona | 5.2 | — | — | — | 7.8 | 13.0 |
| Montana | — | — | 5.0 | — | — | 5.0 |
| Utah | 0.8 | — | — | 2.3 | — | 3.1 |
| Wyoming | 5.3 | 5.0 | 2.5 | — | 5.0 | 17.8 |
| Total | 11.3 | 5.0 | 7.5 | 2.3 | 12.8 | 38.9 |

The major possibility for increasing low-sulfur production growth potential of eastern steam coal sources in the near term is the physical desulfurization or "deep-cleaning" of coal.  The estimated potential increase in low sulfur ($<$1%) bituminous steam coal production based on ⅜-inch topsize crushing and a 90% Btu yield is indicated in Table 3.10.

## TABLE 3.10:  ESTIMATE OF POTENTIAL INCREASE IN LOW SULFUR BITUMINOUS COAL VIA DEEP CLEANING

| | Million Tons |
|---|---|
| Northern Appalachian | 20 |
| Southern Appalachian | 7 |
| Eastern Interior (including Ohio) | 5[*] |

[*]This is an increase of 56% over the present annual mining level of $<$1% sulfur Eastern steam coal.

## PHYSICAL COAL DESULFURIZATION

The major possibility for increasing the low-sulfur production growth potential of Eastern steam coal sources in the near term is the physical desulfurization or "deep-cleaning" of coal.  The physical desulfurization process involves first the crushing of the coal to release the pyritic sulfur and then the physical separation of the pyrite from the coal.

Generally, the physical separation is based on the specific gravity difference between coal and the heavier pyrite particles.  The practice of physically cleaning coal is a well-developed technology which has existed for many years.  In the past the purpose has been to remove ash, thereby upgrading the quality of the coal.  The process can be optimized for pyrite removal (i.e., physical desulfurization) (5).  Since the design and construction of a coal cleaning plant is estimated to be approximately three years, the lead time for opening of a new mine will not materially be affected by this addition.

The desulfurization process (as in most beneficiation processes) results in some loss in total heat content of the processed coal.  Therefore, a balance must be achieved between the acceptable Btu loss and the increase in low sulfur coal. The potential increase in low sulfur coal was computed assuming the coal to be crushed to ⅜-inch top size and washed to an 85% minimum Btu yield.  The effect of desulfurization is to redistribute the coal to lower sulfur categories, giving a net increase in low sulfur coal. Redistribution was based on an examination of several hundred washability analyses performed on Eastern coals for the Office of Air Programs (6). Reserve and production figures were reassigned to the lower categories in the same proportion as were the analyses. The net effect of the desulfurization process on low sulfur coal availability is given in Table 3.11.

### TABLE 3.11: POTENTIAL INCREASE IN LOW SULFUR BITUMINOUS COAL (<1% SULFUR)

| REGION | INCREASE IN LOW SULFUR RESERVES (TONS) | INCREASE IN TOTAL RECOVERABLE LOW SULFUR RESERVES (TONS) | INCREASE IN LOW SULFUR MINEABLE RESERVES OWNED BY OPERATING COMPANIES (TONS) | INCREASE IN LOW SULFUR STEAM COAL PRODUCTION (TONS) |
|---|---|---|---|---|
| N. Appalachian | $34,290 \times 10^6$ | $4,460 \times 10^6$ | $1,480 \times 10^6$ | $20 \times 10^6$ |
| S. Appalachian | $7,100 \times 10^6$ | $780 \times 10^6$ | $320 \times 10^6$ | $7 \times 10^6$ |
| Eastern Interior (including Ohio) | $5,200 \times 10^6$ | $1,300 \times 10^6$ | $0 \times 10^6$* | $5 \times 10^6$ |
| Total | $46,590 \times 10^6$ | $6,540 \times 10^6$ | $1,800 \times 10^6$ | $32 \times 10^6$ |
| % Increase Over Present (<1% S) Raw Coal Level | 58% | 58% | 44% | 56% |

* Ohio only

The general results of the cleanability estimate for steam coal production (of the Eastern Interior, Northern and Southern Appalachian coal regions) are indicated in Figure 3.1.

### FIGURE 3.1: ESTIMATED PHYSICAL DESULFURIZATION POTENTIAL OF EASTERN BITUMINOUS STEAM COAL PRODUCTION

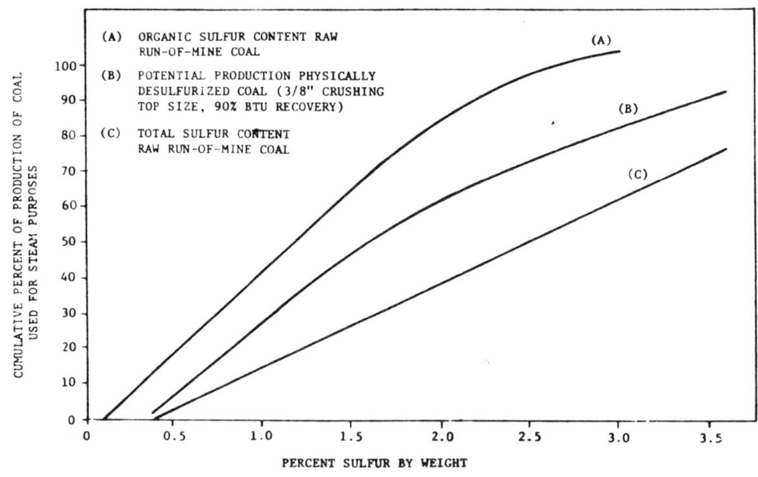


Source: PB-211 505

The figures for total recoverable reserves (Table 3.11) are based on the percent recoverable reserves given in reference (2). In all cases the washed coal has been converted to a Btu equivalent amount assuming the average Btu yield to be 90%.

The estimated maximum annual steam coal production growth of the eastern U.S. with physical desulfurization is given in Table 3.12.

### TABLE 3.12: ESTIMATED MAXIMUM ANNUAL STEAM COAL PRODUCTION GROWTH WITH PHYSICAL DESULFURIZATION

| Weight Percent Sulfur Content (Dry Basis) | Maximum Production with Cleaning | |
|---|---|---|
| | $10^6$ Tons | $10^{12}$ Btu |
| <0.7 | 12 | 314 |
| <1.0 | 30 | 780 |
| <1.5 | 75 | 2,000 |
| All Coal | 180 | 4,900 |

Note: In million tons per year

## CONCLUSIONS

The current annual steam coal production growth of eastern United States steam coals with sulfur contents of 0.7% or less by weight is very modest, in the neighborhood of $1.5 \times 10^6$ tons/yr/yr.

The estimated maximum possible 0.7% sulfur or less by weight production growth of eastern United States steam coals based on existing reserves and a 20-year production lifetime amounts to $9.4 \times 10^6$ tons/yr/yr. The growth period is of course limited by reserves in place.

The estimated maximum possible 0.7% sulfur or less by weight steam coal production growth of Western United States steam coals based on existing reserves and a 20-year production lifetime amounts to $60 \times 10^6$ tons/yr/yr.

The Fort Union region in Montana and the Powder River region in Wyoming have very large strippable subbituminous reserves (in excess of $10 \times 10^9$ tons) with less than 0.7% sulfur content by weight. Coal from this area is attractive for eastern markets due to the quantity of reserves, the geologic features of the area, and general availability of rail lines.

The major cost associated with providing eastern markets with Rocky Mountain coal is the transportation cost. Therefore, major cost reduction efforts should consider ways of reducing the effective transportation cost. In this regard, the total transportation process must be investigated. As a rule water-borne transportation costs less than rail transportation. This advantage is somewhat reduced by the fact that some waterways do not allow a year-round capability because of freeze-ups.

An effective way of reducing transportation cost is to increase the Btu/ton content of coal prior to shipping. In general, western subbituminous coals have substantial moisture contents and a reduction in coal moisture content prior to shipping would increase the as-received Btu content per ton and thereby reduce the effective transportation cost.

The major possibility for increasing the low-sulfur production growth potential of Eastern steam coal sources in the near term is the physical desulfurization or "deep-cleaning" of coal. Statistically the maximum effect of applying the desulfurization process would be an increase in low sulfur ($\leqslant 1\%$ by weight) steam coal production by 32 million tons annually, a 56% increase over present production at this sulfur level. The increase in production of coal with a maximum sulfur content of 0.7%, by weight, would be 17.6 million tons annually, a 107% increase over the present annual production.

Cost-benefit analyses of physically cleaned coal source/user combinations indicate that economic generalizations are impractical. Each case must be evaluated individually to determine the economics associated with physically cleaning coal. A previous study investigating 30 potential clean coal source/user combinations indicated a net cost change to the coal user ranging from a 1¢/MBtu saving to a 9¢/MBtu increase for coal physically cleaned to a 1% or less sulfur content (5).

## REFERENCES


(1)    Averitt, P., "Coal Resources of the United States, January 1, 1967," U.S. Geological Survey Bulletin 1275.

(2)    DeCarlo, J.A., Sheridan, E.T., and Murphy, Z.E., "Sulfur Content of United States Coals," Bureau of Mines Information Circular 8312, 1966.

(3)    Levine, H., "Projection of Usage of Western Coal for Synthetic Fuels, Power Generation and Coking Processes" presented at the Annual Meeting of Rock Mountain Coal Mining Institute, June 28, 1971.

(4)    Draft (1970) – "The Reserves of Bituminous Coal and Lignite for Strip Mining in the United States," by Staff, Bureau of Mines, United States Department of the Interior.

(5)    Yeager, K.E. and Hoffman, L., "The Physical Desulfurization of Coal – Major Considerations for $SO_x$ Emission Control," *Proceedings of the American Power Conference,* 1971.

(6)    Office of Air Programs Coal Washability Program, analyses performed by the Bureau of Mines, Pittsburgh Energy Research Center, Illinois State Geologic Survey, and Commercial Testing and Engineering Company, 1967–1970.


The material in this section is taken from a report by J.A. Cavallaro, et al of the U.S. Department of the Interior, Bureau of Mines (PB-252 965). Further information on this report will be found in the bibliography on page 301. References for this section are given on page 83.

## WASHABILITY STUDIES

The Air Quality Act of 1963 initiated a concerted effort by Federal and local governments to preserve the nation's air quality. Because air pollution from combustion of fossil fuels has long been recognized as a problem by the EPA and its earlier counterparts, major emphasis has been placed on the development of methods for controlling sulfur oxide emissions.

Coal-fired electric utility plants are the major source of sulfur oxide air pollution in the United States today. In 1974 the electric utilities burned 390 million tons of coal with an average sulfur content of 2.2%. The amount of coal consumed by electric utilities is anticipated to reach 500 million tons by 1980 and approximately a billion tons by the year 2000. It is therefore imperative that sulfur oxide emissions be controlled. Available methods for controlling sulfur oxide emissions from stationary combustion sources fall into the following major categories:

(1)  The physical removal (coal cleaning) of pyritic sulfur prior to combustion.

(2)  The removal of sulfur oxides from the combustion flue gas.

(3)  Conversion of coal to a clean fuel by such processes as gasification, liquefaction, and chemical extraction.

Of these three methods, physical removal of pyritic sulfur is the most developed method technologically and is potentially lowest in cost. However, the amount of sulfur reduction that might be obtained is limited, since only pyritic sulfur can be removed.

Sulfur in coal occurs in three forms: organic, sulfate, and pyritic. Organic sulfur, which is an integral part of the coal matrix and which generally cannot be removed by direct physical separation, comprises from 30 to 70% of the total sulfur of most coals.

The sulfate sulfur content is normally an oxidation product which is water-soluble and can be removed readily during coal cleaning. Sulfate sulfur contents are usually less than 0.05% in fresh coal.

Pyritic sulfur is the mineral pyrite which occurs in coal as discrete particles, although often of microscopic size. It is a heavy mineral which has a specific gravity of about 5.0, compared to coal which has a maximum specific gravity of only 1.7. The pyrite content of most coals can be reduced significantly by utilizing coal preparation methods of size reduction and gravimetric separation.

To determine the preparation method and the equipment needed to clean a coal, washability tests must be conducted. A washability analysis is an evaluation of those physical properties of a coal which determine its amenability to improvements in quality by cleaning. This includes stage crushing to release impurities and specific gravity fractionation to show the quality and quantity of the cleaned product. A washability study is made by testing the coal sample at preselected, carefully controlled specific gravities. This is termed "float-sink" analysis or specific gravity separation. Heavy organic liquids are commonly used to obtain the desired specific gravities of separation. Chemical analyses of the various specific gravity fractions of the coal are used to compile the washability data which indicate how well the coal can be prepared.

In 1965, the Federal Government funded a study by the Bureau of Mines to determine the forms of sulfur in the major sources of utility steam coals and the washabilities of these coals. To accelerate this program, in 1967, the Federal

Government funded a similar study of 2-year duration to be carried out by the
Commercial Testing and Engineering Co. The data in this report were compiled
from the work done by these two organizations. This report is an update of an
earlier Bureau of Mines publication which summarized the washability characteristics of 322 raw coal channel samples (1), and presents the washability characteristics of 455 raw coal channel samples.


## TEST PROCEDURE

### Collection of Samples

Face samples were collected from surface and deep mines which were producing
coal primarily for consumption by electric utilities. In general, an attempt was
made to sample the largest utility coal producing mines in the United States;
therefore, the 455 coal mine samples reported in this publication represent mines
which provide more than 70% of the annual utility coal production.

Face samples were collected according to the procedure recommended by Fieldner and Selvig (2) and Holmes (3), except that the dimensions of each sample
cut were expanded to permit 600 pounds of coal to be taken from the face.
Partings and impurities were not removed from the samples unless otherwise
noted. The face was cleared of loose coal or dirt for a width of approximately
5 feet. Loose pieces of roof were also taken down to prevent their falling into
the sample while it was being obtained. Within the cleaned-off area on the face,
the coal was cut from the roof to the floor in a channel 1-inch deep and about
3-feet wide to remove any altered or otherwise inferior coal. The floor was then
cleared and smoothed and a sampling cloth was spread prior to collecting the
sample.

The actual channel sample was cut perpendicular to the lay of the coal bed, approximately 10 inches deep and wide enough to provide a sample of 600 pounds.
For example, for a 4-foot-thick coal bed a channel 30.5 inches wide would be
collected. The exception to this rule would be when a strip mine sample is obtained where the overburden has been removed. In this case, the depth and
width of the channel would be equal. For example, for the 4-foot-thick bed
noted above, the channel would be 17.5 inches deep by 17.5 inches wide. The
collected sample includes all partings and other impurities occurring in the
channel.

### Sample Preparation

The 600-pound channel samples collected in the field are loaded into steel drums
and shipped to the coal preparation laboratory for processing. The sample preparation procedure is outlined in the flowsheet shown in Figure 3.2. Each sample
to be tested is air dried and then crushed to 1½-inch top size using a single roll
crusher. The sample is then coned, long-piled, and shoveled into four pans, according to ASTM specifications, and divided into two portions by combining opposite pans.

One of the 1½-inch by 0 portions is processed as is; the other portion is crushed
in a jaw mill to ⅜-inch top size. This ⅜-inch by 0 material is then riffled into
two portions; one is processed as is (⅜-inch by 0) and the other is crushed to
14 mesh top size in a hammer mill and processed.

A head sample is riffled from the 14 mesh by 0 portion for proximate analysis (moisture, ash, volatile matter, and fixed carbon) and for determination of calorific value, fusibility of ash, free-swelling index, Hardgrove grindability index, and sulfur forms and content (pyritic, organic, and total).  Since the –100 mesh material represents such a small percentage of the weight of the two coarser size fractions analyzed, it is removed prior to float-sink testing and is not presented in this report.

## FIGURE 3.2:  FLOW DIAGRAM SHOWING PREPARATION OF FACE SAMPLES

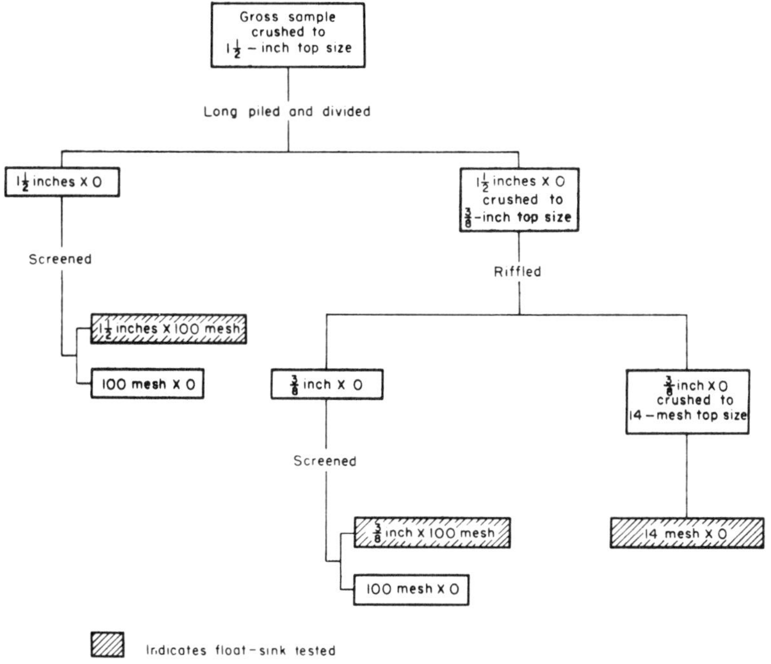


Source:  PB-252 965

The various sized fractions are then float-sink tested at 1.30, 1.40, and 1.60 specific gravities using Certigrav, a commercial organic liquid of standardized specific gravity; the solution tolerance is ±0.001 specific gravity unit and is monitored using a spindle hydrometer.  Those samples processed by Commercial Testing and Engineering Co. were further float-sink tested at 1.90 specific gravity.

For the two coarser sizes the separation is made in a screen-bottom container which is inserted in 10-gallon-capacity vessels containing the organic liquid.  The sample is placed in the 1.30 specific gravity bath, in small quantities to prevent entrapment, and is then stirred and allowed to separate.  The lighter specific gravity coal fraction is removed from the surface of the bath with a screen wire strainer; the heavier specific gravity material settles to the container bottom which is then raised above the liquid level to drain.  The container with the heavier specific gravity material is then placed in the 1.40 specific gravity solution and the process is repeated. This is continued until the sample is separated into the desired specific gravity fractions.

For the 14 mesh by 0 size fraction, the separation is made in glass separatory flasks joined by standard ground taper joints. After the sample separates, a stopper is passed through the float layer and inserted into the neck of the separatory funnel. Both products are filtered; the "floats" are dried and prepared for analysis, while the "sinks" are reintroduced into another separatory flask containing a heavier specific gravity liquid and the float-sink procedure is continued.

Upon completion of the float-sink testing, the specific gravity fractions of the three sized samples are analyzed for ash, pyritic sulfur, and total sulfur content. All chemical analyses are reported on a moisture-free basis unless otherwise noted. Raw coal moisture, as presented in the appendix tables [not included in this book], is the moisture contained in the sample after being air dried at the coal preparation laboratory. The air dry loss is not included in the moisture determination. It is felt that under normal conditions the moisture content as reported would closely simulate the moisture content of the coal burned at the power plant.

Specific gravity separations of fine coal are particularly difficult, especially with coals that are porous and contain high inherent moisture contents, because the heavy liquid used can penetrate the pores and increase the apparent specific gravity of the coal. This explains the unexpectedly low weight recoveries noted occasionally for the float 1.30 specific gravity fraction of the lower rank coal samples crushed to 14-mesh top size.

The float-sink data from the channel samples are not to be construed as representing the quality of the product loaded at the mine where the sample was taken, but rather as indicating the quality of the bed in that particular geographical location. Float-sink data are based upon theoretically perfect specific gravity separations that are approached but not equaled in commercial practice.


## EXPERIMENTAL RESULTS

Four hundred and fifty-five sets of individual washability data have been compiled to show what effect size reduction and specific gravity fractionation have on the liberation and subsequent removal of pyritic sulfur and other impurities. Because of the gross amount of data available, all the individual washability data plus the statistical summations of the data are presented in three appendixes.

### Washability Data

Appendix A of this report [not included in this book, for reasons of space] contains the complete tabulation of the 455 sets of washability data. Cumulative weight and Btu recovery, Btu per pound, ash, pyritic sulfur, total sulfur, and pounds $SO_2$ emission per million Btu levels are given showing theoretical recoveries and product quantities obtained at various specific gravities when the coal samples were crushed to 1½-inch, ⅜-inch, and 14-mesh top sizes. The Btu per pound values for the float 1.60 specific gravity products and the total or raw coal products were obtained by actual analysis; those of the float 1.30, 1.40, and 1.90 specific gravity products were obtained by interpolation from a plot of cumulative ash versus cumulative Btu per pound. The pounds $SO_2$ emission per million Btu were calculated using the corresponding Btu per pound (moisture-free basis) and total sulfur content (moisture-free basis) and assumes that all of the sulfur in the coal goes out of the stack as $SO_2$. Actual emissions may vary since as-fired coals will contain some moisture and all of the sulfur may not go out the stack as $SO_2$.

## Evaluation of Washability Data

Appendixes B and C [not included in this book] are statistical evaluations of the composited washability data presented in Appendix A. The appendixes are broken down by regions.

Data from the various regions are shown composited by coalbeds, giving the number of samples collected from a coalbed by states, and the total number of samples collected from the coalbed.

These data are a composite of all the samples collected for each individual coalbed, or a composite of all the samples collected for all the coalbeds of a region, showing the effect on ash, pyritic sulfur, and total sulfur contents when crushing the coal to three top sizes, 1½ inches, ⅜ inch and 14 mesh. Average values are given plus standard deviation (sigma) values. Average values are the arithmetic means of the data involved in computing any given average. Since the number of pieces of data involved in the computation of an average gives one measure of credence of the average, this number is shown in all output.

Sigma values are given to show the spread of the data about the average. This sigma is the standard deviation of a sample of numbers and is defined as the root mean square deviation. For a normal distribution, 68% of the cases should fall between the average $\overline{X} \pm s$; 95% of the cases between the average $\overline{X} \pm 2s$, and 99.7% of the cases between the average $\overline{X} \pm 3s$. Thus, it is desirable to have N, the number of samples, large; and s, sigma, as small as possible.

Appendix B is a composite, by specific gravity fractions, of all the washability samples of a particular coalbed in a state and a region showing the effects of crushing and gravimetric separation on these coals.

For example, 41 samples were collected from the Pittsburgh bed coal for this study; of these, 1 was from Maryland, 18 were from Ohio, 6 were from Pennsylvania, and 16 were from West Virginia. The data show in detail the average ash content of the Pittsburgh bed coal samples collected for this study to be 13.1%, ± sigma of 6.8%. Thus, in 68% of the samples tested, the raw coal ash content would range from 6.3 to 19.9%. The average pyritic sulfur content was 2.17%, ±1.0%. The average total sulfur content was 3.60%, ±1.3%.

At float 1.30 when the sample is crushed to 1½-inch top size, the average ash is 4.2%, ±0.7%; this represents a reduction in ash content of 63%, ±11%. Therefore, in 68% of the samples tested, the ash varied from 3.5 to 4.9%, representing a reduction in ash content from the raw coal varying from 52 to 76%. The corresponding pyritic sulfur content is 0.53%, ±0.2%, thus providing 71% reduction over the raw coal, ±23%. The total sulfur content at float 1.30 would be 2.06%, ±15%.

As the specific gravity of the product increases, so will the Btu recovery and ash content, but the Btu per pound will decrease. As a general rule, the pyritic sulfur, total sulfur, and pounds $SO_2$ emission per million Btu will also increase. However, there are coals which contain almost no pyrite in the associated rock, most of the pyrite occurring with the coal. Such coals will show lower sulfur contents in the higher specific gravity fractions.

As the sample is reduced in size, impurities are released which are removable by float-sink separation. For example, at float 1.30, the ash content is 4.2% at 1½-inch top size, 4.0% at ⅜-inch top size, and 3.5% at 14-mesh top size.

Appendix C shows the effect of crushing on liberation of impurities by showing the quality of the theoretical products, obtained from cumulative interpolated washability data at 50, 60, 70, 80, 90, and 100% Btu recovery levels. These theoretical products are represented by averages for each of the previously stated levels. The data are arranged the same as for Appendix B.

Generally, these data show that as the recoveries increased, the ash, pyritic sulfur, total sulfur, weight, and pounds of $SO_2$ emission per million Btu also increased. However, the Btu per pound decreased since the ash content increased. As the sample was crushed, more impurities were released and readily separated. That is, the ash, pyritic sulfur, total sulfur, weight recovery, and pounds of $SO_2$ emission per million Btu generally decreased while the Btu per pound increased when the sample was crushed to the finer top sizes and the higher specific gravity material was removed. Figure 3.3 is a nomograph showing the $SO_2$ emissions which will result from burning coals of various sulfur and Btu contents.

**FIGURE 3.3: NOMOGRAPH RELATING SULFUR CONTENT AND CALORIFIC VALUE IN COALS TO POUNDS SO$_2$ EMISSION**

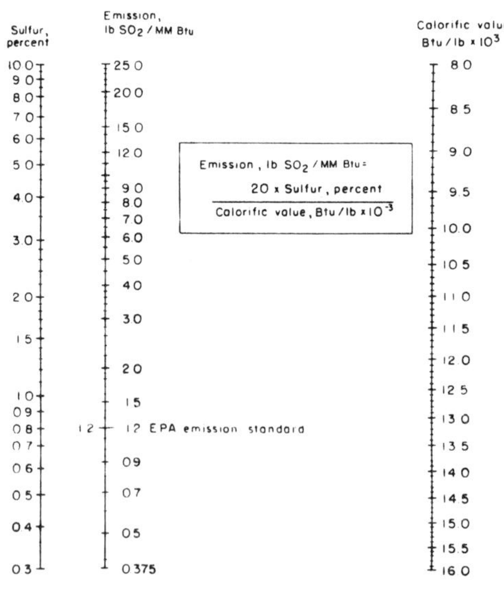


Source: PB-252 965

When using the nomograph or the formula one must be consistent and be sure that both the Btu per pound and sulfur values are on an as-received, moisture-free, or moisture-and-ash-free basis. For example, a coal containing 0.8% sulfur and 13,100 Btu per pound would meet the EPA $SO_2$ emission standard; however, a coal of the same sulfur content but containing only 10,500 Btu per pound would produce 1.5 pounds of $SO_2$/MM Btu and would not be in compliance.

## Computer Programs

For each mine sample, the following data is entered: state, county, town, mine, and bed codes; laboratory number and laboratory identification code; bed-bench and size fraction codes; year of sample collection, company, mine, and county name; the specific gravity fractions of separation and their associated levels with variables, weight and Btu recoveries, ash content, pyritic and total sulfur contents, and pounds of $SO_2$ emission per million Btu. All chemical analyses are reported on a moisture-free basis.

The initial computer program calculates Btu recovery, Btu per pound and pounds of $SO_2$ emission per million Btu for the various cumulated float products. A second computer program was required to make the statistical evaluations. The raw data used consisted of those items calculated in the initial program.

The second computer program calculates the averages and standard deviations of the percent ash, pyritic sulfur, and total sulfur for 100% recovery as determined from the original data. These averages are found for each bed and the required data accumulated for similar averages for the region. In like manner, the average and standard deviation values are calculated for each set of interpolated values at the various yield levels.

## DISCUSSION OF RESULTS

### Northern Appalachian Region Coals

Two hundred twenty-seven coalbed samples collected from Maryland [34], Ohio [58], Pennsylvania [103], and northern West Virginia [32] were evaluated. The average raw coal of the region contained 15.1% ash, 2.01% pyritic sulfur, 3.01% total sulfur, and 12,693 Btu per pound which would produce 4.8 pounds $SO_2$/MM Btu fired at the power plant.

Crushing the raw coal to 14-mesh top-size and removing the sink 1.60 specific gravity material would provide a product analyzing 7.6% ash, 0.66% pyritic sulfur, 1.66% total sulfur and 13,821 Btu per pound, a Btu recovery of 92.1%.

Thus, as shown in Figure 3.4, coal preparation techniques would provide an average pyritic sulfur reduction of 65%, a total sulfur reduction of 41%, and pounds $SO_2$ emission per million Btu reduction of 46%.

Figure 3.5 shows that only 4% of the raw coal samples as mined could meet the EPA emission standard of 1.2 pounds $SO_2$/MM Btu. Only 12% of the samples would comply at a Btu recovery of 90% when crushed to 1½-inch top size and 31% would comply at a Btu recovery of 50% when crushed to 14-mesh top size. However, if the emission standard were raised to 2.0 pounds of $SO_2$/MM Btu, then 15% of the raw coal samples would meet this standard with no preparation, approximately 35% would comply at a 90% Btu recovery when crushed to 1½-inch top size, and about 70% would comply at a 50% Btu recovery when crushed to 14-mesh top size.

**FIGURE 3.4:  THE EFFECT OF CRUSHING**

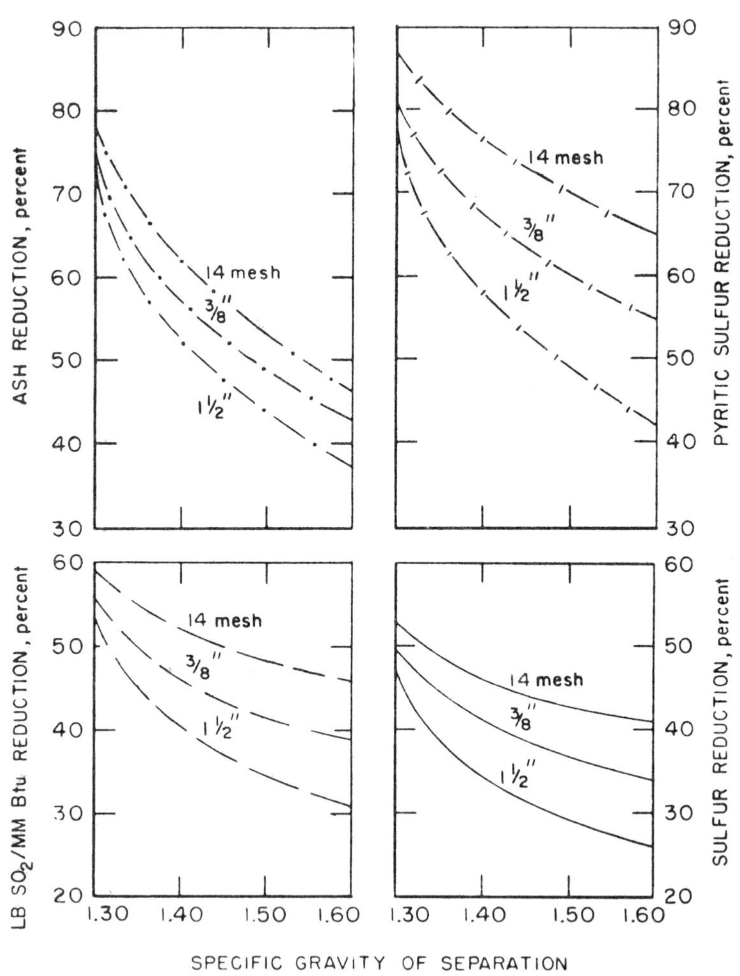


Source:  PB-252 965

## FIGURE 3.5: PERCENT OF COAL SAMPLES MEETING EPA STANDARD

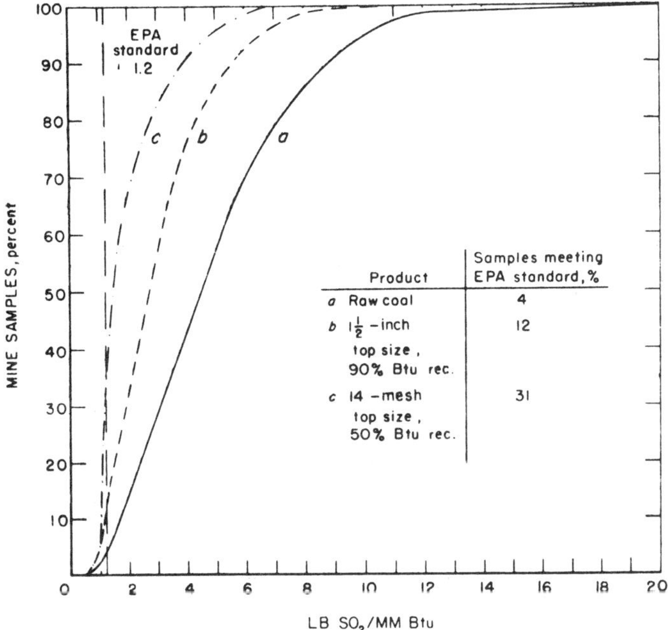


Key: Curve a - no preparation;
Curve b - comparison with those crushed to
1½-inch top-size at a Btu recovery of 90%;
Curve c - crushed to 14-mesh top size at a
Btu recovery of 50%.

Source:  PB-252 965

*Sewickley Coalbed:*  The 15 Sewickley bed coal samples were collected from four Northern Appalachian Region states:  Maryland [2], Ohio [10], Pennsylvania [2], and West Virginia [1].

In general this coal showed significant pyritic sulfur liberation; however, because the organic sulfur content averaged 1.69%, the total sulfur content remained high regardless of the projected gravimetric separation.  The average total sulfur content of this coalbed varied from as high as 4.45% in Ohio to as low as 1.32% in Maryland.

*Pittsburgh Coalbed:*  The 41 Pittsburgh bed coal samples were collected from four states:  Maryland [1], Ohio [18], Pennsylvania [6], and West Virginia [16]. This coal also showed significant pyritic sulfur liberation; however, because the organic sulfur content of this coal averaged 1.43%, the total sulfur content remained high regardless of the projected gravimetric separation.  The average total sulfur content of this coal varied from as high as 4.19% in Ohio to as low as only 0.85% in Maryland.

*Upper Freeport Coalbed:*  The 33 Upper Freeport bed coal samples were collected from three states:  Maryland [7], Pennsylvania [20], and West Virginia [6]. The average pyritic sulfur and total sulfur contents of this coalbed were 1.77 and 2.45%.  However, when this coal is crushed to 14-mesh top-size, it can be gravimetrically upgraded to provide an average product which would meet the EPA $SO_2$ emission standard at a Btu recovery of 70%.  The Upper Freeport coalbed is an excellent example of a high sulfur content coal which can be sufficiently desulfurized to meet the current EPA $SO_2$ emission standard by utilizing appropriate coal preparation techniques.

*Lower Freeport Coalbed:*  The 13 Lower Freeport bed coal samples were collected from three states:  Maryland [1], Ohio [2], and Pennsylvania [10].  The average pyritic and total sulfur content of this coalbed were 1.70 and 2.39%. However, when this coal is crushed to 14-mesh top-size, it can also be gravimetrically upgraded to provide an average product which would meet the EPA $SO_2$ emission standard at a Btu recovery of 80%.  The Lower Freeport coalbed is another excellent example of a high sulfur content coal which can be sufficiently desulfurized to meet the current EPA $SO_2$ emission standard by utilizing appropriate coal preparation techniques.

*Upper Kittanning Coalbed:*  The ten Upper Kittanning bed coal samples were collected from two states:  Maryland [2], and Pennsylvania [8].  The average pyritic and total sulfur contents of this coalbed were 1.62 and 2.21%.  This coalbed was very amenable to upgrading by gravimetric means, providing an average product which would meet the EPA $SO_2$ emission standard at a Btu recovery of 70% when the sample was crushed to a top size of only 1½ inches.

*Middle Kittanning Coalbed:*  The 27 Middle Kittanning bed coal samples were collected from three states:  Ohio [14], Pennsylvania [10], and West Virginia [3].  The average pyritic and total sulfur contents of this coalbed were 1.97 and 3.00%, respectively.  This coal showed significant pyritic sulfur liberation; however, because of the high average organic sulfur content of this coal, the average total sulfur content remained high regardless of the projected gravimetric separation.

*Lower Kittanning Coalbed:*  The 47 Lower Kittanning bed coal samples were collected from four states:  Maryland [3], Ohio [7], Pennsylvania [33], and West Virginia [4].  The average pyritic and total sulfur contents of this coalbed were 2.16 and 3.01%, respectively.  This coal also showed significant pyritic sulfur liberation; however, because of the high average organic sulfur content of this coal, the average total sulfur content remained high regardless of the projected gravimetric separation.  The average total sulfur content of this coalbed was much higher in Ohio, 4.14%, and Pennsylvania, 3.23%, than in Maryland, 0.92%, and West Virginia, 0.78%.

## Southern Appalachian Region Coals

The 35 Southern Appalachian Region coalbed samples were collected from three states:  Kentucky (east) [7], Tennessee [8], southern West Virginia [12], and Virginia [8].  The raw coal of the region averaged 11.0% ash, 0.37% pyritic sulfur, 1.03% total sulfur, and 13,314 Btu per pound; this would produce 1.6 pounds $SO_2$/MM Btu fired at the power plant.

Significant ash and pyritic sulfur reductions are attainable from these coals by gravimetric separations, Figure 3.6, and although the total sulfur reductions are not high, they are adequate to provide an average final product of very low total sulfur content.

## FIGURE 3.6: THE EFFECT OF CRUSHING

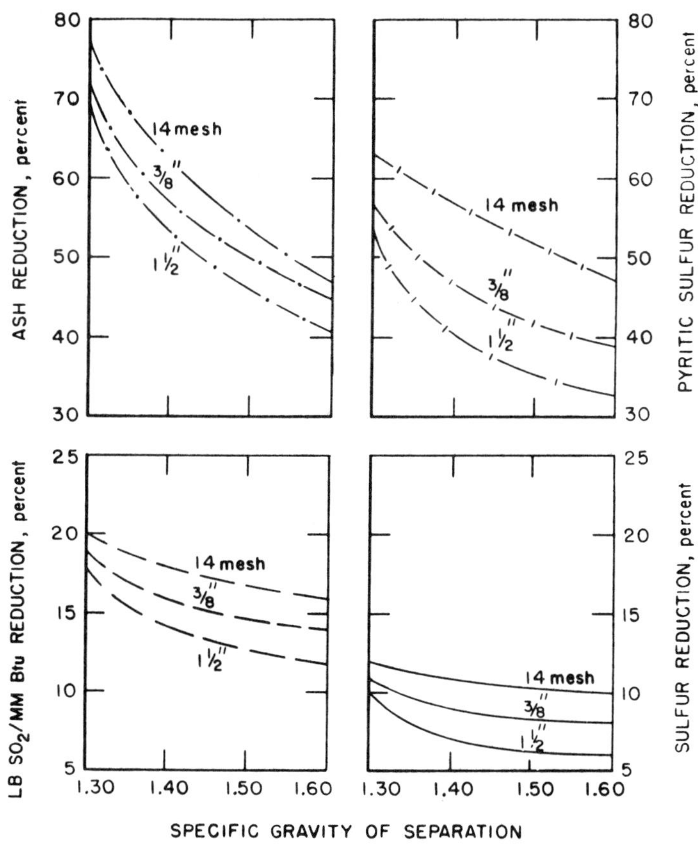



Source:  PB-252 965

Figure 3.7 shows that 35% of the raw coal samples as mined would meet the EPA standard of 1.2 pounds $SO_2$/MM Btu.  50% of the samples would comply at a Btu recovery of 90% when crushed to 1½-inch top size; 63% would comply at a Btu recovery of 50% when crushed to 14-mesh top size.  However, if the emission standard were raised to 2.0 pounds $SO_2$/MM Btu, then more than 80% of the raw coal samples would meet the standard with no preparation and about 90% would comply at a 90% Btu recovery when crushed to 1½-inch top size.

FIGURE 3.7:  PERCENT OF COAL SAMPLES MEETING EPA STANDARD

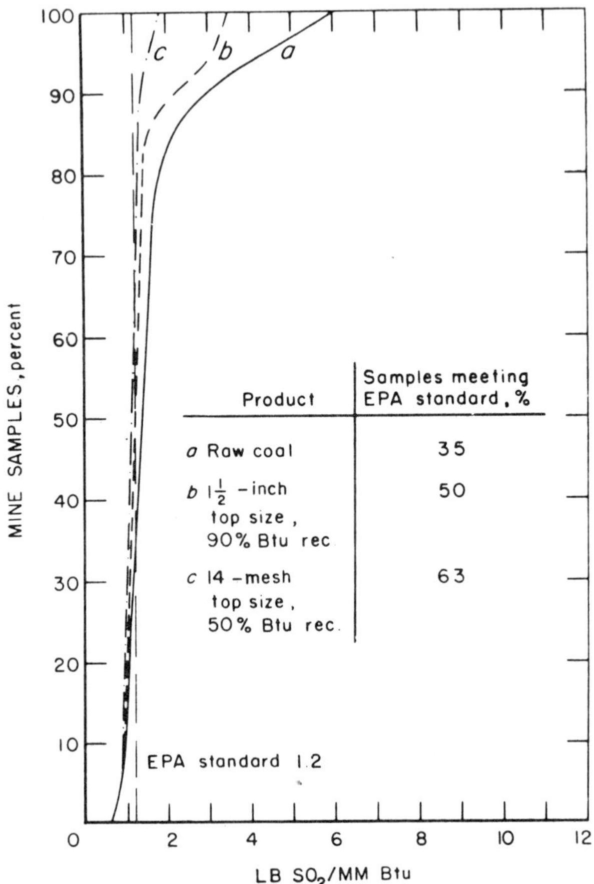



Key: Curve a - no preparation;
Curve b - comparison with those crushed
to 1½-inch top size at a Btu recovery
of 90%;
Curve c - crushed to 14-mesh top size at
a Btu recovery of 50%.

Source:  PB-252 965

## Alabama Region Coals

Only 10 raw coal samples were collected from Alabama.  The average raw coals of the region contained 9.5% ash, 0.69% pyritic sulfur, 1.33% total sulfur, and 13,696 Btu per pound, which would produce 2.0 pounds of $SO_2$/MM Btu fired at the power plant.  The coals of this region, although low in total sulfur content, do not respond to upgrading by crushing and gravimetric separation, Figure 3.8.

FIGURE 3.8:  THE EFFECT OF CRUSHING

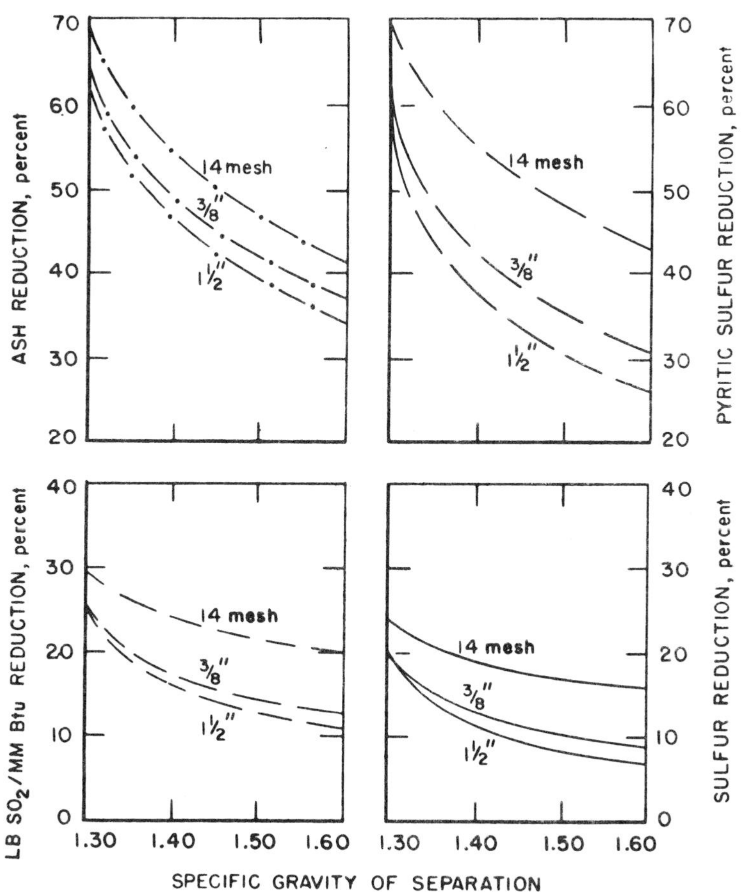



Source:  PB-252 965

Figure 3.9 shows that 30% of the raw coal samples as mined would meet the current EPA standard of 1.2 pounds $SO_2$/MM Btu.  Crushing to 14-mesh top size and accepting only a 50% Btu recovery would result in only 40% of the samples complying.  If the EPA $SO_2$ emission standard were raised to 2.0 pounds $SO_2$/MM Btu, then more than 60% of the raw coal samples would meet the standard with no preparation and all the samples would comply at a 50% Btu recovery when crushed to 14-mesh top size.

## FIGURE 3.9:  PERCENT OF COAL SAMPLES MEETING EPA STANDARD

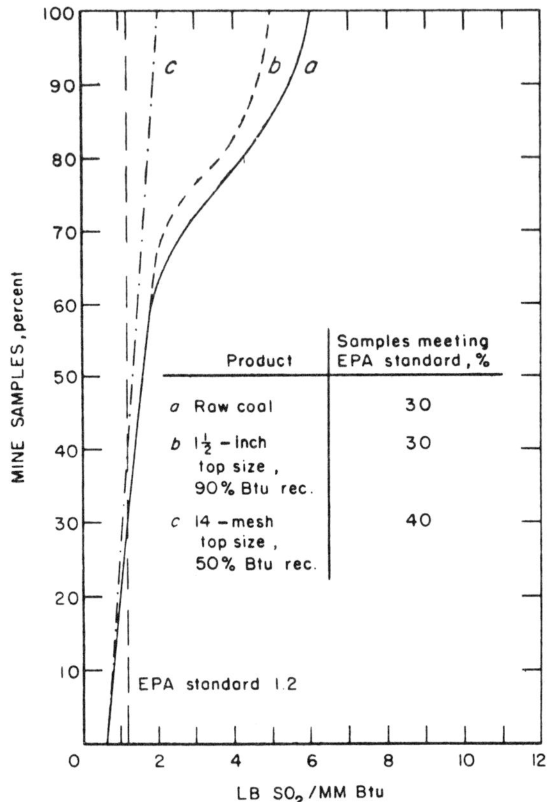


Key: Curve a - no preparation;
Curve b - comparison with those crushed
to 1½-inch top size at a Btu recovery
of 90%;
Curve c - crushed to 14-mesh top size at
a Btu recovery of 50%.

Source:  PB-252 965

## Eastern Midwest Region Coals

The 95 Eastern Midwest Region samples were collected from three states:  Illinois [40], Indiana [20], and Kentucky (west) [35].  The average raw coal from the region contained 14.2% ash, 2.29% pyritic sulfur, 3.92% total sulfur, and 12,189 Btu per pound, which would produce 6.5 pounds $SO_2$/MM Btu fired at the power plant.  The average ash and pyritic sulfur contents of these coals were similar to those of the Appalachian Region coals; however, these coals contained about 0.63% more organic sulfur which makes the problem of producing a low sulfur content product much more formidable.

Figure 3.10 shows that crushing to 14-mesh top size and removing the sink 1.60 specific gravity material would provide a 45% reduction in ash, a 57% reduction in pyritic sulfur, a 32% reduction in total sulfur, and a 37% reduction in the pounds of $SO_2$ emission per million Btu.

## FIGURE 3.10:  THE EFFECT OF CRUSHING

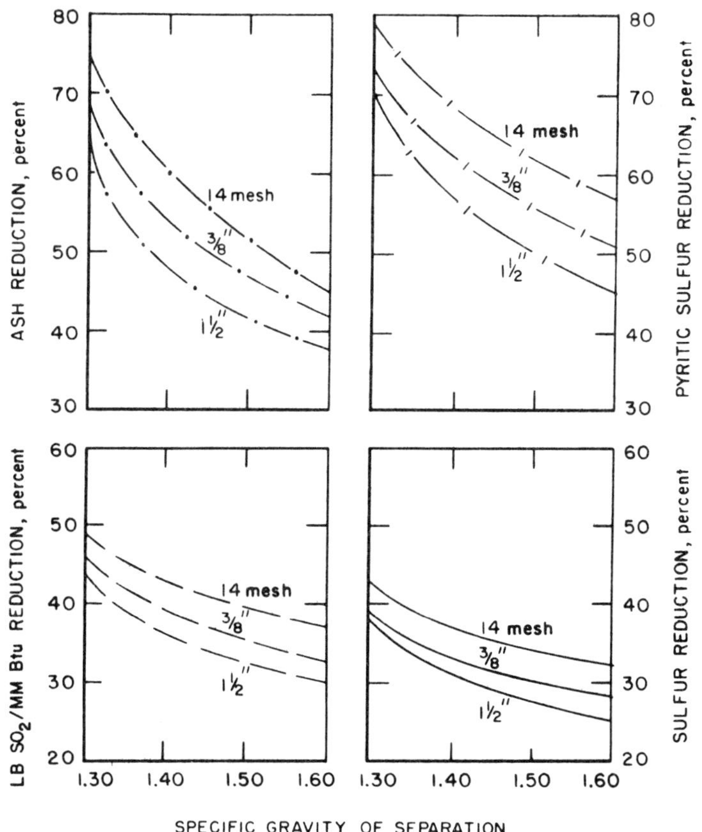



Source:  PB-252 965

The data in Figure 3.11 show that only 1% of the raw coal samples as mined would meet the current EPA $SO_2$ emission standard.  Only 2% of the samples could comply at a Btu recovery of 90% when crushed to 1½-inch top-size, whereas 4% would comply at a Btu recovery of 50% when crushed to 14-mesh top size.  If the emission standard were raised to 2.0 pounds $SO_2$/MM Btu, only 2% of the raw coals would meet the standard with no preparation, and only 12% would meet the standard at a Btu recovery of 50% when crushed to 14-mesh top size.

## FIGURE 3.11:  PERCENT OF COAL SAMPLES MEETING EPA STANDARD

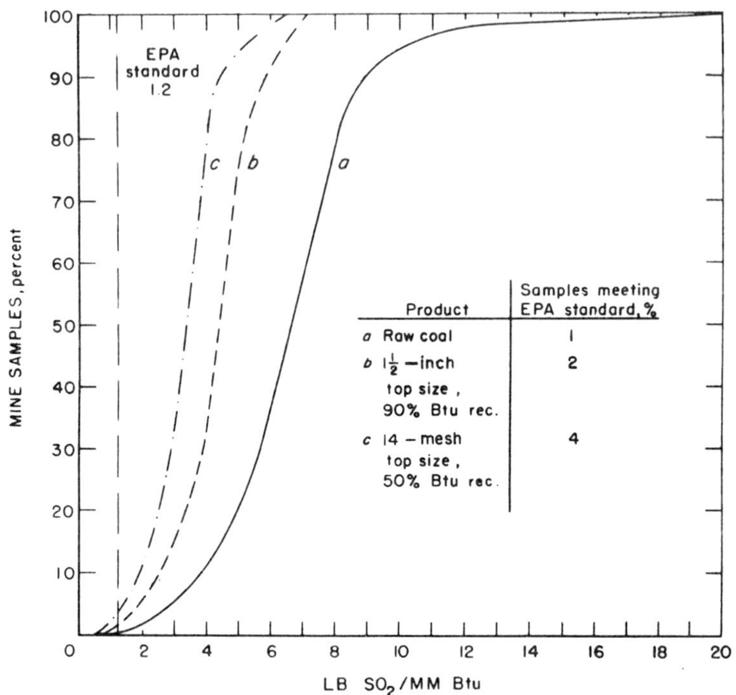


Key: Curve a - no preparation;
     Curve b - comparison with those crushed to 1½-
     inch top-size at a Btu recovery of 90%;
     Curve c - crushed to 14-mesh top-size at a Btu
     recovery of 50%.

Source:  PB-252 965

Although significant numbers of samples were collected from certain coalbeds in the Eastern Midwest Region, individual discussions are not included because of the general similarity in physical characteristics of these coals.

### Western Midwest Region Coals

The 44 Western Midwest Region coalbed samples were collected from five states: Arkansas [3], Kansas [8], Iowa [17], Missouri [9], and Oklahoma [7]. The average raw coal of the region contained 16.2% ash, 3.58% pyritic sulfur, 5.25% total sulfur, and 12,072 Btu per pound, which would produce 9.0 pounds $SO_2$/MM Btu fired at the power plant.

Aside from two of the coals from Oklahoma, one of which could meet the EPA $SO_2$ emission standard as mined and the other which could be upgraded to meet the EPA $SO_2$ emission standard, these coals were generally of very high total sulfur content with high organic sulfur content.

Crushing of these coals to 14-mesh top size and removing the sink 1.60 specific gravity material would provide an ash reduction of 50%, a pyritic sulfur reduction of 55%, a total sulfur reduction of 33%, and a pounds $SO_2$/MM Btu reduction of 40% (Figure 3.12).

## FIGURE 3.12:  THE EFFECT OF CRUSHING

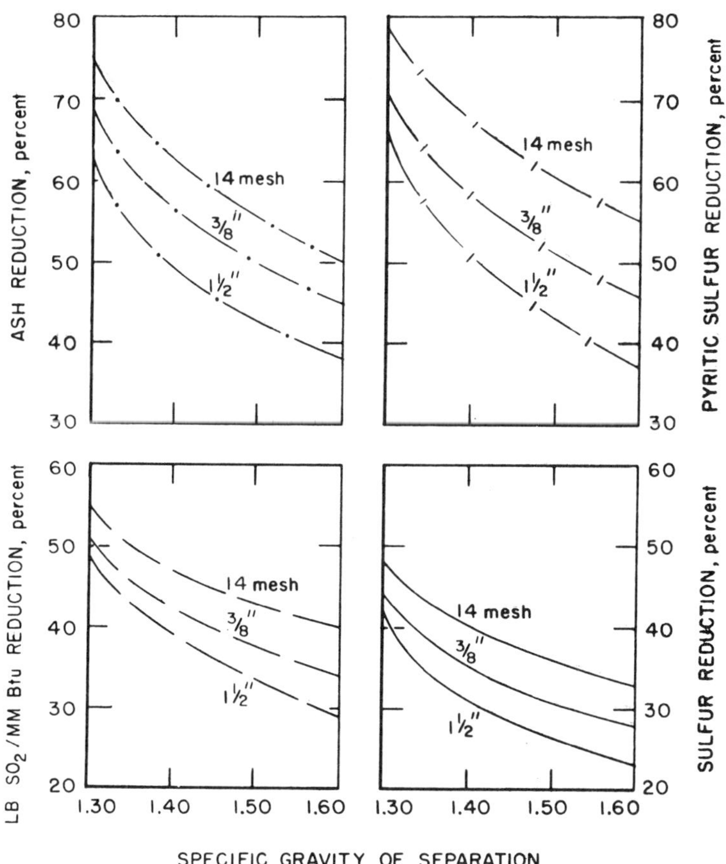



Source:  PB-252 965

Figure 3.13 shows that less than 3% of the raw coal samples as mined could meet the current EPA $SO_2$ emission standard.  Less than 6% of the samples would comply at a Btu recovery of 50% when crushed to 14-mesh top size. Even if the emission standard were raised to 2.0 pounds $SO_2$/MM Btu, only 5% of the raw coal samples would meet the standard with no preparation, whereas approximately 8% would meet the standard at a Btu recovery of 50% when crushed to 14-mesh top size.

## FIGURE 3.13: PERCENT OF COAL SAMPLES MEETING EPA STANDARD

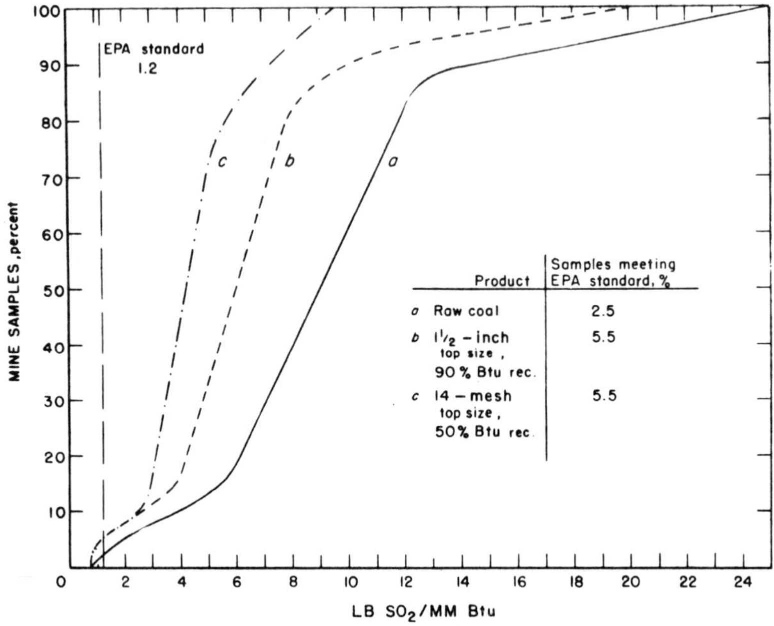



Key: Curve a - no preparation;
   Curve b - comparison with those crushed to 1½-
   inch top size at a Btu recovery of 90%;
   Curve c - crushed to 14-mesh top size at a Btu
   recovery of 50%.

Source: PB-252 965

### Western Region Coals

The 44 Western Region coalbed samples were collected from seven states: Arizona [6], Colorado [11], Montana [5], New Mexico [9], North Dakota [1], Utah [8], and Wyoming [4]. The average raw coal of the region contained 8.9% ash, 0.23% pyritic sulfur, 0.68% total sulfur, and 12,437 Btu per pound, which would produce 1.1 pounds $SO_2$/MM Btu fired at the power plant, thus meeting the current EPA $SO_2$ emission standard.

Figure 3.14 shows that crushing these coals below 1½-inch top size would be of little benefit in reduction of total sulfur content because these raw coals generally contain less than 1% total sulfur. However, crushing to 14-mesh top size and removing the sink 1.60 specific gravity material would provide ash and pyritic sulfur reductions of 33 and 45% respectively.

Figure 3.15 shows that 70% of the raw coal samples as mined would meet the current EPA $SO_2$ emission standard. 94% would comply at a Btu recovery of 90% when crushed to 1½-inch top size, and 98% would comply at a Btu recovery of 50% when crushed to 14-mesh top size.

The coals characteristically are of low sulfur content, often well below 1% total sulfur content.  These coals range in rank from lignite to bituminous, being predominantly subbituminous coals.

However, the inherent moisture content is much higher and the heating value much lower than those of Eastern bituminous coals; therefore, considering also the high transportation costs, extensive use of these coals in eastern utility plants is doubtful.

## FIGURE 3.14:  THE EFFECT OF CRUSHING

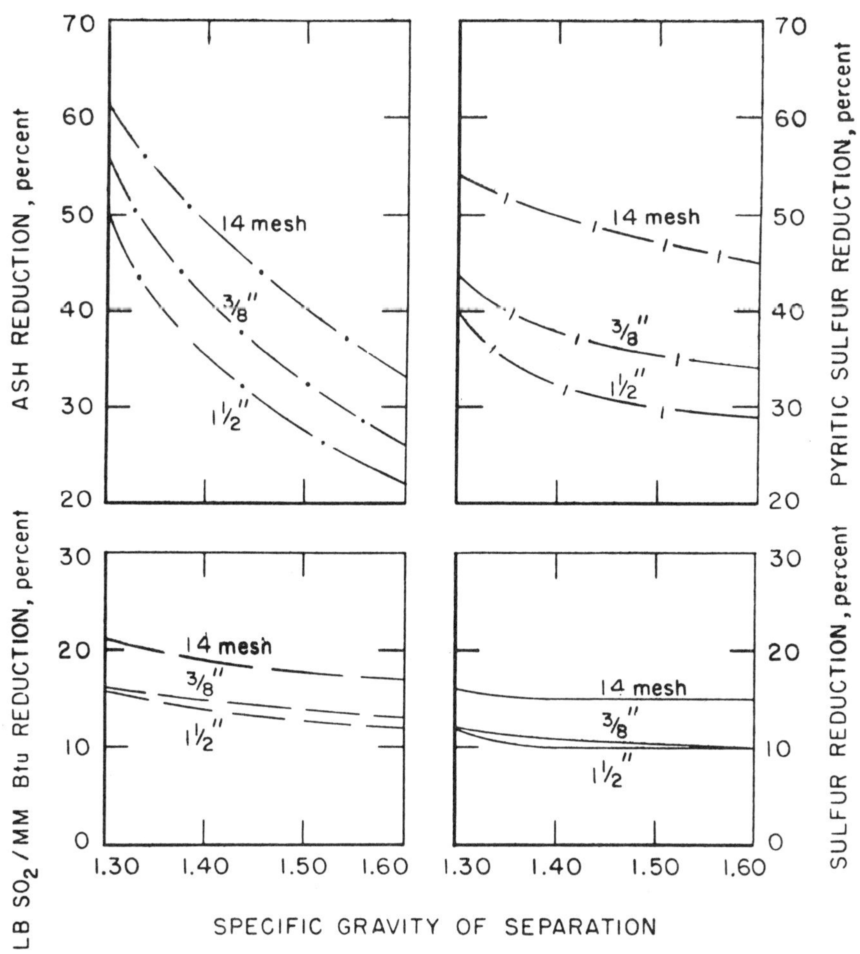


Source:  PB-252 965

**FIGURE 3.15: PERCENT OF COAL SAMPLES MEETING EPA STANDARD**

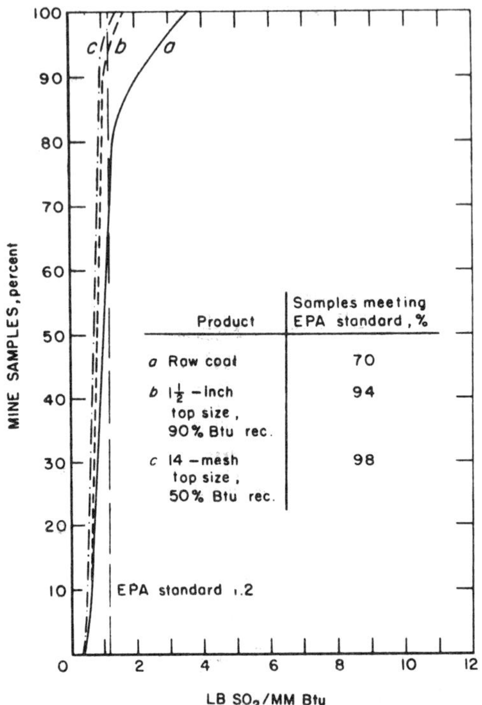


Key: Curve a - no preparation;
Curve b - comparison with those
crushed to 1½-inch top-size at
a Btu recovery of 90%;
Curve c - crushed to 14-mesh top-
size at a Btu recovery of 50%.

Source: PB-252 965

## Summary of All Coals

The 455 raw coal samples averaged 14.0% ash, 1.91% pyritic sulfur, 3.02% total sulfur, and 12,574 Btu per pound, which would produce 4.9 pounds $SO_2$/MM Btu fired at the power plant. The raw coal sulfur contents averaged 63% pyritic sulfur and 37% organic sulfur.

The ash, pyritic sulfur, total sulfur, and heating value contents varied considerably as would be expected when washability data of coals from various regions of the United States are evaluated. This is evidenced by the large sigma values for each of the parameters evaluated.

Figure 3.16 shows that significant reductions of impurities can be obtained, especially ash and pyritic sulfur contents, by crushing and gravimetric separation.

## FIGURE 3.16:  THE EFFECT OF CRUSHING

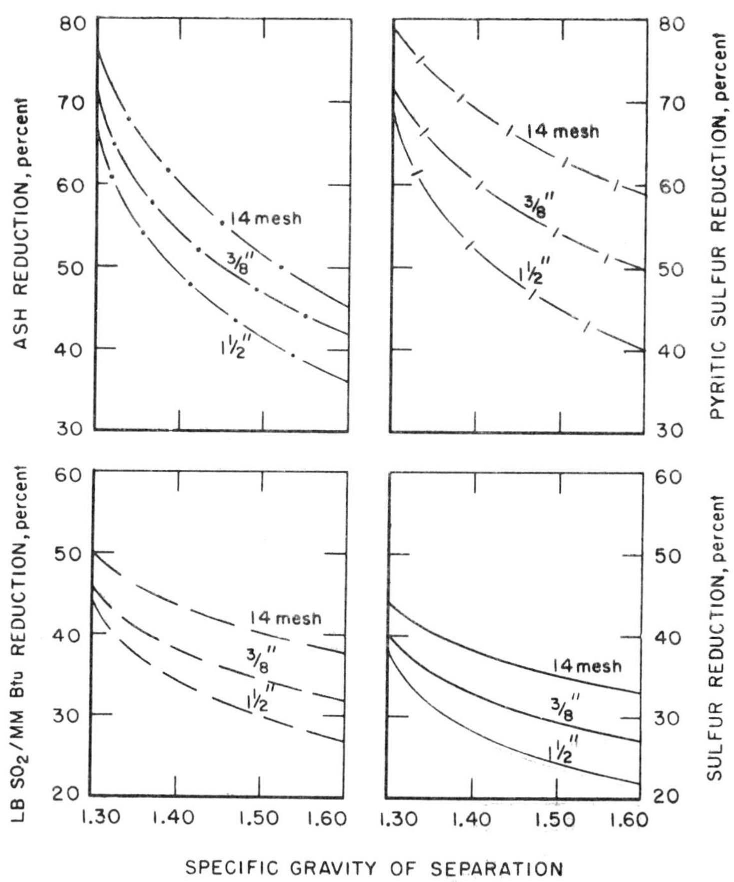


Source:  PB-252 965

Figure 3.17 shows that only 14% of raw coal samples as mined could meet the EPA $SO_2$ emission standard of 1.2 pounds $SO_2$/MM Btu.  24% of the samples would meet the standard at a 90% Btu recovery when crushed to 1½-inch top size, while 32% would meet the standard at a Btu recovery of 50% when crushed to 14-mesh top size.  Even if the standard were raised to 2.0 pounds $SO_2$/MM Btu, only 48% of the samples could meet the standard at a 50% Btu recovery when crushed to 14-mesh top size.

FIGURE 3.17:  PERCENT OF COAL SAMPLES MEETING EPA STANDARD

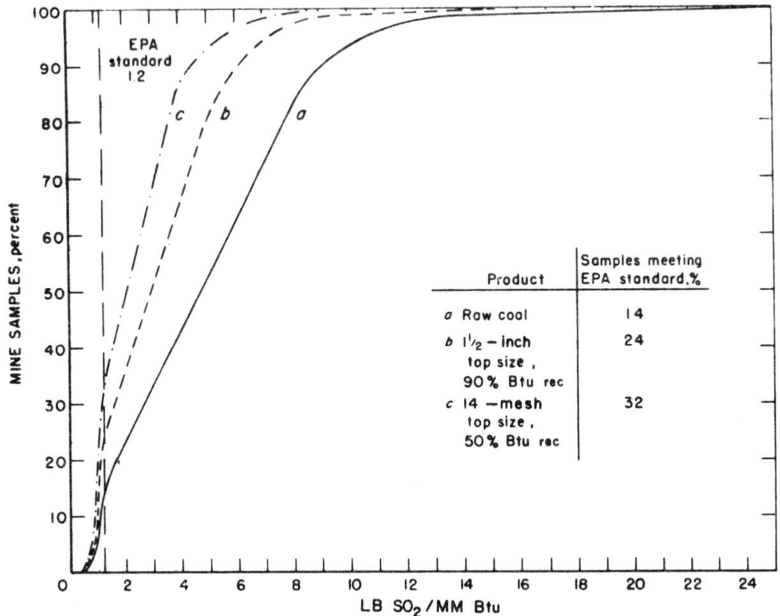


Key:  Curve a - no preparation;
      Curve b - comparison with those crushed to 1½-inch top size
      at a Btu recovery of 90%;
      Curve c - crushed to 14-mesh top size at a Btu recovery of 50%.

Source:  PB-252 965

The composite data show if all the coals were upgraded at a specific gravity of
1.60, the analyses of the clean coal products of the various regions would range
on the average from 5.1 to 8.3% ash, 0.10 to 1.80% pyritic sulfur, 0.56 to 3.59%
total sulfur, 12,779 to 14,264 Btu per pound and would produce 0.95 to 5.5
pounds of $SO_2$/MM Btu recoveries ranging from 91.7 to 97.6%.  The corresponding $SO_2$ removal efficiencies required to comply with the current EPA emission
regulations of 1.2 pounds $SO_2$/MM Btu would range from 0 to 78%.

## CONCLUSIONS

(A)  The 227 Northern Appalachian Region coalbed samples evaluated averaged
2.01% pyritic sulfur and 3.01% total sulfur.  Only 4% of the raw coal samples
would meet the current EPA $SO_2$ emission standard whereas, if a 50% Btu recovery were acceptable, then 31% of the samples would meet the standard when
crushed to 14-mesh top size.  Coals from the Upper and Lower Freeport and
Upper Kittanning coalbeds could be upgraded to meet the current EPA $SO_2$ emission standard utilizing coal preparation techniques.  The coals of this region are
especially amenable to pyritic sulfur liberation, averaging 80% reduction at a 70%
Btu recovery when crushed to 14-mesh top size, and separated gravimetrically.

(B)  The 35 Southern Appalachian Region coalbed samples evaluated contained, on the average, 0.37% pyritic sulfur and 1.04% total sulfur.  63% of the samples would comply with the EPA $SO_2$ emission standard at a Btu recovery of 50% when crushed to 14-mesh top size.

(C)  The 10 Alabama Region coalbed samples evaluated contained, on the average, 0.69% pyritic sulfur and 1.33% total sulfur.  The pyrite does not readily liberate from these coals; therefore, while 30% of the raw coal samples would meet the EPA $SO_2$ emission standard with no preparation, only 40% would meet the standard at a 50% Btu recovery when crushed to 14-mesh top size.

(D)  The 95 Eastern Midwest Region coal samples contained, on the average, 2.29% pyritic sulfur and 3.92% total sulfur.  These coals contained an average of 1.63% organic sulfur and, consequently, could not be upgraded to meet the current EPA $SO_2$ emission standard.  Crushing to 14-mesh top size and removing the sink 1.60 specific gravity material would provide a float product with an average ash reduction of 45% and an average pyritic sulfur reduction of 57%.  Only 1% of the raw coal samples would meet the current EPA $SO_2$ emission standard with no preparation; 4% would meet the standard at a Btu recovery of 50% when crushed to 14-mesh top size.  However, none of the principal coalbeds on the average could be upgraded sufficiently to meet the current EPA $SO_2$ emission standard.

(E)  The 44 Western Mideast Region samples contained an average of 3.58% pyritic sulfur and 5.25% total sulfur.  These coals contained, on the average, 1.67% organic sulfur and thus generally could not be upgraded to meet the EPA $SO_2$ emission standard.  Less than 3% of the raw coal samples would meet the standard with no preparation, whereas only 6% would comply at a Btu recovery of 50% when crushed to 14-mesh top size.  These coals contained on the average more than 1% higher pyritic and total sulfur content than those of the Eastern Midwest Region.

(F)  The 44 Western Region coalbed samples evaluated contained an average of 0.68% total sulfur which would produce 1.1 pounds $SO_2$/MM Btu and would thus meet the current EPA $SO_2$ emission standard as mined.  70% of the raw coal samples would meet the EPA standard with no preparation and 94% would comply at a Btu recovery of 90% when crushed to 1½-inch top size.  These are generally subbituminous coals of high inherent moisture and low Btu content.

(G)  The 455 United States coal samples evaluated contained on the average 1.91% pyritic sulfur and 3.02% total sulfur.  Only 14% of the raw coal samples could meet the current EPA $SO_2$ emission standard of 1.2 pounds $SO_2$/MM Btu with no preparation.  If a 50% Btu recovery was acceptable, then 32% of the samples could be upgraded to meet the standard when crushed to 14-mesh top size.


## REFERENCES

(1)  Deurbrouck, A.W., *Sulfur Reduction Potential of the Coals of the United States,* BuMines RI 7633, 1972, 289 pp.
(2)  Fieldner, A.C. and Selvig, W.A., *Notes in the Sampling and Analysis of Coal,* BuMines Tech. Paper 586, 1938, 48 pp.
(3)  Holmes, J.A., *The Sampling of Coal in the Mine,* BuMines Tech. Paper 1, 1918, 22 pp.

# EPA STUDIES OF COAL QUALITY AND CLEANABILITY

The material in this chapter was taken from a report by L. Hoffman, et al, of the Mitre Corporation, McLean, Virginia which was prepared for the Environmental Protection Agency Office of Research and Development (PB-232 011). Further information on this report will be found in the bibliography on page 301.

## INTRODUCTION

Currently available methods for controlling sulfur oxides emissions from coal-fired stationary combustion sources fall into the following categories:

The use of low sulfur coal either naturally occurring or physically cleaned

Chemical treatment to extract sulfur from coal

Removal of sulfur compounds during the combustion process

Removal of sulfur oxides from the combustion flue gas

Conversion of coal to a clean fuel by such processes as gasification and liquefaction

Of these methods, physical removal of pyritic sulfur (principally $FeS_2$) is the lowest in cost and has the most developed technology.

Sulfur in coal exists in two principal forms, organic and inorganic. The organic sulfur is bound chemically to the coal substance and cannot be physically removed. However, inorganic sulfur (i.e., pyritic sulfur) is not bound chemically with the coal substance and may be removed to varying extents by crushing and physical cleaning. The degree of removal is dependent upon pyrite size and distribution, coal size, and other physical characteristics.

### History of Coal Cleaning

Physical cleaning of coal has been used for many years. Its principal purpose has been to reduce the so-called ash-forming impurities. The process in common use for reducing such impurities is a combination of stage crushing and specific gravity separation. Shale and coal having different specific gravities may be separated. Froth flotation, dependent on differences in surface characteristics of coal and coal impurities, is used for removing ash from fine size coal. Existing cleaning processes, while removing impurities from coal, also reduce the total Btu recovery (i.e., a portion of the heat content of the feed will be lost with the refuse). As a result, the Btu content per unit weight of the processed product increases due to removal of low heat value impurities. In practice an economic balance must be achieved between the Btu loss and the improvement in coal quality. This balance may be further influenced by environmental considerations.

In 1965, EPA sponsored a study to quantify the impact that coal cleaning, optimized for pyrite removal, could have on the control of sulfur oxide emissions. This study found that it was impossible to quantify the impact of coal cleaning on sulfur oxide emissions because of large gaps in available information. The study identified the following areas where required information was either not available or was inadequate for appraising this impact:

> Knowledge of the distribution of sulfur forms in (all) major high-sulfur coalbeds in the United States

> Effectiveness of available commercial coal preparation methods for pyrite separation, together with the development or modification of these techniques to maximize sulfur reduction

> Identification and assessment of processes that could economically utilize coal cleaning reject material for by-product recovery, thereby aiding the overall cleaning economics and reducing potential air, water, and solid pollution

The findings of the 1965 study led EPA to proceed with implementation of a comprehensive program designed to define the potential role of coal cleaning in controlling sulfur oxides emissions from coal-fired sources. The individual program elements were structured to supply the basic information required. The program permitted flexibility for tailoring new program elements in accordance with findings of previous or ongoing studies. An important part of the program was to determine the extent to which the sulfur content of U.S. coals could be reduced by coal cleaning processes based on differences in physical properties.

While pyritic sulfur is amenable to removal by such processes, the other major form of sulfur (organic sulfur compounds) is not. A good indicator of the "cleanability" of pyrites from coal is the float-sink, or washability, test in which crushed coal is tested at various gravities to separate the different specific gravity fractions. The coal fractions float and the heavier fractions sink. Such washability tests form an important element of the total program. The overall study concept included the following elements:

> A data base of coal cleanability, containing results of pyrite (and ash) washability studies, to determine the sulfur-release potential of coals from principal utility-coal-producing beds of the United States

> In-depth studies of existing physical cleaning techniques to obtain maximum pyrite removal

The modification of existing methods and evaluation of new methods for sulfur reduction

Studies covering pyrite utilization and related economics

A feasibility study (with costs) of a prototype cleaning plant to evaluate both economic and technical factors associated with maximizing sulfur reduction

A feasibility study with costs of a high-sulfur combustor for utilizing reject material from coal cleaning

The overall basic program concept (Figure 4.1) has been to develop a means for achieving full utilization of all values in the coal.  The deep-cleaning of coal to maximize sulfur reduction could result in reject material high in coal and sulfur values.  This refuse material could be reprocessed into various products:  nonpolluting refuse for disposal; a pyrite concentrate that could be converted into sulfuric acid; and coal products of various sulfur levels for utilization with flue gas cleaning systems to generate electrical energy.  The full utilization of these values could result not only in lower cost fuels, but also in the conservation of the nation's fossil fuel resources.

## FIGURE 4.1:  FUEL UTILIZATION

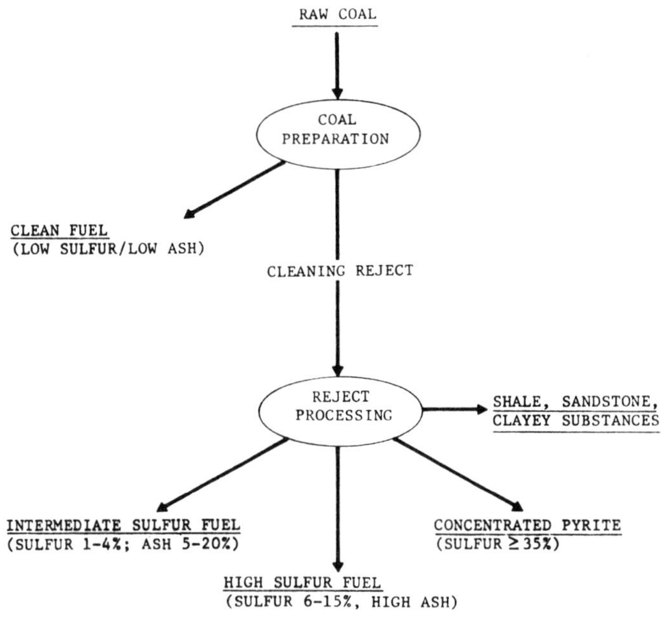


Source:  PB-232 011

A diagram identifying the major elements of the program sponsored by EPA (in whole or part) to develop the information necessary to quantify the potential role of coal cleaning in the control of sulfur oxides emissions and to define the relative attractiveness of cleaning options is given in Figure 4.2. This figure indicates the interrelations among the various program elements, together with the beginning date for each funded effort and the identification of future efforts required to reach demonstration of commercial applicabilities of considered techniques.

## FIGURE 4.2: PYRITE COAL PROGRAM

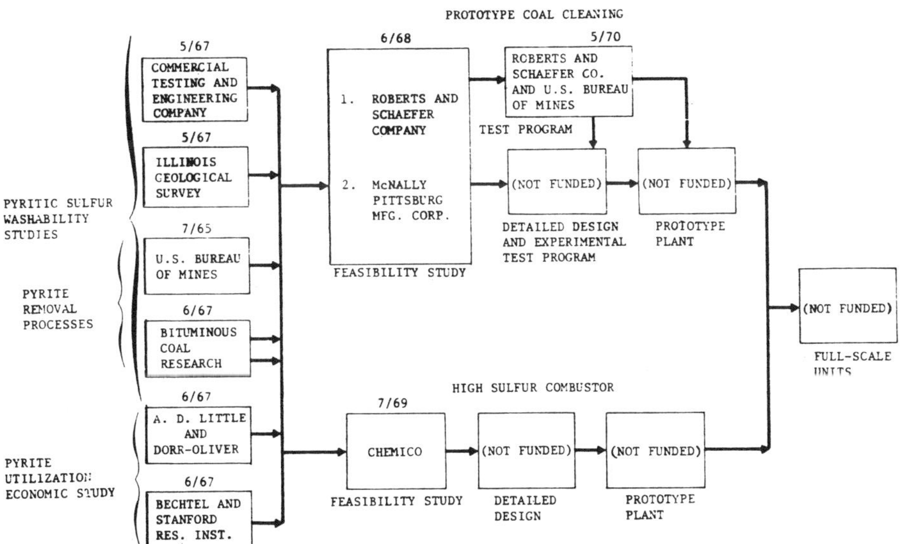


Source:　PB-232 011

*Washability Studies:*　The initial EPA-sponsored washability study began in 1965.　Representative samples were collected from mines of major producers of coal primarily for use by the utilities.　The samples were stage crushed to three top sizes:　coarse—1½ inches, intermediate—⅜ inch, and fine—14 mesh. Washability studies were performed on the various sized fractions with organic liquids of standardized specific gravities to determine the effects of top size and specific gravity parameters upon pyrite liberation and separation.

Because of the large number of coals to be tested and the need to accelerate their evaluation, EPA expanded the testing program by supporting three additional organizations.　The first of these concentrated on regions where preliminary data indicate the presence of coals cleanable for major sulfur reduction.

In another study, EPA cooperatively supported a two-year investigation to evaluate all actively mined coal beds in the state of Illinois, which has the largest bituminous coal reserves in the United States.　This effort determined the important physical and chemical properties of Illinois coal, including its washability and distribution of sulfur forms.

To further develop basic data on the effects of size reduction on pyrite liberation, selected samples (in a separate effort) were further reduced in top size to
30 mesh and 60 mesh and evaluated by float-sink evaluation.  Although the
cleaning of this size material by specific gravity methods is not normal commercial practice, this information would define the maximum effects of size reduction on pyrite liberation and separation.

The four funded washability studies contributed to the development of a substantial body of data on the cleanability of U.S. coals.  This information is
stored in a computerized data bank to facilitate retrieval, updating, and analysis.

*Pyrite Separation:*  Concurrent with the washability studies, EPA sponsored two
separate investigations of techniques for separating and concentrating pyrites
from fine coal.  One of these evaluated the effectiveness of commercially available equipment; the wet concentrating table, the compound water cyclone, and
the concentrating spiral.  The other evaluated conventional techniques to define
operating parameters, modification of these techniques to maximize sulfur reduction, and the development of new techniques.

*By-Product Recovery:*  In 1967, EPA sponsored two preliminary studies that
investigated the economics of recovering by-products from the reject material
resulting from coal cleaning.  One of these included a review and analysis of
commercially applicable processes for recovering sulfur and the applicability of
these processes utilizing coal cleaning refuse.  The second was a detailed feasibility study of fluid bed roasting with conversion to sulfuric acid, plus a market
analysis for its principal product, commercial grade sulfuric acid.  The study
findings indicated that the high-sulfur combustor had the highest economic via-
bility potential.

After the preliminary economic studies were completed, EPA funded a feasibility study of a high-sulfur combustor capable of burning the reject material
to recover its energy and sulfur values.  This study produced an engineering
design and cost analysis associated with the utilization of high sulfur reject
material to generate power and recover sulfuric acid or elemental sulfur.

*1968 EPA Studies:*  In order to determine the scope and cost of conducting
investigations and demonstrations of pyrite removal using commercial equipment on a scale approximating commercial capacity, EPA in 1968 funded two
independent and competing engineering studies.  These efforts, predicated on
findings of the washability investigations, were for designing a prototype coal
cleaning plant and for estimating capital and operating cost for the plant.  Two
studies were required to identify the most acceptable design/cost option due to
plant and operational complexities.

These various studies are described and their results are summarized in the following chapters of this report.  The economics, technologies, etc., provided in this
report are those of the addressed studies.  It should be noted that since the
reported studies were performed, there are many conditions that have changed
and the economic findings and conclusions of the various efforts may not be
currently valid.

## WASHABILITY STUDIES

This section outlines the procedures employed, details the number of samples obtained from various locations, summarizes the results of the washability studies, and reports on the findings of an effort that examined the comparability of coal core samples and face samples. The comparability of core samples and face samples effort was to determine whether useful washability data could be obtained from 2-inch diameter core samples. [A more detailed description of sampling and washability procedures and an in-depth discussion of results can be found in References (3), (4), (5), (6), (7), and (8).]

The principal purpose of the effort was to determine the sulfur forms and distribution of sulfur in coal samples from mines which were principally utility coal producers. Coal samples were obtained from the following coal regions:

> Northern Appalachian region: the states of Maryland, Pennsylvania (bituminous), and Ohio and counties located in central and northern West Virginia.

> Southern Appalachian region: the states of Tennessee, Virginia, and eastern Kentucky and some of the southern West Virginia counties.

> Alabama region: the state of Alabama.

> Midwest region: the states of Illinois, Indiana, and western Kentucky.

> Western region: all states west of the Mississippi.

Generally, the washability data of the coals indicated that the total sulfur could be reduced by using some combination of stage crushing and specific gravity separation. The washability data were analyzed with the use of a specifically designed computer program.

### Experimental Procedure – Collection of Coal Mine Samples

Coal samples were collected either from a fresh face in the mine or from the tipple (raw R.O.M.). The face samples were obtained by cutting channels approximately ten inches deep and wide enough to provide a sample of approximately 600 lb (8). Each channel was selected so as to be as representative as possible of coal being mined. Tipple samples were collected by taking increments at fixed time and/or weight intervals so as to insure that all working faces were represented in the gross sample.

### Procedure for the Conduct of the Washability Studies

The sample preparation procedures by investigations with specific gravity test points for the washability studies are provided in Figure 4.3 and Table 4.1. Each gross sample was air-dried and crushed to 1½ inch top size using a single roll tooth crusher. The coal was then divided into required portions by means of cone and quartering. The preparation of each coal portion for float and sink tests was as indicated in Figure 4.3. The 100 mesh x 0 material, separated by screening from the coal fraction used for float and sink tests, was analyzed for ash, pyritic sulfur, and total sulfur.

The float and sink analyses were conducted on the various fractions at specific gravities as indicated by Table 4.1.

FIGURE 4.3:  FLOW DIAGRAM SHOWING PREPARATION OF GROSS SAMPLE

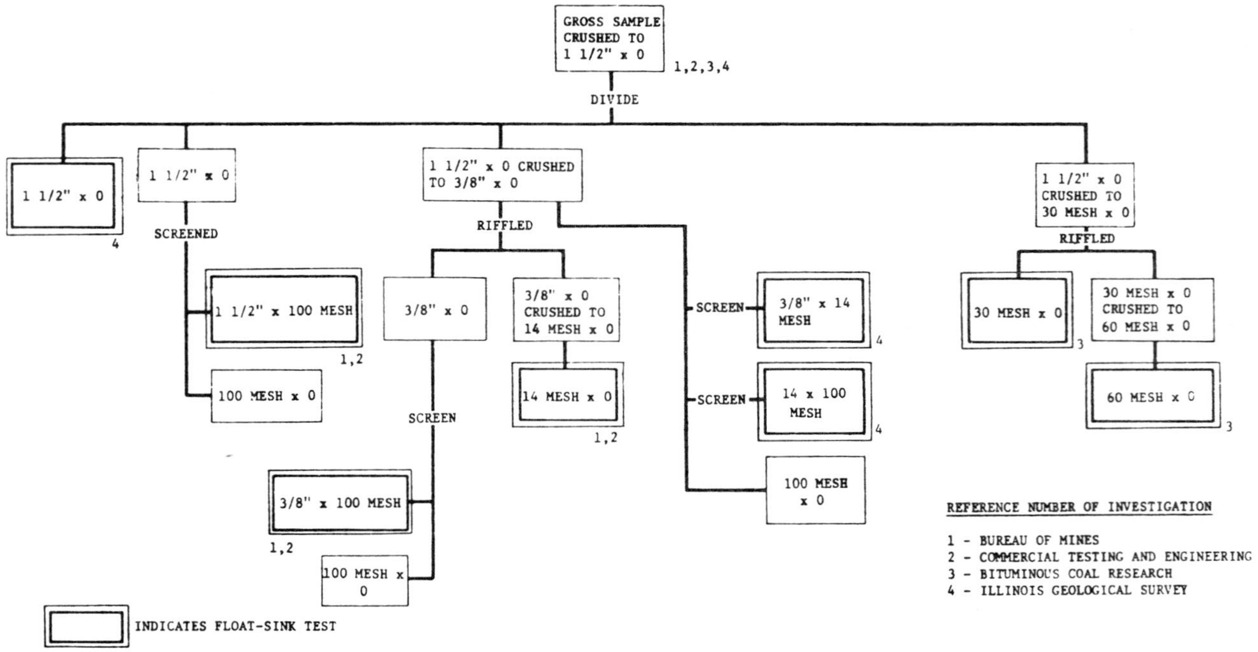

Source:  PB-232 011

## TABLE 4.1:  TABLE OF FLOAT-SINK TEST SIZES WITH SPECIFIC GRAVITIES

| Investigator | Sizes Washed | Test Points (Specific Gravities) | | | | | |
|---|---|---|---|---|---|---|---|
| | | 1.25-1.29 | 1.30 | 1.30-1.39 | 1.40 | 1.60 | 1.90 |
| Bureau of Mines | 1 1/2 in. x 100 mesh | | x | | x | x | |
| | 3/8 in. x 100 mesh | | x | | x | x | |
| | 14 mesh x 0 | | x | | x | x | |
| Commercial Testing and Engineering Company | 1 1/2 in. x 100 mesh | | x | | x | x | x |
| | 3/8 in. x 100 mesh | | x | | x | x | x |
| | 14 mesh x 0 | | x | | x | x | x |
| Bituminous Coal Research, Inc. | 30 mesh x 0 | | | | | x | |
| | 60 mesh x 0 (p.c.) | | | | | x | |
| Illinois State Geological Survey | 1 1/2 in. x 0 | x | | x | x | x | |
| | 3/8 in. x 14 mesh | x | | x | x | x | |
| | 14 mesh x 100 mesh | x | | x | x | x | |

Source:  PB-232 011

After completing the float-and-sink analysis, the specific gravity fractions of the samples were analyzed for ash, pyritic sulfur, and total sulfur content.  [Additional details of the float-sink tests are provided in Reference (8).]

**Washability Computer Program**

A computer program was used to interpolate the washability data to find the theoretical specific gravity of separation, yield, ash content, and pyritic and total sulfur contents for specific levels of yield or total sulfur.  The values of the sulfur and yield interpolants were read into the computer and stored.  In addition, the following data were entered for each mine sample:  "state, county, town, mine, and bed codes; laboratory number, and laboratory identification code, bed-bench and size fraction codes; year of sample collection; company name, mine name, and county name; the specific gravity fractions of separation and their associated levels with the variables, ash, pyritic and total sulfur" (8).

Using these data, three tables are printed out for each mine.  These tables are: (1) the original cumulative washability data; (2) values of specific gravity of separation, yield, ash, and pyritic sulfur for requested levels of total sulfur; and (3) values of specific gravity of separation, yield, ash, pyritic sulfur, and total sulfur for the requested levels of yield.

A second computer program was written to make a statistical evaluation of the washability data.  This program calculated averages and the standard deviation (sigma) values.

The computer program was used to calculate the percentages of total recovery of ash, pyritic sulfur, and total sulfur for the desired levels of interpolation. Next, the program was used to calculate the averages and standard deviations of the ash, pyritic sulfur, and total sulfur.  These averages and standard deviation values were found for each bed and the required data accumulated for similar averages for the region. Also, the averages and standard deviation values were calculated for each set of interpolated values at the various yield levels.

## Program Results

The average composition of the coals evaluated in the Bureau of Mines study (data from Commercial Testing and Engineering Co.) was 14.4% ash, 2.05% pyritic sulfur and 3.23% total sulfur (8). The results generally indicated that the largest reduction in ash and sulfur through washing was obtained by reducing coal top size from 1½ inches to ⅜ inch. Crushing from ⅜ inch to 14 mesh showed relatively lower ash and sulfur reduction. Summary data covering the 322 coal samples evaluated by the Bureau of Mines are as follows:

| Percent Yield | No. of Samples | Ash, Percent (Raw Avg.: 14.4, Sigma: 5.8) | | Pyritic Sulfur, Percent (Raw Avg.: 2.05, Sigma: 1.35) | | Total Sulfur, Percent (Raw Avg.: 3.23, Sigma: 1.75) | |
|---|---|---|---|---|---|---|---|
| | | Average | Sigma* | Average | Sigma | Average | Sigma |
| | | | | 1 1/2 Inch Top Size | | | |
| 60 | 260 | 5.8 | 2.3 | 0.72 | 0.57 | 1.95 | 1.06 |
| 70 | 290 | 6.4 | 2.7 | 0.79 | 0.59 | 2.03 | 1.07 |
| 80 | 284 | 7.1 | 2.8 | 0.89 | 0.65 | 2.13 | 1.12 |
| 90 | 168 | 7.2 | 2.4 | 1.00 | 0.68 | 2.27 | 1.21 |
| | | | | 3/8 Inch Top Size | | | |
| 60 | 25. | 5.1 | 2.2 | 0.57 | 0.48 | 1.82 | 1.04 |
| 70 | 287 | 5.6 | 2.4 | 0.61 | 0.48 | 1.83 | 1.00 |
| 80 | 277 | 6.3 | 2.4 | 0.68 | 0.49 | 1.92 | 1.04 |
| 90 | 145 | 6.5 | 2.2 | 0.75 | 0.56 | 1.95 | 1.12 |
| | | | | 14 Mesh Top Size | | | |
| 60 | 270 | 4.6 | 2.1 | 0.46 | 0.40 | 1.70 | 0.98 |
| 70 | 298 | 5.2 | 2.3 | 0.50 | 0.39 | 1.72 | 0.96 |
| 80 | 274 | 5.9 | 2.2 | 0.57 | 0.42 | 1.79 | 0.99 |
| 90 | 136 | 6.0 | 2.0 | 0.59 | 0.48 | 1.79 | 1.07 |

*Sigma ($\sigma$) is the standard deviation value.

Source:   PB-232 011

The sulfur reductions of these coals attained by washing are further summarized in Figure 4.4. This figure indicates that significant total sulfur reductions are attainable by removing heavy impurities. At 60% yield, 20 to 25% of the mine samples showed a 1% or less total sulfur product.

An evaluation of the data from the 322 mines sampled showed that approximately 30% of these mines have coal that washed to a total sulfur content of 1% or less. On crushing to 1½ inch top size, significant reductions in sulfur content were attainable by removing 10 or 20% of the heaviest material. Furthermore, the total sulfur content was reduced by approximately 50% in more than half of the samples collected by crushing of the coal prior to removing the heaviest impurities. Both the distribution and release potential of the sulfur in coals vary between the regions and the coals within the region.

FIGURE 4.4:  WASHABILITY SUMMARY OF ALL COALS

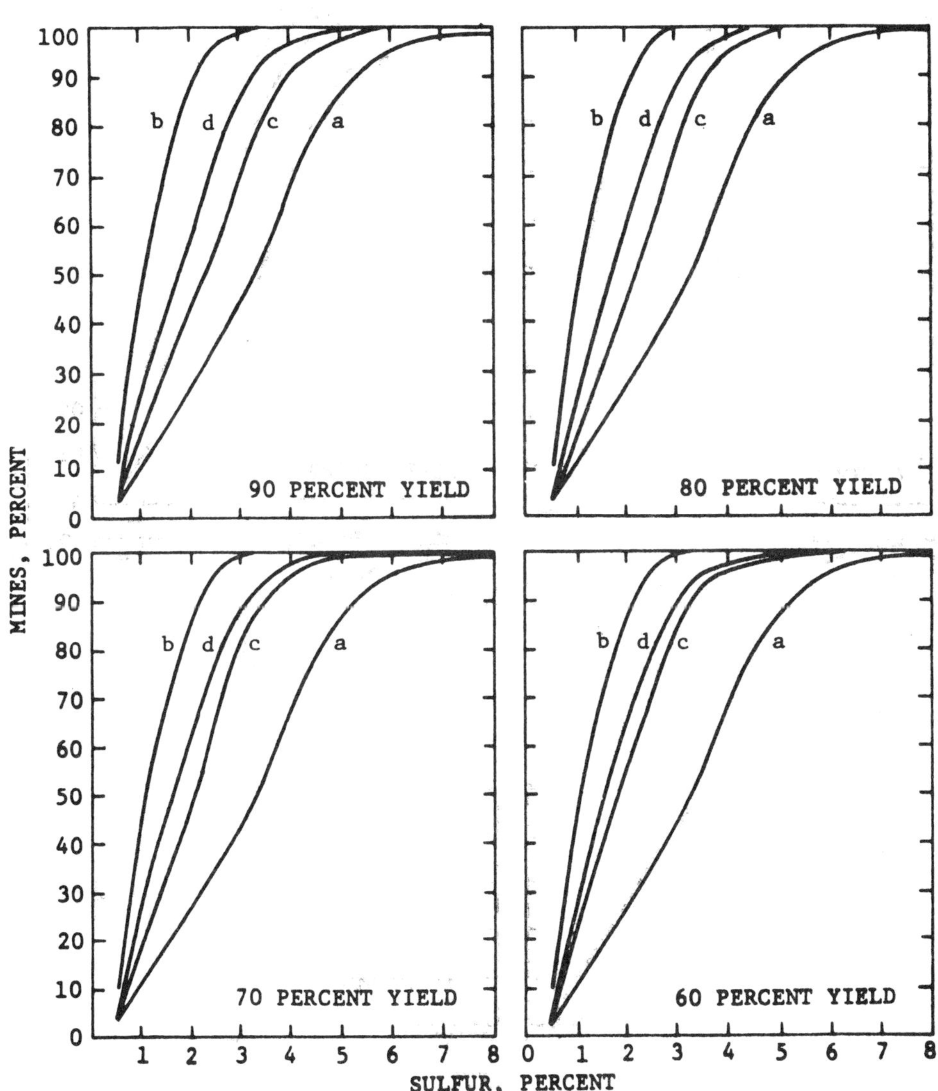


(a)    Raw coal total sulfur content curve
(b)    Raw coal organic sulfur content curve
(c)    Washed coal total sulfur content coal curve when raw coal was crushed
         to minus 1½ inches
(d)    When raw coal was crushed to minus 14 mesh

Source:  PB-232 011

The Illinois Geological Survey washability studies indicated that most Illinois coals have a total sulfur content of 3 to 5% and there are only a few Illinois coals whose sulfur content can be reduced to 1.5% or less. The coals capable of this amount of sulfur reduction had only 2% or less sulfur in the raw coal samples.

With regard to the forms of sulfur in the coals, it was found that approximately 50% of the sulfur in the average Illinois coal was in the pyritic form. The average pyritic sulfur reduction was approximately 60% with an 80% yield and 75% with a 40% yield. Usually, the float coal fractions had less sulfur when the coal was crushed to finer sizes. In many of the coals tested, however, the differences were not great enough to warrant grinding the coals to finer sizes for sulfur reduction. [Note: Complete data on washability studies for each region's coal samples are given in this report, but are omitted here for reasons of space.]

## LABORATORY AND PILOT STUDIES EVALUATING PYRITE REMOVAL PROCESSES

This section describes the EPA-sponsored laboratory and pilot study projects conducted by the Bureau of Mines and Bituminous Coal Research, Incorporated. The objectives were to determine the operating parameters for maximum pyrite separation, to modify existing techniques and/or develop new techniques to separate pyrite from fine coals, and to evaluate methods to process coal cleaning rejects so as to concentrate the pyrite and reclaim clean coal products.

The equipment or processes were evaluated by the two contractor organizations as indicated below. The Bureau of Mines efforts concentrated on equipment operational parameters and the development of new techniques while the Bituminous Coal Research efforts concentrated on the applicability of equipment to various coals.

|  | Investigated by | |
| --- | --- | --- |
| Equipment/Process | Bureau of Mines | Bituminous Coal Research, Inc. |
| Wet concentrating table (Deister) | X | X |
| Concentrating spiral (Humphry) | X | X |
| Hydrocyclones | X | X |
| Air classifiers |  | X |
| Electrokinetic techniques | Y |  |
| Agglomo-separation | X,Y |  |
| Froth flotation | X,Y |  |

    X designates pilot plant tests
    Y designates laboratory-scale tests

These processes are described briefly in the following sections and results of their performance are summarized.

### Wet Concentrating Table

The wet table is a rectangular or rhomboid-shaped normally riffled deck operated in essentially a horizontal plane. A drive mechanism imparts a differential motion to the deck along its long axis while water flows by gravity along the short axis of the table. The rapid shaking motion causes particles of different densities to migrate to different zones on the table's periphery.

Commercial tables have a deck approximately 16 feet long on the clean coal side and 8 feet long on the refuse side. In these studies, a quarter-size deck, 8 feet long on the clean coal side and 4 feet long on the refuse side, was used. The direction of the feed coal mixture and the reporting of the clean coal, middlings, and refuse to the various sections of the table are shown in Figure 4.5. As the specific gravity of the coarse particles increases, the material deposits further around the table toward the refuse end.

In tests by the Bureau of Mines and Bituminous Coal Research, Inc., the wet concentrating table proved effective in separating liberated pyrite and ash from coal. Coals with a high proportion of pyritic sulfur relative to organic sulfur are particularly amenable to sulfur reduction by this process.

**FIGURE 4.5: DISTRIBUTION OF CLEAN COAL, MIDDLINGS, AND REF- USE ON THE WET CONCENTRATING TABLE**

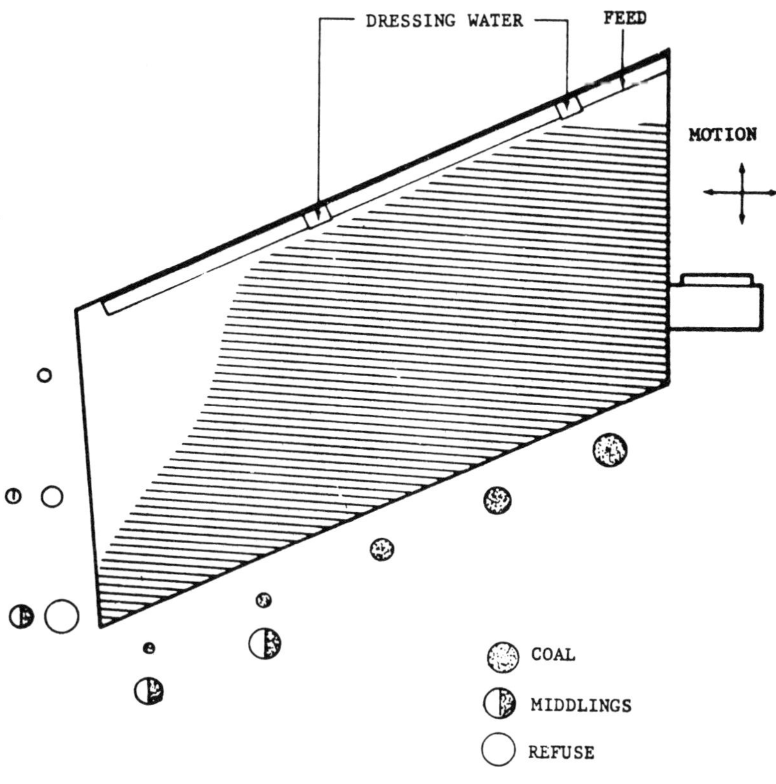


Source: PB-232 011

Two series of wet concentrating table tests were performed by the Bureau of Mines, one on a normal size table feed and the second on a rather fine size feed. The results of cleaning minus ¼ inch coal are shown in Table 4.2.

**TABLE 4.2: PRODUCT ANALYSIS OF A ¼ INCH BY 0 SIZE COAL WASHED ON A WET CONCENTRATING TABLE**

| Products | Percent | | | |
| --- | --- | --- | --- | --- |
| | Weight | Ash | Pyritic Sulfur | Total Sulfur |
| *1/4 inch by 14 mesh, 41.6 percent* | | | | |
| Clean coal | 87.3 | 2.6 | 0.56 | 1.13 |
| Refuse | 12.7 | 36.9 | 19.19 | 19.60 |
| Feed | 100.0 | 7.0 | 2.92 | 3.47 |
| *14 by 48 mesh, 38.9 percent* | | | | |
| Clean coal | 86.1 | 2.6 | .45 | 1.08 |
| Refuse | 13.9 | 40.9 | 17.15 | 17.66 |
| Feed | 100.0 | 7.9 | 2.77 | 3.38 |
| *48 by 200 mesh, 15.2 percent* | | | | |
| Clean coal | 83.8 | 5.9 | .56 | 1.17 |
| Refuse | 16.2 | 68.0 | 17.69 | 17.79 |
| Feed | 100.0 | 15.8 | 3.32 | 3.85 |
| *200 mesh by 0, 4.3 percent* | | | | |
| Clean coal | 86.1 | 28.4 | 2.51 | 3.08 |
| Refuse | 13.9 | 71.9 | 27.51 | 27.64 |
| Feed | 100.0 | 34.6 | 6.00 | 6.51 |
| *1/4 inch by 0, 100.0 percent* | | | | |
| Clean coal | 86.3 | 4.2 | .60 | 1.20 |
| Refuse | 13.7 | 45.6 | 18.49 | 18.86 |
| Feed | 100.0 | 9.9 | 3.07 | 3.63 |

Source:  PB-232 011

The pyritic sulfur content was reduced from 3.07 to 0.60% and the total sulfur from 3.63 to 1.20% at a clean coal recovery of 86.3%. Further analyses on the coal (first series of tests) were made by screening the minus ¼ inch coal into four size fractions as indicated by Table 4.2. The combined pyritic sulfur reduction for the three larger size fractions was approximately 84% each. Also, the efficiency of the separation deteriorated in the minus 200 mesh size range where the pyritic sulfur reduction was only 58%.

For the Bureau's second series of tests (Table 4.3) conducted on 35 mesh x 0 size coal, the 35 by 200 mesh fraction shows the pyritic sulfur content reduced by 75%, from 2.05% in the feed to 0.50% in the clean coal. The pyritic sulfur in the minus 200 mesh by 0 size fraction was reduced by 23%, or from 1.69% in the feed to 1.30% in the clean coal. The results of the second series of wet concentrating table tests are shown in Table 4.3. When the sample was broken down into the plus and minus 200 mesh size fractions, the results showed that the efficiency of the table for separating pyrite from coal deteriorated below the 200 mesh size range.

### TABLE 4.3: PRODUCT ANALYSIS OF A 35 MESH BY 0 SIZE COAL WASHED ON A WET CONCENTRATING TABLE

| Product | Percent | | | |
|---|---|---|---|---|
| | Weight | Ash | Pyritic sulfur | Total sulfur |
| **35 by 200 mesh, 63.9 percent** | | | | |
| Clean coal | 63.7 | 8.7 | 0.50 | 1.25 |
| Refuse | 36.3 | 53.2 | 4.77 | 5.18 |
| Feed | 100.0 | 24.8 | 2.05 | 2.67 |
| **200 mesh by 0, 36.1 percent** | | | | |
| Clean coal | 92.4 | 16.1 | 1.30 | 1.98 |
| Refuse | 7.6 | 27.6 | 6.51 | 7.08 |
| Feed | 100.0 | 17.0 | 1.69 | 2.36 |
| **35 mesh by 0, 100.0 percent** | | | | |
| Clean coal | 74.0 | 12.0 | .87 | 1.58 |
| Refuse | 26.0 | 50.5 | 4.95 | 5.39 |
| Feed | 100.0 | 22.0 | 1.92 | 2.56 |

Source:  PB-232 011

In addition to the above, two-stage cleaning tests, in which a coarse sample (3/8 inch x 0) was tabled on a quarter-sized unit and the clean fraction pulverized to a finer size (30 mesh x 0) which was retabled, were performed by Bituminous Coal Research, Inc. Equipment for testing the coarse sample had a capacity of 2½ to 3 tons per hour, and that for the fine sample of 1 to 1¼ tons per hour. These tests indicated that pyrite removal can be improved by use of the concentration table in two stages. A total of eight coals, five eastern and three (mid)western, were subjected to two-stage tabling.

These coals are identified and their sulfur contents are shown in Table 4.4. The results are given in Table 4.5 which also shows feed rate during testing, percent total sulfur in the feed, and percent sulfur in clean coal. Reductions in total sulfur range from 29 to 68%.

**Concentrating Spiral**

The concentrating spiral, although widely used for ore dressing, has never been used for coal preparation even though it had been previously considered for coal cleaning purposes. It was evaluated because of its low capital and operating costs.

# TABLE 4.4: ANALYSIS OF COALS SELECTED FOR CONCENTRATING TABLE TESTS

| Coal Identification Seam County, State | BCR Lot No. | Weight Percent, Dry Basis | | | | | Organic Sulfur as Percent of Total Sulfur |
|---|---|---|---|---|---|---|---|
| | | Ash | Total Sulfur | Sulfate Sulfur | Pyritic Sulfur | Organic Sulfur | |
| **Eastern Coals** | | | | | | | |
| No. 6-A Harrison, Ohio | 1735 | 10.4 | 2.50 | 0.01 | 1.86 | 0.63 | 25.2 |
| Upper Freeport Westmoreland, Pa. | 1750 | 22.5 | 3.74 | 0.03 | 3.22 | 0.49 | 13.1 |
| No. 8 Jefferson, Ohio | 1768 | 19.4 | 4.71 | 0.07 | 3.69 | 0.95 | 20.2 |
| No. 6 Columbiana, Ohio | 1745 | 10.4 | 2.44 | 0.05 | 1.79 | 0.60 | 24.6 |
| Lower Kittanning Indiana, Pa. | 1755 | 20.1 | 4.66 | 0.05 | 4.00 | 0.61 | 13.1 |
| **Midwestern Coals** | | | | | | | |
| Lower Cherokee | 2415 | 12.8 | 5.12 | 0.26 | 4.24 | 0.62 | |
| Upper Spandra | 2416 | 6.5 | 1.59 | 0.08 | 0.77 | 0.74 | |
| Tebo Seam | 2417 | 29.8 | 6.92 | 0.68 | 4.44 | 1.80 | |

Source:  PB-232 011

## TABLE 4.4: CONCENTRATING TABLE TESTS—EFFECT OF TWO-STAGE CLEANING

| Coal Identification Seam / County — State | BCR Lot No. | Feed Size | Feed Rate tph | Total Sulfur, Feed to Table* | Percent Recovery, Clean Coal | Total Sulfur, Clean Coal | Total Sulfur Reduction |
|---|---|---|---|---|---|---|---|
| **Eastern Coals** | | | | | | | |
| No. 6-A | 1950 | 3/8 in x 0 | 2.39 | 2.82 | 91.8 | 2.12 | 24.8 |
| Harrison, Ohio | 1950 | 30 Mesh x 0 | 1.13 | — | 87.5** | 1.40 | 50.4 |
| Upper Freeport | 2012 | 3/8 in x 0 | 2.66 | 3.34 | 84.9 | 1.76 | 47.3 |
| Westmoreland, Pa. | 2012 | 30 Mesh x 0 | 1.09 | — | 81.4** | 1.07 | 68.0 |
| No. 8 | 2013 | 3/8 in x 0 | 2.85 | 3.83 | 92.4 | 3.29 | 14.1 |
| Jefferson, Ohio | 2013 | 30 Mesh x 0 | .99 | — | 87.4** | 2.72 | 29.0 |
| No. 6 | 2026 | 3/8 in x 0 | 2.62 | 4.38 | 94.4 | 2.57 | 41.3 |
| Columbiana, Ohio | 2026 | 30 Mesh x 0 | 1.00 | — | 90.7** | 1.65 | 62.3 |
| Lower Kittanning | 2031 | 3/8 in x 0 | 2.55 | 4.77 | 83.5 | 1.98 | 58.5 |
| Indiana, Pa. | 2031 | 30 Mesh x 0 | .82 | — | 78.8** | 1.54 | 67.7 |
| **Midwestern Coals** | | | | | | | |
| Lower Cherokee | 2415 | 3/8 in x 0 | | *** | 89.3 | 3.53 | 33.7 |
| | | 30 Mesh x 0 | | | 95.5** | 3.14 | 42.8 |
| Upper Spadra | 2416 | 3/8 in x 0 | | *** | 96.4 | 1.21 | 24.7 |
| | | 30 Mesh x 0 | | | 96.5** | 0.97 | 40.5 |
| Tebo Seam | 2417 | 3/8 in x 0 | | *** | 77.6 | 4.57 | 42.9 |
| | | 30 Mesh x 0 | | | 93.9** | 4.29 | 50.8 |

*    Composite of Table Product
**   Based on 100 percent 3/8 inch x 0 raw coal feed.
***  See Table 4.4 for approximation

Source: PB-232 011

The concentrating spiral (see Figure 4.6) consists of a spiral conduit of modified semicircular cross section.

**FIGURE 4.6:  HUMPHREY'S SPIRAL CONCENTRATOR CLOSED CIRCUIT TEST UNIT**

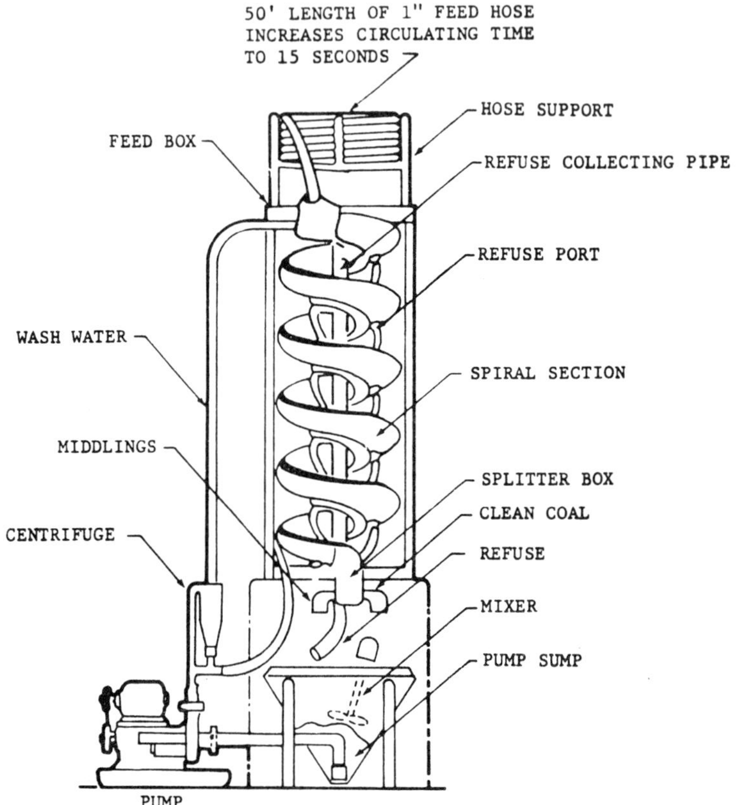


Source:  PB-232 011

In operation, pulp is fed from the top of the spiral and as it flows downward centrifugal force causes the heavier particles to concentrate in a band along the inner side of the spiral.  They are removed through adjustable ports located on each turn of the spiral at the lowest point in the cross section of the conduit. As the spiral stream is discharged from the lower end of the spiral, an adjustable splitter divides the stream into two products:  a clean coal product (outer coal) and a middling product (inner coal).

The Bureau of Mines evaluated spiraling on coals of top sizes ranging from 4 mesh down to 35 mesh and on double-sized fractions.  It was found that the best separation occurred when the top size of the spiral feed was minus 14 mesh or less.  Surprisingly, the capacity of the spiral remained about constant, regardless of the feed top size.  Consequently, the capacity of the spiral for washing fine-size coal was very high.

The results of spiraling 35 mesh by 0 Middle Kittanning bed coal are shown in Table 4.6.

**TABLE 4.6: PRODUCT ANALYSIS OF 35 MESH BY 0 MIDDLE KITTANNING BED COAL WASHED OVER THE COAL SPIRAL**

| | | Percent | | |
|---|---|---|---|---|
| Product | Weight | Ash | Pyritic Sulfur | Total Sulfur |
| *35 by 200 mesh, 62.2 percent of feed* | | | | |
| Clean coal | 88.3 | 7.2 | 0.29 | 0.92 |
| Refuse | 11.7 | 42.8 | 13.51 | 13.79 |
| Feed | 100.0 | 11.4 | 1.85 | 2.43 |
| *200 mesh by 0, 37.8 percent of feed* | | | | |
| Clean coal | 94.7 | 12.0 | 1.71 | 2.34 |
| Refuse | 5.3 | 21.4 | 8.48 | 8.96 |
| Feed | 100.0 | 12.5 | 2/07 | 2.69 |
| *Composite 35 mesh by 0, 100.0 percent of feed* | | | | |
| Clean coal | 90.7 | 9.1 | .86 | 1.48 |
| Refuse | 9.3 | 38.2 | 12.43 | 12.75 |
| Feed | 100.0 | 11.8 | 1.94 | 2.53 |

Source:  PB-232 011

A feed coal having a total sulfur content of 2.53% was reduced to 1.48% by spiraling at a clean coal recovery of 90.7%.  The final sulfur content of the washed 35 by 200 mesh coal was 0.92% sulfur, while the sulfur content of the washed minus 200 mesh coal was 2.34% sulfur.  This rapid deterioration of the efficiency of the washing technique with size reduction is a normal characteristic of specific gravity separators.  The excellent results achieved by the Bureau of Mines with the spiral on fine-size coal suggest it should be considered along with hydrocyclones and the wet concentrating tables as a "rougher" cleaning unit to remove pyrite from high sulfur coals prior to flotation.

Five different coals were used by Bituminous Coal Research, Inc., in their evaluation of the concentrating spiral for cleaning coal.  These coals were from the following seams:  No. 6-A seam, Harrison County, Ohio; Lower Kittanning seam, Indiana County, Pennsylvania; No. 6 seam, Columbiana County, Ohio; Upper Freeport seam, Westmoreland County, Pennsylvania; and No. 8 seam, Jefferson County, Ohio.

**TABLE 4.7: CONCENTRATING TABLE AND SPIRAL CONCENTRATOR TESTS—EFFECTS OF TWO-STAGE CLEANING**



| Coal Identification Lot | Seam | Location County, State | Total Sulfur, Feed to Table* Percent | Clean Coal, Two-stage Cleaning Percent | Total Sulfur Reduction, Two-stage Cleaning Percent |
|---|---|---|---|---|---|
| 2031 | Lower Kittanning | Indiana, Pennsylvania | 3.66 | 78.3 | 65.3 |
| 2026 | No. 6 | Columbiana, Ohio | 2.90 | 89.3 | 48.3 |
| 2012 | Upper Freeport | Westmoreland, Pennsylvania | 3.30 | 77.9 | 54.5 |
| 2013 | No. 8 | Jefferson, Ohio | 3.98 | 87.0 | 28.4 |

* Based on Composite of Table Products, Rough Cleaning

Source:  PB-232 011

Their sulfur contents are given in Table 4.4. Coal from the No. 6-A Ohio seam was tested at the 30 mesh x 0 size, while the other four were precleaned on the wet concentrating table at the ⅜ inch x 0 size and the clean-coal fraction then crushed to 30 mesh x 0. The fine coal was then cleaned on the spiral.

A 15.3% reduction in total sulfur, 94.8% recovery, was obtained by single stage cleaning of the 30 mesh x 0, Ohio 6-A seam coal. This result was obtained by combining the screen fractions of the middling material and the clean coal. For both the single and double-stage cleaning, it was found that all the clean coal fractions contained a significant quantity of high-ash, high-sulfur sink material. Results of the double-stage cleaning appear in Table 4.7. Comparison of these sulfur reductions with those obtained by cleaning the same coals with the wet concentrating table (Table 4.5) indicate that the spiral produces similar reductions and yields for two of the coals tested and somewhat lower reductions for the other two.

**Water Cyclone**

Water cyclone performance evaluation tests were conducted by the Bureau of Mines on five units in commercial plants. Preliminary results of these tests showed excellent pyritic sulfur reductions were attainable. The clean coal recoveries were highly dependent on the size of the material. Clean coal recovery of the plus 48 mesh material was low, while most of the minus 100 mesh material reported to the clean coal product.

Water cyclone tests were also conducted by Bituminous Coal Research, Inc. (6) (10). Feed coals were pulverized to minus 30 mesh size prior to test runs. Ohio 6-A seam coal was used in the first set of tests. For the remaining four tests, feed material was prepared by pulverizing the minus ⅜ inch clean fraction from the concentrating table to minus 30 mesh. This approach is analogous to the two-stage cleaning with the concentrating table.

In two-stage cleaning of the minus 30 mesh Ohio 6-A coal, the total sulfur was reduced from 2.90 to 1.59% (see Table 4.8).

**TABLE 4.8: COMPARISON OF CONCENTRATING TABLE RUNS AND COM-
POUND WATER CYCLONE RUN WITH 30 MESH x 0,
R.O.M., OHIO NO. 6-A SEAM**

| Cleaning Unit | Total Sulfur in Feed, Weight Percent | Recovery, Weight Percent | Total Sulfur Reduction, Weight Percent |
|---|---|---|---|
| Compound Water Cyclone (Test No. 3, Run No. 6) | 2.47 | 91.4 | 21.1 |
| Compound Water Cyclone (Test No. 2, Run No. 2) | 2.39 | 90.9 | 27.6 |
| Concentrating Table Run | 2.44 | 92.0 | 36.5 |

Source: PB-232 011

A comparison of the compound water cyclone run with the concentrating table run is shown in Table 4.9. This table indicates that the concentrating table performed better than the compound water cyclone in the cleaning of this coal. This finding is supported by the results of cyclone cleaning the other four coals (at 30 mesh x 0 size) which had been precleaned on the concentrating table. Table 4.8 gives results under one set of test conditions. These tests show smaller sulfur reductions for three of the four coals tested than were achieved by the wet concentrating table (Table 4.5). Bituminous Coal Research, Inc. believes that improved results in total sulfur reduction could have been obtained by using a closer-sized feed or feeds to the compound water cyclone, i.e., 30 mesh x 200 mesh and 200 mesh x 0 (6).

A diagram of McNally Visman Tricone of the type evaluated by Bituminous Coal Research, Inc., is shown in Figure 4.7.

**FIGURE 4.7: COMPOUND WATER CYCLONE**

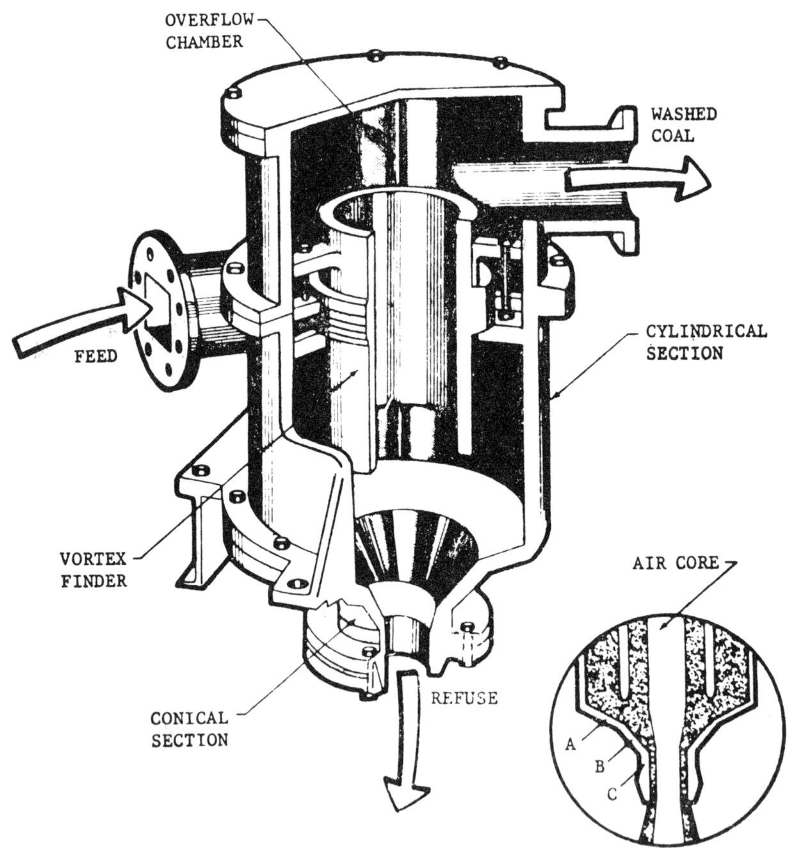


Source: PB-232 011

**TABLE 4.9: CONCENTRATING TABLE AND COMPOUND WATER CYCLONE TESTS—EFFECTS OF TWO-STAGE CLEANING**

Run No. 2 Operating Conditions

| Coal Identification Lot | Seam | Location County, State | Total Sulfur Feed to Table* Percent | Clean Coal Two-stage Cleaning Percent | Total Sulfur Reduction, Two-stage Cleaning Percent |
|---|---|---|---|---|---|
| 2031 | Lower Kittanning | Indiana, Pennsylvania | 3.66 | 73.2 | 57.9 |
| 2026 | No. 6 | Columbiana, Ohio | 2.90 | 80.3 | 44.8 |
| 2012 | Upper Freeport | Westmoreland, Pennsylvania | 3.30 | 69.9 | 62.7 |
| 2013 | No. 8 | Jefferson, Ohio | 3.98 | 78.8 | 34.7 |

*Based on Composite of Table Products, Rough Cleaning.

Run No. 2 Operating Conditions:
Cone Type "M", Vortex Finder Clearance: 1", Inlet Pressure: 8 psi, Feed Solids Concentration: 8.0%.

Source: PB-232 011

It consists of a cylindrical section, a compound conical section and a vortex finder.  In operation, coal/water tangentially enters near the top of the cylindrical section, forming a strong vertical flow.  Refuse moves along the wall of the cyclone and is discharged through the apex.  The washed coal passes through the vortex finder to the overflow chamber.  Washed coal is discharged from this chamber through the tangential outlet.  This particular design of the water-only cyclone has three conical sections.  Particles of different sizes and specific gravities form a hindered settling bed in the first conical section.  Light coarse particles are removed from the first conical section and light middlings are removed from the second conical section.  In the last conical section, the bed is destroyed and the heavy fractions pass through the apex.  The central current of the department water is weak at this point and only the fine, light particles are discharged through the vortex finder.

### Air Classifier

This is a two-stage process.  First air is used to separate the relatively coarse grind of pulverized coal into a fine fraction and a coarse, pyrite-rich fraction using the BCR-Majac and Alpine Zigzag air classifiers.  Pyrite can then be removed from the coarse fraction by wet-coal preparation techniques.

Float-and-sink analyses at 1.60 specific gravity were conducted on both the coarse and fine fractions.  The composite of the coarse and fine clean coal fractions was also analyzed.

Ten samples from the group of coals used for float-sink washability studies were selected for testing.  (Note that this group is not from the same sample lots as used for testing other types of cleaning equipment, e.g., the spiral.)  The reduction of total sulfur by use of the Majac unit ranged from 20.1 to 51.7%.  The results of the ten coals are given in Table 4.10.  One coal was also processed on the Alpine Zigzag unit (Table 4.11).

### Electrokinetics

The electrokinetic process for the physical cleaning of coal uses the technique of electrophoresis for separating the pyrite from the coal.  Electrophoresis is defined as the migration of electrokinetically charged particles in a liquid toward an electrode of opposite charge in a dc electrical field.  The speed of migration is directly proportional to the magnitude of the electrokinetic charge of the particles and the applied voltage and inversely proportional to the distance between the electrodes.  The electrokinetic process was investigated on a laboratory scale.

The application of electrophoresis to physically wash coal was based on earlier studies which showed that, while coal and pyrite were both negatively charged above a pH of 4, the coal was more negatively charged than the pyrites.  In the use of the electrophoretic technique for the separation of pyrites from coal, however, it was found that the electrophoretic mobilities of coal and pyrite were never sufficient to overcome the effects of the specific gravity and the size variance in particles.  Therefore, a modified cell was built to take into account not only the electrophoretic mobilities of coal and pyrite, but also their specific gravities (13).

## TABLE 4.10: PYRITIC SULFUR REDUCTION IN MAJAC AIR CLASSIFICATION TESTS

| BCR Lot No. | Seam | Total Sulfur in Raw Coal Percent | Pyritic Sulfur in Raw Coal, Percent | Recovery Majac Product Percent | Total Sulfur in Majac Product Percent* | Total Sulfur Reduction Percent | Pyritic Sulfur in Majac Product* Percent | Pyritic Sulfur Reduction, Percent |
|---|---|---|---|---|---|---|---|---|
| 1752 | Upper Kittanning | 2.71 | 2.31 | 90.4 | 1.31 | 51.7 | 0.94 | 59.3 |
| 1750 | Upper Freeport | 3.74 | 3.22 | 83.6 | 1.98 | 47.1 | 1.48 | 54.0 |
| 1771 | Lower Freeport | 2.54 | 1.89 | 92.6 | 1.49 | 41.3 | 0.98 | 48.1 |
| 1745 | No. 6 | 2.44 | 1.79 | 94.3 | 1.46 | 40.2 | 1.01 | 43.6 |
| 1770 | Thick Freeport | 2.08 | 1.68 | 90.9 | 1.25 | 39.1 | 0.92 | 45.2 |
| 1735 | No. 6-A | 2.50 | 1.86 | 93.2 | 1.65 | 34.0 | 0.98 | 47.3 |
| 1730 | No. 6 | 4.72 | 3.70 | 92.2 | 3.16 | 33.1 | 2.17 | 41.4 |
| 1747 | Lower Kittanning and Lower Freeport | 2.42 | 1.84 | 90.7 | 1.65 | 31.8 | 1.08 | 41.3 |
| 1757 | Freeport | 2.52 | 2.16 | 77.5 | 1.90 | 24.6 | 1.39 | 35.6 |
| 1733 | No. 8 | 4.58 | 2.84 | 87.6 | 3.66 | 20.1 | 1.95 | 31.3 |

* Majac product composed of raw fine coal plus cleaned coarse fraction (1.60 specific gravity)

Source: PB-232 011

## TABLE 4.11: SULFUR REDUCTION IN ALPINE ZIGZAG CLASSIFIER TESTS

A--Total Sulfur Reduction

| BCR Lot No. | Seam | Rank | Total Sulfur in Raw Coal, Percent | Zigzag Product, Percent | Total Sulfur in Zigzag Product,* Percent | Total Sulfur Reduction, Percent |
|---|---|---|---|---|---|---|
| 1733 | No. 8 | HVC | 4.58 | 84.1 | 3.82 | 16.6 |

* Zigzag product composed of raw fine coal plus cleaned coarse fraction (1.60 specific gravity)

B--Pyritic Sulfur Reduction

| BCR Lot No. | Seam | Rank | Pyritic Sulfur in Raw Coal, Percent | Zigzag Product, Percent | Pyritic Sulfur in Zigzag Product,** Percent | Pyritic Sulfur Reduction, Percent |
|---|---|---|---|---|---|---|
| 1733 | No. 8 | HVC | 2.84 | 84.1 | 1.90 | 33.1 |

** Zigzag product composed of raw fine coal plus cleaned coarse fraction (1.60 specific gravity)

Source:  PB-232 011

The electrophoretic cell consisted of a rectangular plastic column with inside dimensions of ³⁄₈ inch by ³⁄₈ inch by 24 inches. Platinum electrode wires were used to produce a dc field. The column was longitudinally inclined at an angle of 45 degrees and laterally inclined at angles varying from 4 to 11 degrees. The anode was located on the high side of the laterally inclined column and the cathode was on the lower side. During operation, the coal slurry was fed into the top of the column and, as the slurry moved down the length of the column with an applied dc field, the heavier and less negatively charged pyrites tended to slide toward the cathodic side and the coal was drawn toward the anode.

Initial tests were conducted with the electrophoresis column using synthetic coal-pyrite mixtures. Next, a series of tests were conducted on a high-pyritic-sulfur coal. Results of three test runs are given in Table 4.12.

## TABLE 4.12: FRACTIONATION OF A SAMPLE OF 100 BY 325 MESH LOWER FREEPORT BED COAL IN AN ELECTROPHORESIS COLUMN

| TEST | APPLIED VOLTAGE VOLTS, DC | AVERAGE CURRENT MILLIAMPS | RECEPTACLE | RECOVERY, PERCENT | ASH, PERCENT | PYRITIC SULFUR, PERCENT | TOTAL SULFUR, PERCENT |
|---|---|---|---|---|---|---|---|
| 1 | 500 | 5.0 | Coal | 94.3 | 13.2 | 1.01 | 1.61 |
|  |  |  | Reject | 5.7 | 45.0 | 19.73 | 19.95 |
| 2 | 400 | 2.0 | Coal | 55.6 | 17.0 | .34 | 1.07 |
|  |  |  | Reject | 44.4 | 14.8 | 4.63 | 5.58 |
| 3 | 250 | 1.5 | Coal 1st pass | 13.8 | 44.5 | .38 | .61 |
|  | 500 | 5.0 | Coal 2nd pass | 77.3 | 8.9 | .62 | 1.38 |
|  |  |  | Reject | 8.9 | 44.8 | 21.00 | 22.65 |

Source: PB-232 011

The laboratory-scale investigation indicated that this was an effective method to remove pyrite and other impurities such as silica from coal. Economic assessments, unfortunately, indicated that the technique would be too costly for commercial application. Therefore further development was not pursued. The electrophoresis column could probably best be used as a laboratory instrument for investigating the physical desulfurization of coal.

**Agglomo-Separation (Oil)**

The Agglomeration-Separation process developed by the National Research Council of Canada was investigated by the Bureau of Mines. The process was reported to recover a product low in ash approaching the clean coal product. It was felt that a similar sulfur reduction might be obtained. In addition, the process had the capability to agglomerate fine-size coal to a larger size and eventually to provide for pelletization.

High energy mixing is employed to disperse kerosine or similar light oils in a coal-water slurry. The oil selectively coats the coal particles causing the particles to agglomerate and then, with an assist from entrained air, rise to the surface, while the impurities remain in the water.  The investigation was conducted on both 35 mesh x 0 and 325 mesh x 0 sized coals using kerosine as the agglomerating oil in slurries containing 10% of coal by weight.  Experiments were conducted on both the laboratory and bench scale.  A blender-type mixer was used in the laboratory experiments, and a 4-gallon cell was used for bench-scale testing.

Four variables—kerosine dosage, retention time, slurry pH, and mixer speed—were evaluated in the laboratory.  Kerosine dosage had the most significant effect.  Generally, a concentration of 8% kerosine, based on the weight of dry coal in the slurry, produced the best results.  At lower kerosine levels the recovery of clean coal decreased, while using larger amounts of kerosine provided minimal increases in recovery of clean coal.  Increasing the retention time from 1 to 2 minutes increased recovery.  Beyond 2 minutes, however, there was no increase in recovery.  Neither the pH of the slurry nor the speed of mixing had a significant effect on the product recovery.

Tables 4.13 and 4.14 show the results of two samples of 325 mesh x 0 size coals evaluated in the laboratory.

## TABLE 4.13:  EVALUATION OF THE AGGLOMO-SEPARATION PROCESS FOR THE LOWER FREEPORT BED COAL USING FOUR PROCESS VARIABLES

```
Coal Size                       Minus 325 mesh
Total Sulfur, percent           2.5
Ash, percent                    19.3
Agglomerating Oil               Kerosine
Slurry, percent solids          10
```

| TEST NO. | KEROSINE DOSAGE, PERCENT | AGITATOR SPEED, RPM | RETENTION TIME, MIN. | SLURRY pH | CLEAN COAL PERCENT | | |
|---|---|---|---|---|---|---|---|
| | | | | | YIELD | SULFUR | ASH |
| 1 | 2 | 10,000 | 1 | 6.0 | 58.5 | 1.9 | 12.8 |
| 2 | 4 | 10,000 | 1 | 6.0 | 70.5 | 2.2 | 11.2 |
| 3 | 8 | 10,000 | 1 | 6.0 | 81.8 | 2.1 | 11.2 |
| 4 | 2 | 10,000 | 2 | 6.0 | 60.4 | 2.0 | 12.8 |
| 5 | 4 | 10,000 | 2 | 6.0 | 75.0 | 1.9 | 11.2 |
| 6 | 8 | 10,000 | 2 | 6.0 | 82.5 | 2.2 | 10.5 |
| 7 | 2 | 10,000 | 1 | 10.0 | 68.5 | 1.9 | 12.1 |
| 8 | 4 | 10,000 | 1 | 10.0 | 78.5 | 1.9 | 12.0 |
| 9 | 8 | 10,000 | 1 | 10.0 | 82.1 | 2.1 | 10.1 |
| 10 | 2 | 10,000 | 2 | 10.0 | 77.4 | 1.8 | 10.0 |
| 11 | 4 | 10,000 | 2 | 10.0 | 81.5 | 2.0 | 8.7 |
| 12 | 8 | 10,000 | 2 | 10.0 | 82.7 | 2.2 | 9.2 |
| 13 | 2 | 14,000 | 1 | 6.0 | 62.0 | 1.8 | 11.6 |
| 14 | 4 | 14,000 | 1 | 6.0 | 70.1 | 1.9 | 11.6 |
| 15 | 8 | 14,000 | 1 | 6.0 | 82.6 | 2.1 | 10.6 |
| 16 | 2 | 14,000 | 2 | 6.0 | 70.3 | 1.8 | 10.6 |
| 17 | 4 | 14,000 | 2 | 6.0 | 75.4 | 1.9 | 12.3 |
| 18 | 8 | 14,000 | 2 | 6.0 | 83.9 | 2.0 | 9.2 |

(continued)

**TABLE 4.13:  (continued)**

| TEST NO. | KEROSINE DOSAGE, PERCENT | AGITATOR SPEED, RPM | RETENTION TIME, MIN. | SLURRY pH | CLEAN COAL PERCENT | | |
|---|---|---|---|---|---|---|---|
| | | | | | YIELD | SULFUR | ASH |
| 19 | 2 | 14,000 | 1 | 10.0 | 68.6 | 1.7 | 10.7 |
| 20 | 4 | 14,000 | 1 | 10.0 | 76.3 | 2.0 | 11.9 |
| 21 | 8 | 14,000 | 1 | 10.0 | 85.0 | 2.1 | 9.3 |
| 22 | 2 | 14,000 | 2 | 10.0 | 74.5 | 1.8 | 10.2 |
| 23 | 4 | 14,000 | 2 | 10.0 | 78.5 | 1.9 | 8.8 |
| 24 | 8 | 14,000 | 2 | 10.0 | 83.6 | 2.1 | 8.5 |

Source:  PB-232 011


**TABLE 4.14:  EVALUATION OF THE AGGLOMO-SEPARATION PROCESS FOR THE PITTSBURGH BED COAL USING FOUR PROCESS VARIABLES**

| Coal Size | Minus 325 mesh |
|---|---|
| Total Sulfur, percent | 2.3 |
| Ash, percent | 16.1 |
| Agglomerating Oil | Kerosine |
| Slurry, percent solids | 10 |

| Test no. | Kerosine dosage, percent | Agitator speed, rpm | Retention time, min. | Slurry pH | Clean Coal Percent | | |
|---|---|---|---|---|---|---|---|
| | | | | | yield | sulfur | ash |
| 25 | 2 | 6,000 | 1 | 3.0 | 18.1 | 1.7 | 10.9 |
| 26 | 4 | 6,000 | 1 | 3.2 | 54.1 | 1.6 | 11.7 |
| 27 | 8 | 6,000 | 1 | 2.9 | 60.4 | 1.7 | 9.9 |
| 28 | 2 | 6,000 | 2 | 3.1 | 17.1 | 1.5 | 8.8 |
| 29 | 4 | 6,000 | 2 | 3.0 | 49.5 | 1.7 | 10.5 |
| 30 | 8 | 6,000 | 2 | 3.0 | 63.5 | 1.7 | 10.0 |
| 31 | 2 | 6,000 | 1 | 10.0 | 39.8 | 1.8 | 12.0 |
| 32 | 4 | 6,000 | 1 | 10.0 | 54.5 | 1.6 | 11.0 |
| 33 | 8 | 6,000 | 1 | 10.0 | 65.8 | 1.8 | 10.6 |
| 34 | 2 | 6,000 | 2 | 10.0 | 36.3 | 1.6 | 10.7 |
| 35 | 4 | 6,000 | 2 | 10.0 | 53.3 | 1.8 | 10.5 |
| 36 | 8 | 6,000 | 2 | 10.0 | 60.6 | 1.6 | 9.0 |
| 37 | 2 | 12,000 | 1 | 3.2 | 35.9 | 1.7 | 9.7 |
| 38 | 4 | 12,000 | 1 | 3.1 | 52.9 | 1.7 | 9.8 |
| 39 | 8 | 12,000 | 1 | 3.4 | 68.1 | 1.7 | 10.1 |
| 40 | 2 | 12,000 | 2 | 3.0 | 24.4 | 1.4 | 8.0 |
| 41 | 4 | 12,000 | 2 | 3.1 | 48.8 | 1.6 | 9.7 |
| 42 | 8 | 12,000 | 2 | 3.2 | 64.7 | 1.6 | 10.3 |
| 43 | 2 | 12,000 | 1 | 10.0 | 40.1 | 1.6 | 10.8 |
| 44 | 4 | 12,000 | 1 | 10.0 | 53.4 | 1.6 | 10.2 |
| 45 | 8 | 12,000 | 1 | 10.0 | 62.6 | 1.6 | 9.0 |
| 46 | 2 | 12,000 | 2 | 10.0 | 42.3 | 1.5 | 10.3 |
| 47 | 4 | 12,000 | 2 | 10.0 | 56.3 | 1.7 | 10.7 |
| 48 | 8 | 12,000 | 2 | 10.0 | 67.5 | 1.4 | 9.0 |

Source:  PB-232 011

Reductions in total sulfur range from about 12 to 39% while yields of clean coal range from 18 to 85%, depending on test conditions and the type of coal. Under the same conditions ash reduction ranges from 25 to 60%.

The agglomo-separation process provided only limited sulfur reduction but a substantial reduction in ash. The oil coats the pyrite as well as the coal particles, resulting in agglomerates containing both the clean coal and the pyrite. There is usually some sulfur reduction accompanying low recoveries of clean coal due to low concentrations of agglomerating oil but, as the oil concentration is increased to obtain maximum clean coal recovery, the percentage of sulfur in the product is usually equal to or greater than that of the raw coal.

The laboratory work was scaled up to a 500 pound per hour pilot plant consisting of two 4-gallon capacity cells equipped with turbine blade impellers, a Bird dewatering screen bowl centrifuge, and a Dravo 39 inch pelletizing disc. With this arrangement, a 10% slurry of 35 mesh x 0 size coal was agglomerated with 8% kerosine. The agglomerated clean coal was dewatered to about 16% moisture content in the centrifuge and then pelletized on the disc-pelletizer using an asphalt binder. This step was performed to study pelletization as a means of overcoming some of the problems encountered in drying and handling fine-sized coals. Pellets made with approximately 5% asphalt binder averaged ½ inch in diameter and contained about 16% moisture. The pellets were firm when wet and had good strength and water resistance when dried.

The quality of the clean coal produced by the pilot plant was similar to that obtained in the laboratory unit. The sulfur reduction was negligible, but the recovery of low ash coal was good.

## Froth Flotation

"Froth flotation is a process for separating fine-size particles by selective attachment of air bubbles to coal particles, causing them to be buoyed up into a froth while leaving the refuse particles in the water. This process deals with fine particles in a rather turbulent and foamy aqueous system where specific gravity is not as significant as the surface properties of the particles; although much heavier than the coal, due to surface characteristics the pyrite remains with the coal in the froth product" (12).

The process consists of thoroughly mixing the fine raw coal particles with water. A frothing agent is added, and air is bubbled through the mixture, causing a froth to be produced at the surface. This froth, containing primarily the hydrophobic components, is skimmed away, thereby separating the hydrophobic from the hydrophilic fractions which remain suspended in the bulk of water. The froth flotation process is reasonably new for coal processing, although it has long been used in the mineral industry to effect separations between various minerals often of a very complex nature.

The research program to determine the pyrite reduction potential of the froth flotation process pursued two approaches. These are the depression of coal, and the depression of pyrite. The depression of pyrite approach, involving a one-stage process, was not encouraging. The detailed procedures, results, and conclusions are presented in Reference (15).

The second approach was concerned with a two-stage flotation process for separating the pyrite remaining in the "initial" froth flotation product. This product from a conventional single-stage flotation process is repulped in fresh water; treated with a coal flotation depressant, a pyrite collector, and a frother; and is rewashed in a second bank of froth flotation cells. The pyrites are removed with the froth, leaving a depressed fraction of clean coal containing less pyrites. This two-stage process can be reversed, i.e., the pyrite is floated (and the coal is depressed) in the first stage and the coal is conventionally floated in the second stage. The two-stage froth flotation process was developed because in the one-stage process the fine-sized pyrites reported with the froth product. Work on the two-stage flotation process in a half ton per hour capacity pilot plant has been completed.

Detailed results of the froth flotation research are presented in Tables 4.15 and 4.16 and in References (15) and (16). A more complete description of the process may be found in References (12) and (17).

## TABLE 4.15: TWO-STAGE FLOTATION RESULTS WITH LOWER FREEPORT BED COAL SLURRY

| | | Analysis, percent | | | Distribution, percent | | |
| Product | Weight, percent | Ash | Total sulfur | Pyritic sulfur | Ash | Total sulfur | Pyritic sulfur |
|---|---|---|---|---|---|---|---|
| **Test 1:** First stage; 40 g/ton MIBC<br>Second stage; 250 g/ton Aero Depressant 633,<br>150 g/ton potassium amyl xanthate,<br>40 g/ton MIBC | | | | | | | |
| Clean coal 2 | 64.2 | 10.0 | 1.11 | 0.51 | 22.2 | 30.8 | 20.5 |
| Reject 2 | 9.1 | 15.5 | 7.06 | 6.54 | 4.9 | 27.8 | 37.2 |
| Clean coal 1 | 73.3 | 10.7 | 1.85 | 1.26 | 27.1 | 58.6 | 57.7 |
| Reject 1 | 26.7 | 79.8 | 3.57 | 3.51 | 72.9 | 41.4 | 42.3 |
| Feed | 100.0 | 28.9 | 2.31 | 1.60 | 100.0 | 100.0 | 100.0 |
| **Test 2:** First stage; 40 g/ton MIBC<br>Second stage; 500 g/ton Aero Depressant 633,<br>200 g/ton potassium amyl xanthate,<br>40 g/ton MIBC | | | | | | | |
| Clean coal 2 | 63.5 | 9.3 | 1.03 | .46 | 20.4 | 28.8 | 16.2 |
| Reject 2 | 8.1 | 17.0 | 7.93 | 7.35 | 4.8 | 28.3 | 33.1 |
| Clean coal 1 | 71.6 | 10.2 | 1.81 | 1.24 | 25.2 | 57.1 | 49.3 |
| Reject 1 | 28.4 | 76.2 | 3.43 | 3.22 | 74.8 | 42.9 | 50.7 |
| Feed | 100.0 | 28.9 | 2.27 | 1.80 | 100.0 | 100.0 | 100.0 |
| **Test 3:** First stage; 40 g/ton MIBC<br>Second stage; 350 g/ton Aero Depressant 633,<br>250 g/ton potassium amyl xanthate,<br>40 g/ton MIBC | | | | | | | |
| Clean coal 2 | 59.7 | 9.8 | .94 | .32 | 20.4 | 24.3 | 10.4 |
| Reject 2 | 12.7 | 13.4 | 5.84 | 5.06 | 5.9 | 32.1 | 35.1 |
| Clean coal 1 | 72.4 | 10.4 | 1.80 | 1.15 | 26.3 | 56.4 | 45.5 |
| Reject 1 | 27.6 | 77.2 | 3.61 | 3.60 | 73.7 | 43.6 | 54.5 |
| Feed | 100.0 | 28.7 | 2.31 | 1.83 | 100.0 | 100.0 | 100.0 |

Source: PB-232 011

**TABLE 4.16: TWO-STAGE FLOTATION RESULTS WITH MIDDLE KITTANNING BED COAL SLURRY**

| Product | Weight Percent | Analysis, Percent | | | Distribution, Percent | | |
|---|---|---|---|---|---|---|---|
| | | Ash | Total Sulfur | Pyritic Sulfur | Ash | Total Sulfur | Pyritic Sulfur |
| Test 1: First stage; 300 g/ton kerosine, 80 g/ton MIBC | | | | | | | |
| Second stage; 250 g/ton Aero Depressant 633, 150 g/ton potassium amyl xanthate, 40 g/ton MIBC | | | | | | | |
| Clean coal 2 | 74.5 | 3.2 | 1.22 | 0.69 | 32.2 | 31.6 | 21.9 |
| Reject 2 | 10.1 | 5.6 | 3.23 | 2.68 | 7.6 | 11.3 | 11.5 |
| Clean coal 1 | 84.6 | 3.5 | 1.46 | .93 | 39.8 | 42.9 | 33.4 |
| Reject 1 | 15.4 | 28.8 | 10.65 | 10.18 | 60.2 | 57.1 | 66.6 |
| Feed | 100.0 | 7.4 | 2.88 | 2.35 | 100.0 | 100.0 | 100.0 |
| Test 2: First stage; 300 g/ton kerosine, 80 g/ton MIBC | | | | | | | |
| Second stage; 350 g/ton Aero Depressant 633, 250 g/ton potassium amyl xanthate, 40 g/ton MIBC | | | | | | | |
| Clean coal 2 | 76.1 | 2.7 | .95 | .33 | 27.4 | 25.2 | 10.6 |
| Reject 2 | 9.0 | 10.5 | 6.22 | 6.22 | 12.6 | 19.5 | 23.5 |
| Clean coal 1 | 85.1 | 3.5 | 1.51 | .95 | 40.0 | 44.7 | 34.1 |
| Reject 1 | 14.9 | 30.1 | 10.68 | 10.53 | 60.0 | 55.3 | 65.9 |
| Feed | 100.0 | 7.5 | 2.87 | 2.38 | 100.0 | 100.0 | 100.0 |
| Test 3: First stage; 300 g/ton kerosine, 80 g/ton MIBC | | | | | | | |
| Second stage; 500 g/ton Aero Depressant 633, 350 g/ton potassium amyl xanthate, 40 g/ton MIBC | | | | | | | |
| Clean coal 2 | 78.8 | 2.6 | .89 | .34 | 30.1 | 24.4 | 11.4 |
| Reject 2 | 9.9 | 12.2 | 7.83 | 7.23 | 17.8 | 27.0 | 30.5 |
| Clean coal 1 | 88.7 | 3.7 | 1.66 | 1.11 | 47.9 | 51.4 | 41.9 |
| Reject 1 | 11.3 | 31.7 | 12.30 | 12.13 | 52.1 | 48.6 | 58.1 |
| Feed | 100.0 | 6.8 | 2.87 | 2.35 | 100.0 | 100.0 | 100.0 |

Source:  PB-232 011

Of the innovative processes for pyrite separation investigated by the Bureau of Mines under the EPA Clean Coal Program, the two-stage froth flotation technique appears to show the greatest potential for scale-up to commercial operation.  The importance of this process results from the fact that no other way has been identified to remove pyrite from fine coals.

## SUMMARY AND CONCLUSIONS

### Washability Studies

Sulfur is found in raw coal in two major forms:  organic and inorganic (pyritic) sulfur.  Organic sulfur is chemically bound to the organic structure of the coal and cannot be removed by use of conventional washing methods.  The pyritic sulfur occurs in particles of varying size mixed with the coal.  By crushing (to liberate the pyritic sulfur) and washing (to remove the liberated particles), the total amount of sulfur in the coal may be reduced.  The degree of removal of the pyritic sulfur is dependent on the size of the particles of pyrites; the finer the particles, the more difficult to remove by washing.  The total sulfur reduction, therefore, is dependent on the amount of organic sulfur and the effects of crushing to liberate the pyritic sulfur.  The desulfurization process results (as do most beneficiation processes) in some loss in total heat content of the processed coal.

A summary of 322 washability tests reported by the Bureau of Mines shows that the organic sulfur (see Figure 4.4) and the effects of crushing (Figure 4.8) vary widely, as would be expected, when washability data of coals throughout the United States are combined.

The average total sulfur content of all coal studied in the Bureau of Mines reported investigation of coal washability is 3.23% at a raw coal top size of ⅜ inch (Figure 4.9).  When washed to a 90% yield, this average was reduced to 1.95%; reducing the yield to 60% results in a 1.82% average value.  Figure 4.10 shows the average pyritic sulfur content of the raw coal is 2.05%.  At 90% yield this average was reduced to 0.75%; at 60% yield it was reduced to 0.51%.  On the average, regardless of yield, the organic sulfur content of the coals was 1.2%. The average percent total sulfur reduction and the average percent pyritic sulfur reduction at the varied yields are shown in Figures 4.11 and 4.12.  Figure 4.4 shows that at a 60% yield only 20 to 25% of the coals sampled could produce a 1% or less total sulfur washed coal.

The production of 1% or less total sulfur coal by means of physical cleaning is governed by the amount of organic sulfur and particle sizes of the pyrite in the coals (the fine, intimately mixed pyrite particles are impossible to remove by conventional washing processes), regardless of cost.  In general, the lower the level of total sulfur required in the washed coal, the higher the cost of cleaning due to the lower recovery of the washed coal (i.e., greater loss of coal values during washing).

The three regions indicating a potential for washing to a total sulfur content of 1% or less are the Northern Appalachian, the Southern Appalachian, and Eastern Interior or Midwest Region.  The remaining coal producing regions given less attention or excluded are:  the Western Interior (the coals in this region are characteristically high in sulfur content regardless of the physical treatment used) and Northern and Southern Rocky Mountain (coals are of low sulfur content).

As shown by Figures 4.8 and 4.9, coal crushed to ⅜ inch top size and cleaned to yield in the order of 90% will provide the predominant benefit.  The figures show that, on the average, sulfur content will decrease by less than 10% with an additional yield sacrifice of 30% (i.e., 90 to 60%).  Even so, it must be emphasized that individual cases may vary and must be independently assessed.

In addition to the Bureau of Mines reported washability studies, 67 washability tests on Illinois coals were reported by the Illinois Geological Survey. The test samples were taken from mines located in significant mining areas in the state. These tests indicated that most Illinois coals have 3 to 5% total sulfur and only in those coals having a relatively low-sulfur content as mined could the total sulfur content be reduced to 1.5% or less by washing techniques. Of the 67 coal samples tested, 6 could be washed to a total sulfur level of 1% and 10 could be washed to 1.5% with an 80% minimum recovery.

**FIGURE 4.8: EFFECT OF CRUSHING ON LIBERATION OF TOTAL SULFUR, SUMMARY OF ALL COALS**

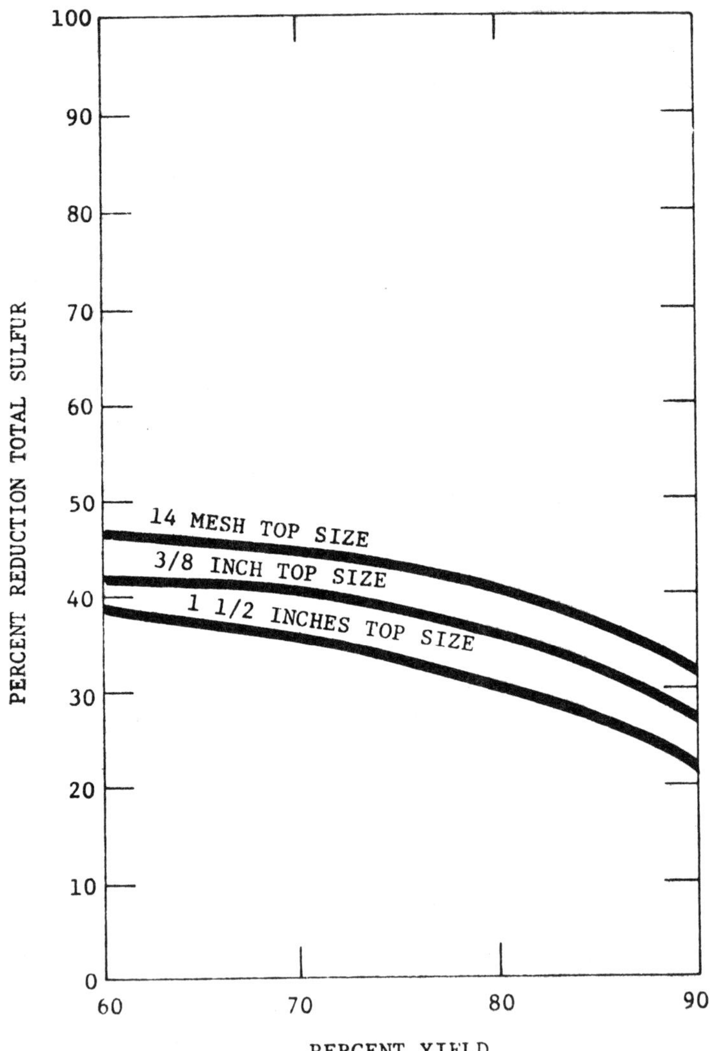


Source: PB-232 011

## FIGURE 4.9: AVERAGE TOTAL SULFUR CONTENT, ±1 STANDARD DEVIATION AT ⅜ INCH TOP SIZE, SUMMARY OF ALL COALS

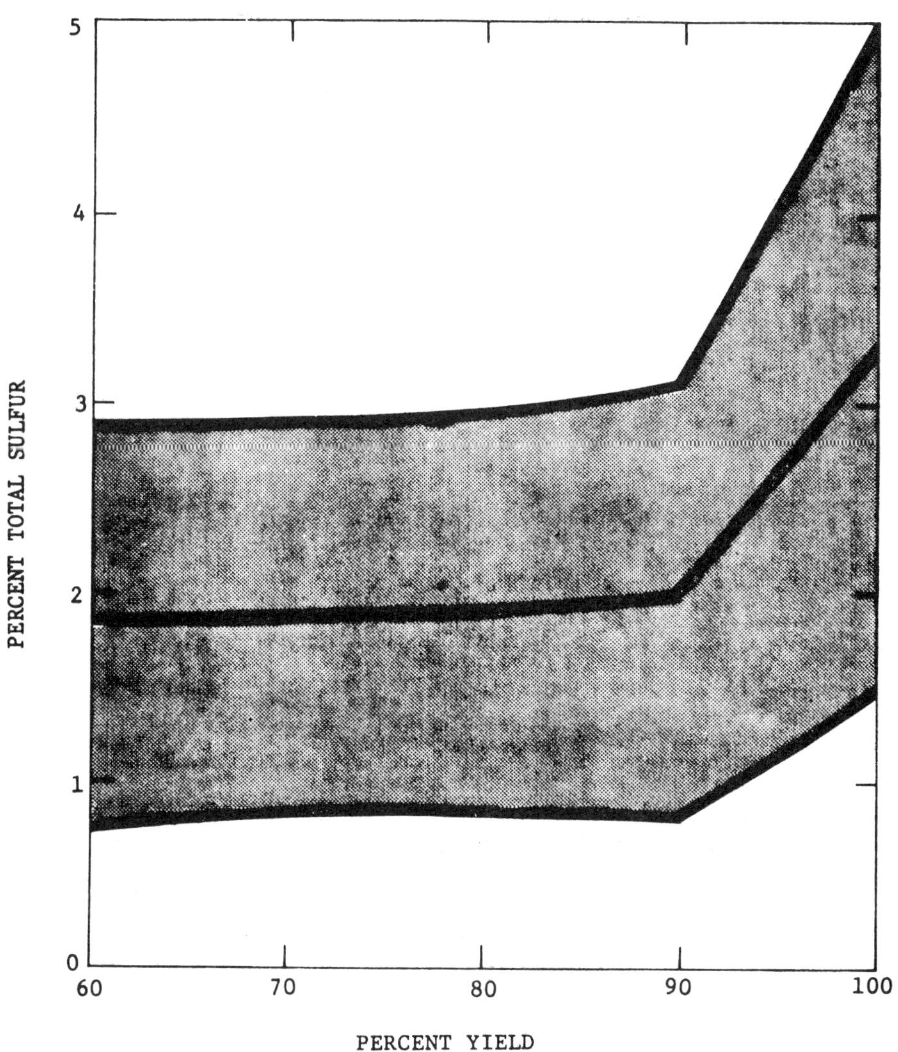


Source:  PB-232 011

FIGURE 4.10:  AVERAGE PYRITIC SULFUR CONTENT, ±1 STANDARD DEVIATION AT ⅜ INCH TOP SIZE, SUMMARY OF ALL COALS

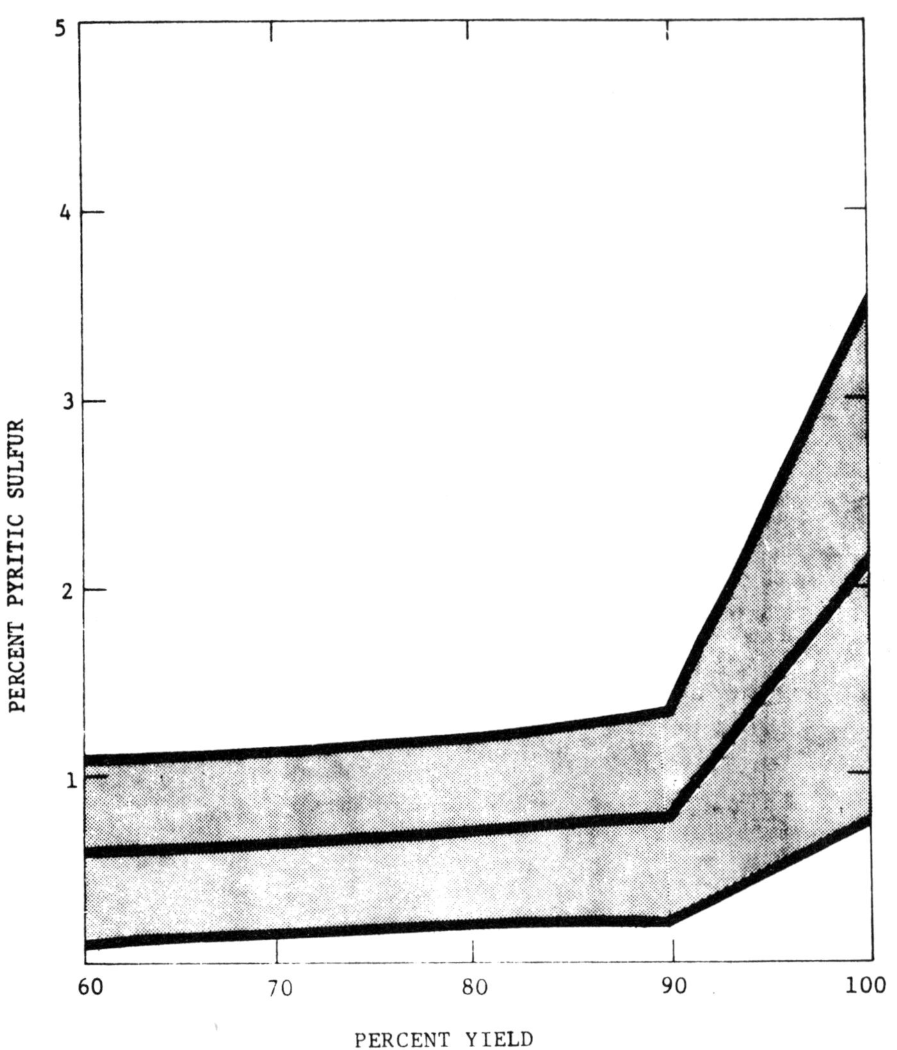


Source:  PB-232 011

**FIGURE 4.11:   AVERAGE PERCENT TOTAL SULFUR REDUCTION
±1 STANDARD DEVIATION AT ⅜ INCH TOP SIZE,
SUMMARY OF ALL COALS**

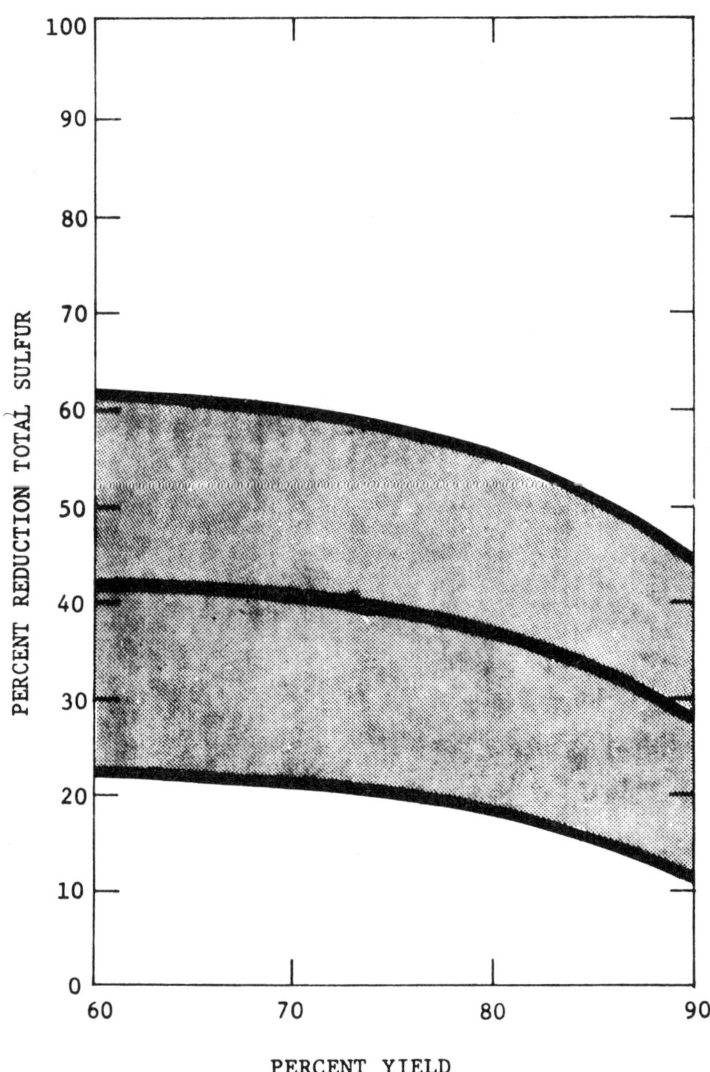


Source:   PB-232 011

**FIGURE 4.12: AVERAGE PERCENT PYRITIC SULFUR REDUCTION,**
**±1 STANDARD DEVIATION AT ⅜ INCH TOP SIZE,**
**SUMMARY OF ALL COALS**

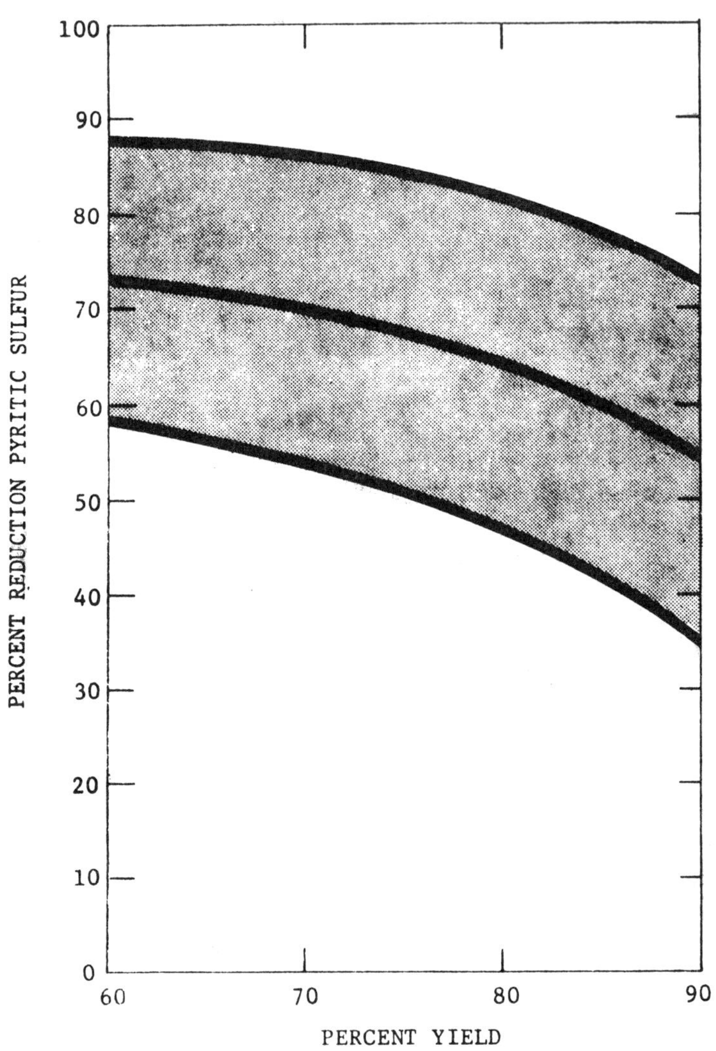



Source: PB-232 011

## Summary of Pyrite Removal Studies

The goals of EPA-sponsored laboratory and pilot study projects conducted by the Bureau of Mines and Bituminous Coal Research, Incorporated were to determine the operating parameters for maximum pyrite separation, to modify existing techniques and/or develop new techniques to separate pyrite from fine coal, and to evaluate methods to process coal cleaning rejects so as to concentrate the pyrite and reclaim clean coal products.

*Wet Concentrating Table:*  In tests, the wet concentrating table proved effective in removing free pyrite from coal.  Coals with a high proportion of pyritic sulfur relative to organic sulfur are particularly amenable to sulfur reduction by this process.  Two-stage cleaning tests, in which a coarse sample was tabled and the clean fraction pulverized and retabled, indicated that pyrite removal can be enhanced by use of the concentrating table in two stages.

*Concentrating Spiral:*  The concentrating spiral experienced a rapid deterioration of washing efficiency with feed size below 35 mesh, this being a normal characteristic of specific gravity separators.  Overall, the excellent results the concentrating spiral provided on fine-size coal suggest that it should be considered as a "rougher" cleaning unit to remove pyrite from high sulfur coals prior to flotation.  Double-stage cleaning indicate that for some coals the spiral process produces similar reductions and yields to those obtained with the wet concentrating table.

*Water-Only Cyclone:*  Preliminary results of the Bureau of Mines water cyclone performance evaluation tests showed excellent pyritic sulfur reductions were obtainable.  The clean coal recoveries were highly dependent on the size of the material.  Clean coal recovery of plus 48 mesh material was low while most of the minus 100 mesh material reported to the clean coal product.

Water cyclone tests were also conducted by Bituminous Coal Research, Inc. (BCR, Inc.).  Feed coals were pulverized to minus 30 mesh prior to test runs. BCR, Inc., results indicated that in general the compound water cyclone provided smaller sulfur reductions than were achieved by the concentrating table. However, BCR, Inc., believes that improved results in total sulfur reduction could be obtained by using a closer-sized feed to the water cyclone.

*Air Classifier:*  Air is used to separate a relatively coarse grind of pulverized coal into a fine fraction and a coarse, pyrite-rich fraction using the BCR-Majac and Alpine Zigzag air classifiers.  Pyrite could then be removed from coarse fractions by wet-coal preparation techniques.  The reduction of total sulfur through use of the Majac unit ranged from 20.1 to 51.7% for 10 samples.  Only one coal was processed on the Alpine Zigzag unit.

*Electrokinetics:*  The electrokinetic process for the physical cleaning of coal uses the technique of electrophoresis for separating the pyrite from the coal.  The laboratory-scale investigation indicated that pyrite and other impurities such as silica could be separated from the coal.  The technique, unfortunately, was deemed uneconomical for scaling up to the pilot stage as a coal cleaning procedure.

*Agglomo-Separation (Oil):* An oil agglomeration process for the recovery of clean coal from slime-size material and the sulfur and ash reduction potential of this process was investigated by the Bureau of Mines. This process provided negligible reduction in sulfur; however, the reduction in coal ash content was good.

*Froth Flotation:* Froth flotation is a process for separating fine-size particles by selective attachment of air bubbles to coal particles, causing the coal to be buoyed up into a froth while leaving the refuse particles in the water. Of the several innovative processes for pyrite separation investigated under the EPA Clean Coal Program, the froth flotation technique appears to show the greatest potential for scale-up to commercial operation.

**Prototype Coal Cleaning Plants**

The washability studies and pilot plant investigations indicated the amenability of a wide range of coals to physical desulfurization under carefully controlled conditions. Although testing for pyrite removal could be conducted to a limited degree in some existing preparation plants, the equipment and cleaning circuits of such plants are not normally arranged to maximize sulfur reduction but to meet ash and Btu specifications. Ideally, a plant to be used for investigating and demonstrating desulfurization methods at high through-put capacities would provide flexibility of equipment configurations, cleaning circuits, and operating conditions.

In order to determine the scope and cost of conducting investigations and demonstrations of pyrite removal using commercial cleaning equipment on a scale approximating commercial capacity, EPA funded two independent and competitive studies for designing a prototype coal cleaning plant of 50 to 100 tons per hour capacity and for estimating the capital and operating cost for the plant. The study objectives were to produce a prototype plant design, together with cost estimates for building, equipping and operating a plant capable of reducing the sulfur of coals having widely variable washability characteristics.

The prototype plant design and cost studies were conducted by the McNally Pittsburg Manufacturing Corporation and the Roberts and Schaefer Company of Chicago. The major difference in the two studies was plant design; Roberts and Schaefer's design was the more complex of the two; however, it offered the most flexibility.

The Roberts and Schaefer cost estimate [1974 figures] for a 52-month program covering plant cost, operating cost, laboratory services, and contractor's fee came to $16,457,965. In addition, a cost estimate was provided for a modified, alternate, lower price program that would not unduly sacrifice program objectives. The cost estimate for the alternate 44-month program (instead of the 52-month basic program) amounted to $8,923,200.

The McNally Pittsburg cost estimate [1974 figures] for a 52-month program covering plant cost and operating cost, which include laboratory services, amounted to $11,406,775. Even though not specifically stated, it is assumed that the contractor's fee charge is included in the above amount.

## Conclusions

Information was developed on the sulfur forms and distribution of sulfur in coal samples from mines which were producers of utility coal. Related washability studies were performed to assess the effects of specific gravity and size reduction on the liberation and separation of pyritic sulfur and ash from coals. Data on 322 samples were reported by the Bureau of Mines and data on 67 samples were reported by the Illinois Geological Survey. This effort is continuing.

An evaluation of the data from 322 mines samples showed that approximately 30% of these mines have coal that washed to a total sulfur content of 1% or less. Both the distribution and release potential of the sulfur in coals varies between the regions and coals within the region.

The Illinois Geological Survey data indicated that there are only a few Illinois coals whose sulfur content can be reduced to 1.5% or less. Coals capable of this amount of sulfur reduction had only 2% or less sulfur in the raw coal samples.

An evaluation of data obtained from 195 coal samples from the Appalachian region coal bed showed that significant total sulfur reductions are obtainable at yields of 70 to 80%.

An evaluation of data obtained from 13 coal samples from the Southern Appalachian region showed percentage reductions in pyritic sulfur contents (via washing) similar to those for Northern Appalachian coals. However, due to the low pyritic sulfur content of the southern coals, the attractiveness for cleaning is not as great.

The Illinois Geological Survey concluded that, for the (Illinois) coals tested for washability, two-inch-diameter core sample data correlated very closely with face sample data.

All existing commercial coal cleaning processes with the exception of froth flotation proved effective in separating liberated pyrite from coal. Of the several innovative processes for pyrite liberation investigated, the two-stage flotation technique appears to show the greatest potential for scale-up to commercial operation. The degree of sulfur reduction was dependent on equipment, operating characteristics, and, of course, the characteristics of the coal being tested. The studies defined how sulfur reduction could be optimized.

Based on float-sink test results, an estimate was made of the physical desulfurization potential of eastern bituminous steam coal (9). The estimate was based on washing at $\frac{3}{8}$ inch top size with a 90% Btu recovery. The resulting estimates of cumulative availability by sulfur content for both clean and raw coal are given in Figure 4.13.

The technology exists for utilizing reject and middling products to recovery sulfur values and generate energy. The variables are such (e.g., market for sulfur values, cost of coal, cleaning cost, transportation economics, plant cost) that economic generalizations are not possible and each case must be individually examined.

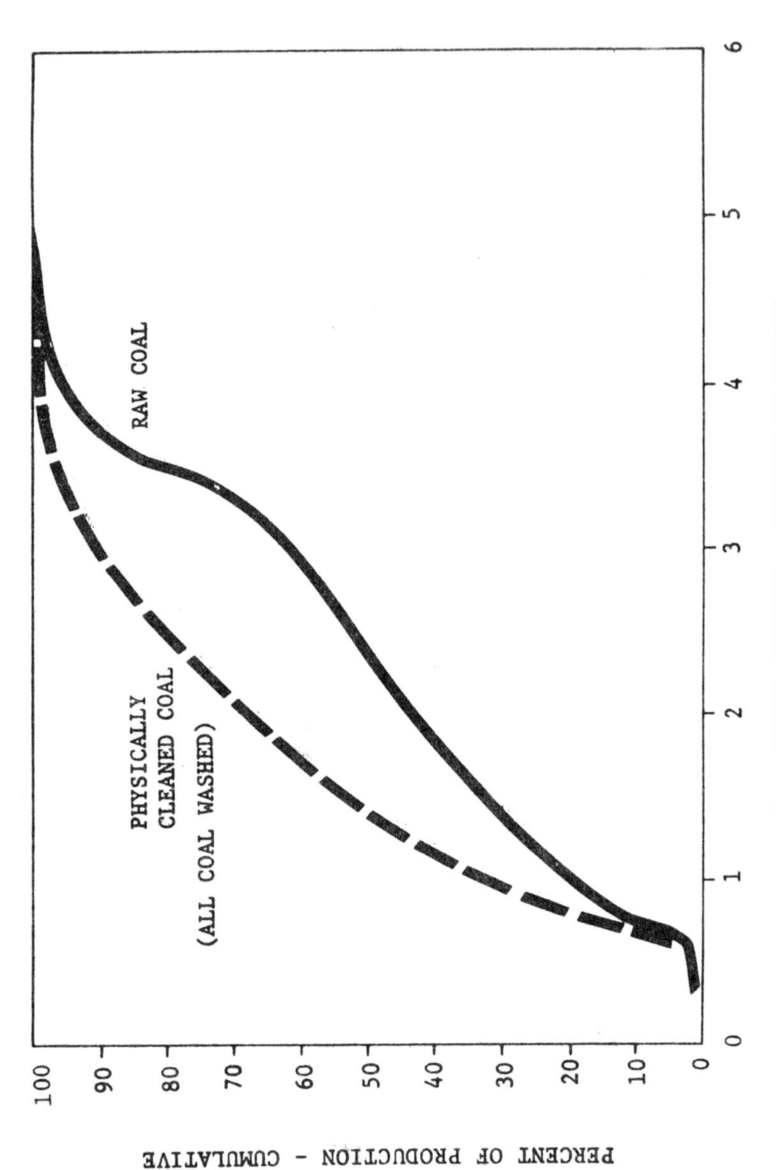


FIGURE 4.13:  ESTIMATED PHYSICAL DESULFURIZATION POTENTIAL OF EASTERN BITUMINOUS STEAM COAL PRODUCTION

Source:  PB-232 011

Physical cleaning, in general, will provide a higher Btu, lower ash, more uniform product. Monetary benefits will result from using physically cleaned coal. These benefits are attributable to higher Btu content (of cleaned coal), an effective transportation cost saving, an ash disposal cost saving, a grinding cost saving, and a plant maintenance cost saving (9).

Physical cleaning of the higher sulfur content coals, thereby providing a higher Btu and lower ash product at high yield and reasonable cost, could provide fuel for a combustor employing a moderate-cost/moderate-capability supplemental stack gas cleaning system. This use of physical cleaning and flue gas cleaning could reduce the overall $SO_2$ emission control cost.

Physical cleaning of coal is essentially a proven technique with only degree (levels and amounts) to be quantified. The process is capable of providing a near-term, even though limited, benefit.

Physical cleaning of coal as a means to control sulfur oxides emissions will have the least impact on the economy. Wide-spread commercial application of physical cleaning will help maintain a viable coal industry and will permit increased use of coal, by far the most abundant fossil fuel resource.

Studies performed by the Bechtel Corporation and A.D. Little, Incorporated indicate that adequate technology exists, in the form of several commercially used and proven processes, for recovery of sulfur values of pyrite materials as elemental sulfur or sulfuric acid and energy values as steam or electric power. These processes have been applied to mined pyrites but have not yet been applied to utilizing pyrite-bearing refuse from coal preparation plants as feedstock.


## REFERENCES

(1) W.E. Morrison, *Disaggregated Energy Consumption by Source, Form, and Sector*, Office of Economics, Federal Power Commission, February 1973.

(2) Paul Weir Company, Inc., "An Economic Feasibility Study of Coal Desulfurization," Volumes I and II, A Study for the Division of Air Pollution, Public Health Service, U.S. Department of Health, Education and Welfare, Contract No. PH 86-65-29, October 1965.

(3) "Report of Results of Washability Tests on Raw Run of Mine Coal Samples," for U.S. Public Health Service, National Center for Air Pollution Control, Cincinnati, Ohio, Commercial Testing and Engineering Company, Contract No. PH27-00079, December 1967.

(4) "Report of Results of Washability Tests on Raw Run of Mine Coal Samples," for U.S. Public Health Service, National Center for Air Pollution Control, Cincinnati, Ohio, Commercial Testing and Engineering Company, Contract No. CPA-69-645, November 1969.

(5) "Report of Results of Washability Tests on Raw Run of Mine Coal Samples," for U.S. Public Health Service, National Center for Air Pollution Control, Cincinnati, Ohio, Contract No. CPA-69-530, April 1969.

(6) "An Evaluation of Coal Cleaning Processes and Techniques for Removing Pyritic Sulfur from Fine Coal," Bituminous Coal Research, Inc., Public Health Service, for U.S. Department of Health, Education and Welfare, Contract No. PH-86-67-139, September 1969.

(7) R.J. Helfinstine, N.F. Shimp, and J.A. Simon, "Sulfur Varieties in Illinois Coals, Float-Sink Tests," Report of Study Phase I, supported in part by U.S. Public Health Service, Department of Health, Education and Welfare, Contract No. PH 86-67-206, Illinois State Geological Survey, Urbana, Illinois, August 10, 1969.

(8) A.W. Deurbrouch, "Sulfur Reduction Potential of the Coals of the United States," Report of Investigation 7633, Bureau of Mines, U.S. Department of Interior, 1972.

(9) L. Hoffman, et al. "Survey of Coal Availabilities by Sulfur Content," Mitre Corporation Technical Report MTR-6086, May 1972 [see Chapter 3 of this book].

(10) "An Evaluation of Coal Cleaning Processes and Techniques for Removing Pyritic Sulfur from Fine Coal," Bituminous Coal Research, Inc., Monroeville, Pennsylvania, February 1970.

(11) R.J. Helfinstine, N.F. Shimp, J.A. Simon, and M.E. Hopkins, "Sulfur Reduction in Illinois Coals—Washability Studies," Report of Study Phase II, supported in part by U.S. Public Health Service, Department of Health, Education and Welfare, Contract No. PH86-67-206, Illinois State Geological Survey, Urbana, Illinois, July 28, 1971.

(12) J.A. Cavallaro, A.W. Deurbrouck, and A.F. Baker, "Physical Desulfurization of Coal," Paper D28, Bureau of Mines, United States Department of the Interior.

(13) K.J. Miller and A.F. Baker, "Electrophoretic-Specific Gravity Separation of Pyrite from Coal," Bureau of Mines Report of Investigations, October 1970.

(14) A.W. Deurbrouck, Private Communication.

(15) A.F. Baker and K.J. Miller, "Hydrolyzed Metal Ions as Pyrite Depressants in Coal Flotation: A Laboratory Study," Report of Bureau of Mines Investigations 7518, May 1971.

(16) K.J. Miller and A.F. Baker, "Flotation of Pyrite from Coal," Bureau of Mines, Technical Progress Report-51, February 1972.

(17) J.A. Cavallaro and A.W. Deurbrouck, "Froth Flotation Washability Data of Various Appalachian Coals Using Timed Release Analysis Technique," U.S. Bureau of Mines Report No. RI6652, 1965.

(18) Roberts and Schaefer Company, "Design and Cost Analysis for a Prototype Coal Cleaning Plant," Volume 1, July 15, 1969, Revised August 12, 1969.

(19) Roberts and Schaefer Company, "Design and Cost Analysis for a Prototype Coal Cleaning Plant," Volume 2, July 15, 1969, Revised August 12, 1969.

(20) Roberts and Schaefer Company, "Supplemental Report to Design and Cost Analysis Study for a Prototype Coal Cleaning Plant," January 7, 1970.

(21) McNally, Pittsburg Mfg. Corp. "A Study on Design and Cost Analysis of a Prototype Coal Cleaning Plant," prepared for National Air Pollution Control Administration, Contract PH22-68-59, Parts I-VI, November 1969; Part VII, "Coal Cleaning Plant: Prototype Plant Specifications" and Part VIII, "Coal Cleaning Plant: Prototype Plant Design Drawings."

(22) Roberts and Schaefer Company, "Research Program for the Prototype Coal Cleaning Plant," Contract No. CPA 70-157, Program Element No. IA2013, January 1973.

# PROCESSES FOR DESULFURIZING COAL

## HYDROGENATION

### Lime as a Getter for Hydrogen Sulfide

*B.S. Lee and F.C. Schora, Jr.; U.S. Patent 3,640,016; February 8, 1972; assigned to Institute of Gas Technology* developed a method for desulfurizing caking coal which specifically removes the equilibrium limitation in the reaction

$$FeS + H_2 \longrightarrow Fe + H_2S$$

by hydrogenating in the presence of a getter for the hydrogen sulfide. The getter chemically combines with the hydrogen sulfide formed. Introducing the getter strongly shifts the equilibrium conversion toward greater hydrogen utilization by permitting more desulfurization to take place. Examples of cheap and effective getters are lime, calcined dolomite, or other alkaline earth metal oxides. In the following discussions lime is used as the example with the understanding that the others are just as suitable.

This process may be practiced as follows: Raw coal from storage is dried, crushed and sized to obtain a particle size suitable for fluidization, typically minus one-fourth inch. Particle size is not critical but should be chosen depending upon reactor volume and configuration and gas flow rates. Limestone or other source of alkaline earth metal oxide material, such as dolomite, is also dried, crushed and sized to approximately the same particle size as the coal and fed to the calciner to produce the oxide. The coal and lime are then fed into the desulfurizer which may be a reactor of any conventional type having a reaction chamber and means to inject gas into it.

A fluidized bed may be used and would be particularly advantageous for good temperature control. The material in the fluidized bed may comprise fine coal particles with coarse particles of lime raining down through the bed, thus permitting subsequent separation of lime and coal particles by screening. An alternative is a fluidized bed of coal and lime of essentially the same size, the mixture being separated later by differences in density. A third possible arrangement is a fluidized packed bed where small coal particles are fluidized in the interstices of a fixed bed of large pieces of lime. Broadly speaking, any reactor arrangement where the coal and lime particles are placed in a relatively close degree of juxtaposition permitting hydrogen gas to flow therethrough can be used in this method. The process is effected at substantially atmospheric pressure and at temperatures between about 600° and 800°F.

Steam-laden hydrogen gas exiting from the reactor is directed through a heat exchanger and condenser to remove water vapor and then recycled with makeup hydrogen from supply to the desulfurizer.

The reacted lime-coal mixture is removed from the desulfurizer and the coal and lime fractions are separated by techniques well-known in the art which utilize either the size or the density difference between the materials. The unreacted lime-CaS mixture is then regenerated in a regenerator where steam and $CO_2$ are fed through the mixture to produce $Ca(OH)_2$ or $CaCO_3$ and $H_2S$. The resultant $H_2S$ gas is then sent to sulfur recovery where elemental sulfur is produced by well-known techniques such as the Claus process.

The regenerated calcium carbonate is then fed to the calciner along with a makeup limestone feed. The carbon dioxide is used as feed for the lime regenerator.

*Example:* A mixture containing equal volumes of coal and calcined limestone in the size range of –16 to +80 mesh was reacted at 750°F with hydrogen in a fluidized bed reactor. The reactor was a 2-inch diameter stainless steel cylinder about 6 feet long. Flow rate of hydrogen through the coal-limestone mixture was 53 scf/hour (1.5 linear feet/second). The coal employed as the starting material contained 1.92% pyritic sulfur, 1.78% organic sulfur and 0.18% sulfate sulfur. All percentages are based on the weight of coal. Analysis of the coal after the treatment showed substantially no pyritic sulfur. The organic sulfur content was reduced by about 34%.

Although the sulfate content was also substantially reduced, this item is not particularly significant in view of the low percentage present in the starting material. No hydrogen sulfide was detected in the effluent gas, indicating that substantially all hydrogen sulfide has reacted with the lime.

**Formation of Coal or Coke Pellets or Balls**

The process developed by *E.B. Mancke; U.S. Patent 3,756,791; September 4, 1973; assigned to Bethlehem Steel Corporation* relates to a continuous method for simultaneously calcining and desulfurizing agglomerates of coal and/or coal derivatives by heated particles of a desulfurizing agent flowing countercurrently to the agglomerates in a rotary kiln. The agglomerates so manufactured have a medium or low reactivity rate when exposed to carbon dioxide at elevated temperatures and a major portion of their sulfur has been removed.

These agglomerates are suitable for charging into blast furnaces, cupolas and the like. The agglomerates are heated and then are calcined and a major portion of the sulfur that is contained therein is removed. The heat and material necessary for calcination and sulfur removal are provided by heated relatively fine particles of a desulfurizing agent. The agglomerates which are within a temperature range of about ambient temperature to about 850°F and particles of a desulfurizing agent heated to at least about 1400°F are charged into opposite ends of a rotary kiln.  The agglomerates and desulfurizing agent pass through the furnace countercurrently to each other.  During the countercurrent flow of the charged materials, a heat exchange occurs at a controlled rate between the agglomerates and the heated particles of the desulfurizing agent.  The agglomerates are heated to a temperature of at least about 1400°F at discharge and the heated particles of the desulfurizing agent are reduced to any temperature below about 1100°F at discharge.

Simultaneously with the heat exchange, the sulfur in the agglomerates reacts with hydrogen formed in the furnace as the coal agglomerates are heated to elevated temperatures.  The hydrogen reacts with sulfur present as organic sulfur and pyritic sulfur in the agglomerates to produce hydrogen sulfide.  The hydrogen sulfide contacts and reacts with the heated particles of the desulfurizing agent to form a sulfide with the material therein.  The sulfur is thereby removed from the gas.  Free hydrogen gas is re-formed which can then react with sulfur in the agglomerates.  It can be seen that the hydrogen acts as a vehicle to transport a portion of the sulfur originally in the agglomerates to the heated particles of the desulfurizing agent.  Of course, it is not possible to remove all the sulfur from the agglomerates since sulfur is difficult to remove, but the amount removed is sufficient to reduce the sulfur content in the agglomerates to an acceptable amount.

By agglomerates are meant briquettes, extrusions or pellets of coal and/or coal derivatives formed by charging coal particles and/or coal char to a press and applying pressure thereto at ambient temperature or temperatures of about 300° to 950°F, or forming balls thereof on a balling drum and the like.  The method is particularly adapted to the processing of balls of coal which are formed by a hot balling technique.  The technique includes heating the particles of coal and char separately so that a mixture thereof attains a temperature within the range of about 700° to 850°F and forming the particles into balls in a balling drum or the like.  The heated particles of the desulfurizing agent can be either solid or porous so long as the particles are of a size smaller than the agglomerates and have sufficient surface area available for reacting with hydrogen sulfide to thereby remove sulfur from the agglomerates.

As a suitable apparatus for accomplishing the process, a rotary kiln, having two concentrically aligned cylinders, an outer shell and an inner apertured tube, may be used.  A worm conveyor is attached to the inner surface of the outer shell. Lifter bars whose free ends are curved to form a cup-like part are attached to the inner surface of the outer shell, parallel with the inner surface.  The apertures in the inner tube must be sufficiently large to allow the small particles of desulfurizing agent to pass through them, but sufficiently small to retain the agglomerates inside the tube.  The agglomerates are charged into one end of the inner tube, pass down the tube, and are discharged into a chamber where they are cooled.  The heated particles of the desulfurizing agent are charged into one end of the outer shell and, as the rotary kiln slowly rotates, are carried up toward the apex of the outer shell by the lifter bars.  They fall downward through the apertures in the inner tube, but a sufficient number are retained in the inner tube to form a relatively deep bed.

The agglomerates flow down through this bed.  A controlled heat exchange occurs between the agglomerates and the heated particles of the desulfurizing agent so that the agglomerates are heated and the particles cooled. Simultaneously, as the agglomerates lose volatile matter with increasing temperature, hydrogen is formed and desulfurization of the agglomerates occurs. A portion of the heated particles fall down through the apertures and are transported by the worm conveyor and discharged from the kiln.

The worm conveyor is in spaced relationship with the outer surface of the inner tube to allow the passage of gases fed to the kiln or formed by heating the agglomerates therein. The heated particles of a desulfurizing agent, for example, manganese ores in which a major part of the managanese is manganous oxide, calcined dolomite, calcium oxide, or magnesium oxide, smaller in size than the agglomerates, are heated to a temperature above about 1400°F and preferably above about 1950° to 2200°F prior to being charged into the kiln. The agglomerates are within a temperature range of about ambient temperature to about 850°F when charged into the kiln. As the kiln rotates, the agglomerates and the particles of the heated desulfurizing agent flow in countercurrent directions to one another.

Generally, all the heat necessary to calcine the agglomerates is provided by the hot particles of the desulfurizing agent. However, if necessary, additional heat can be provided by blowing a limited amount of air into the kiln. Since the hydrogen-containing atmosphere is hot, the oxygen in the air will burn, thereby providing heat.

Simultaneously with the transport of the sulfur from the agglomerates to the heated particles of the desulfurizing agent, the agglomerates are heated to a temperature of not less than about 1400°F and generally from 1650° to 1950°F and preferably above 1950°F by heat transferred from the hot particles of the desulfurizing agent.

The calcined agglomerates can be cooled to as low as about 200°F in an inert atmosphere, which can be hydrogen, nitrogen or relatively nonoxidizing gases and mixtures thereof, in a chamber outside the kiln and are then passed to storage or use. A portion of the desulfurizing agent is regenerated on any given cycle by any one of several known methods. The regenerated portion can be recharged into the kiln for reuse.

*Example:* By the use of the described apparatus it is possible to charge about 1.1 tons of agglomerates of coal containing about 2% sulfur at a temperature of about 700°F into the inner tube at one end of a rotary kiln and about 2.6 tons of heated particles of manganous oxide at a temperature of about 1950°F into the outer shell at the other end of the kiln. As the kiln rotates, the agglomerates and heated particles of manganous oxide will flow countercurrently to each other. The temperature of the agglomerates will be raised to a temperature approaching 1950°F and the heated particles of manganous oxide will be cooled to about 850° to 950°F. The agglomerates will be charged into a cooling chamber where an inert atmosphere will be maintained. The heated particles of manganous oxide will be discharged from the other end of the kiln. The total time for the agglomerates to pass through the kiln where they are calcined and desulfurized could be about 4 hours. The sulfur in the calcined agglomerates will be decreased to acceptable levels. All the hydrogen required for desulfurization can generally be formed during the heating of the agglomerates.

### Treatment with Hydrogen plus Methane

In the process developed by *C.M. Whitten, V. Mansfield, L.J. Petrovic and J.C. Agarwal; U.S. Patent 3,759,673; September 18, 1973; assigned to Peabody Coal Company* coal is treated in a reducing atmosphere in a fluid bed reactor to partially devolatilize it, remove all moisture, and remove some $H_2S$ and $SO_2$. The then charred coal is treated in a multistage contactor with hot gas composed of hydrogen and methane to remove most of the remaining sulfur.

Referring to Figure 5.1, the starting material, for example bituminous coal sized to ½ inch x 0 dried and preheated, is fed into a fluid bed reactor for complete drying and light heat treatment at from 700° to 800°F to prevent subsequent agglomeration in the multistage contactor. In the fluid bed reactor, partial devolatilization occurs and some sulfur, particularly the pyritic type, is removed. It is essential that the hot gas feed as indicated at **6** to the fluid bed reactor be as oxygen-free as possible so as to provide a reducing atmosphere in the reactor. These are recycle gases from the remainder of the system consisting essentially of hydrogen and methane.

Only enough air is added to provide heat via a partial combustion which is required to heat and maintain the bed at temperature. Accordingly, the fluidized bed remains in a reducing atmosphere. The gases exhausted as indicated at **8** from the fluid bed reactor can be utilized for combustion or recycled through the recovery system.

From fluid bed reactor **2** the coal, then char, is fed as indicated by the feed line **10** into a multistage contactor **12**. About 85% of the coal input to fluid bed reactor **2** is reported into the char entering multistage contactor **12**, which consists of an enclosure bearing a hopper input through which the char is deposited upon a perforated endless belt which runs over sprockets at each end of the enclosure. The char is spread to form a bed by means of a spreader gate. Beneath the upper belt run is a zoned airbox with at least six zones having individual gas input pipes. Over the upper belt run are partitions which cooperate with the airbox zones to form individual plenum chambers. Pipes exhaust the gases from between the partitions. After being treated in contactor **12**, the desulfurized char drops off the end of the belt run into an outlet **32**.

Treatment of the coal in the multistage contactor is differentiated from the treatment in the fluid bed reactor in that in the contactor the bed remains static while being exposed to the hydrogen containing gases. In effect, each zone becomes a separate reactor where the temperature may be varied as desired. The contactor is operated at elevated temperatures (i.e., 1000° to 1700°F) and elevated pressures, 14.7 to 500 psia, which could result from placing a compressor in line **44** before the preheater.

The temperatures are maintained by the exothermic reactions of carbon with small amounts of oxygen and carbon monoxide that appear in the gases or are introduced into the gases. The incoming hydrogen containing gas is essentially $H_2S$ free and within the contactor, the ratio of $H_2S$ to $H_2$ is held at less than 1 to 100 parts. If needed, a small amount of air may be introduced into the incoming gas line as indicated at **31** so as to produce sufficient reactions in the individual plenum chambers within the contactor as to maintain the desired elevated temperatures, or the air may be fed individually to the airbox zones in

FIGURE 5.1:  CHAR TREATED WITH HYDROGEN PLUS METHANE

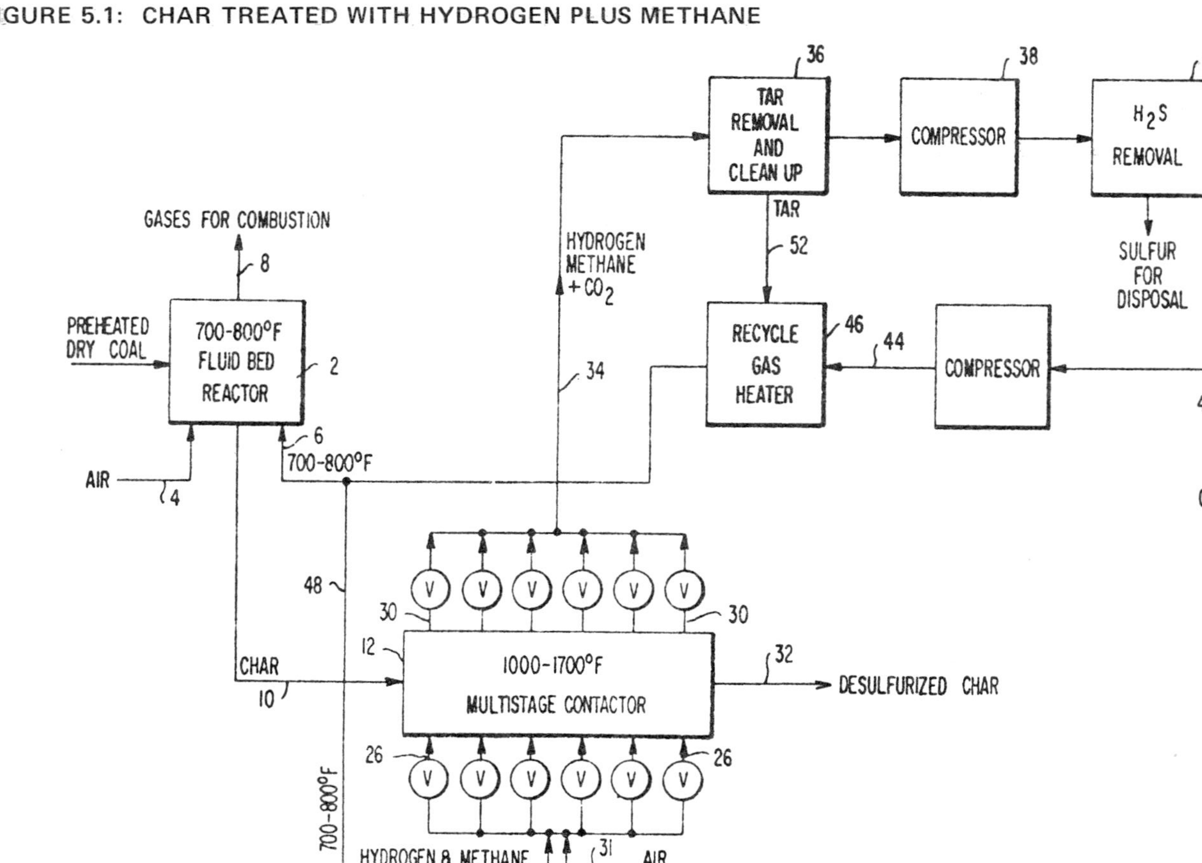



Source:  U.S. Patent 3,759,673

controlled amounts. An added benefit to the treatment in the multistage contactor is that the coal is continuously devolatilized thermally, thereby adding $H_2$ and methane to the exhausted gas and providing the capacity for further sulfur removal. All hydrogen produced within the process is by thermal devolatilization.

The volume of hydrogen containing gas required for desulfurization in the multistage contactor may be very substantially reduced by adding a $H_2S$ acceptor to the char fed to the contactor. Acceptor technology has been described by others. This may be preferred because it simplifies gas handling and the gas cleanup facilities. It has been demonstrated that the quantity of hydrogen needed in the presence of an acceptor, such as calcined dolomite, is only one to two times stoichiometric. The quantity of an acceptor required with hydrogen is also one to two times stoichiometric. While regeneration and recycle of the acceptor are not a simple matter, it can be accomplished with known equipment.

The char fed into multistage contactor **12** produces more hydrogen than is lost by reaction or leakage; therefore, gas containing hydrogen, methane, and CO is continuously withdrawn through output line **34** from the contactor. This gas is low in Btu value (100-250), but it may be burned for its heating value. The contactor has the capability of providing the residence time that enables operation at lower pressures and temperatures and the recycle of large volumes of gases. The coal used to illustrate the concept of desulfurization is presented in the table below. The char from the fluidized bed reactor prior to entry of the multistage contactor and after the multistage contactor is also shown in the table. The composition of each material is given in weight percent, dry basis.

|  | Coal, (Ill. #6) | Char from Fluidized Bed | Char from Desulfurizer |
|---|---|---|---|
| Ash | 17.73 | 20.42 | 24.11 |
| Fixed carbon | 44.08 | 50.77 | 66.95 |
| Volatile | 38.19 | 28.79 | 8.94 |
| Sulfur | 4.20 | 4.00 | 1.20 |

The above case represents about 70% by weight recovery of the input coal as char. Residence time in the contactor should be about 120 minutes. The processing conditions in the multistage contactor for these results need to be from 1000° to 1700°F, with the temperature increasing by about 200°F during passage over each of the zones of the airbox until from 1600° to 1700°F is reached and there is obtained a hydrogen partial pressure of about one atmosphere. Under these conditions, Illinois #6 seam test coal is about 75 to 80% desulfurized. Retention time of less than 120 minutes will result in less sulfur removal. If hydrogen partial pressure is increased to 5 atmospheres, 75 to 80% desulfurization may be accomplished in a shorter length of time.

It has been determined, however, that coals with higher percentages of the organic sulfur are more difficult to desulfurize and an $H_2S$ acceptor has to be utilized if the sulfur is to be reduced to less than 1% in the desulfurized char.

The gases withdrawn from the multistage contactor through outlet pipes **30** and common outlet line **34**, essentially hydrogen, methane and $CO_2$, are fed through a tar removal and clean-up apparatus **36**, then to a compressor **38**, then to $H_2S$ removal apparatus **40** from which sulfur can be removed for disposal. Excess gas from H$_2$S removal apparatus can be withdrawn for combustion as indicated at **42** and the remainder is recycled through recycle gas heater **46**. From heater **46** the steam is split, some being recycled back to fluid bed reactor **2**, and the other being fed via line **48** back into multistage contactor **12**.

**Treatment with Oxygen and Steam to Generate Hydrogen**

According to the process described by *W.C. Schroder; U.S. Patent 3,909,212; September 30, 1975* the sulfur content of solid carbonaceous fuels, such as coal or lignite, is reduced by reacting a portion of the fuel with oxygen and steam without complete carbonization of the fuel, so as to generate nascent hydrogen at the surface and within the fuel particles for reaction with the sulfur in the fuel to form hydrogen sulfide. The resulting sulfur-containing gases are removed and a low sulfur, solid fuel is recovered.

This process differs from gasification as practiced heretofore by reacting only the minimum amount of coal with steam and air (or oxygen) which is necessary to reduce the sulfur in the coal to the desired level and all the remaining coal is removed from the reaction zone. The desulfurized coal still contains sufficient volatile matter so that it is a satisfactory fuel for combustion purposes. The product is not a coke or char. It has been carbonized only to the extent necessary to remove a substantial amount of the sulfur.

Desulfurization of coal as practiced according to this process reacts on only 5 to about 14% of the coal. Consequently, the capital cost of the equipment and facilities to carry out this process are approximately one-tenth of the cost for gasification. The purification step is also reduced accordingly. Operating costs are greatly reduced and it is estimated that they will increase fuel costs only by 15 to 20% over the cost of the coal.

This process can be used on strongly caking bituminous coals, noncaking subbituminous coals, or lignite. The coal particle, without previous thermal treatment of any kind, is introduced into a bed of particles at operating temperature, which heats the coal particle rapidly. Combustion with oxygen at the particle surface provides further heat for rapid expansion of the particle and creation of large surface areas. The increased porosity of the particle as well as its large surface area make the hydrogen more effective in sulfur removal.

The coal particle heats and expands in the presence of steam which reacts with the hot carbon to produce nascent hydrogen which is highly effective in reacting with sulfur or sulfur compounds to produce H$_2$S. The H$_2$S is carried out with the other gaseous constituents and removed from the gas by well-known purification methods.

Temperatures, nascent hydrogen production and coal throughput are all controllable, and are so controlled that they provide maximum sulfur removal, minimum change in the coal itself and minimum heat loss. The result is a process which is low in capital investment and operating cost and gives minimum increase in fuel cost over the cost of the coal itself. The coal product can be used in existing boilers or in new boilers without the addition of special facilities to the boiler itself and without loss of steam output and without change in the method of operating the boiler.

*Example:* Crushed or pulverized coal is fed continuously at a controlled rate to a vessel by gravity, screw conveyor, or other suitable means, and partly devolatized coal is withdrawn from the bottom. The vessel is steel, refractory lined to withstand internal temperatures up to about 1700°F. It is also well insulated on the inside to reduce heat loss. A portion of the coal in the vessel is burned continuously by passage of steam and air (or oxygen) through the coal bed. Recycle gas is also fed to the coal bed.

Air, steam and recycle gas may be blown upwardly through the pulverized coal bed. For a bed of crushed coal the air, steam and recycle gas may be blown upward or downward, since crushed coal may be held on a grate near the bottom of the vessel. Steam is fed as necessary to produce hydrogen.

For strongly caking bituminous coals it is necessary to recycle coal from near the bottom of the coal bed to the top, to prevent agglomeration of the bed by the introduction of the fresh coal. For this purpose, the coal is withdrawn from the bottom and is lifted to the top of the bed by a small amount of steam and air. The gas generated leaves the reactor and passes through a cyclone separator. Solid material from the bottom of the separator is returned to the coal bed.

The most suitable temperature to be maintained in the vessel depends on the coal, rate of throughput and degree of desulfurization desired. Bituminous coal requires higher temperatures and longer retention times than subbituminous coals or lignite. In most cases bituminous coals will require retention times from 8 to 10 minutes and temperatures in the range from 1100° to 1500°F. Retention times for subbituminous coal and lignite are from about 4 to 8 minutes at temperatures of 900° to 1400°F. The pressure in the reaction vessel is in the range of from atmospheric up to about 10 atmospheres. Optimum pressures for desulfurization are in the range of 2 to 6 atmospheres.

To heat one ton of coal to 1200° to 1500°F, decompose steam, and react the sulfur and some of the nitrogen compounds with hydrogen requires the combustion of approximately 180 to 250 pounds of coal. The resulting volume of combustion gas is generally between 20,000 and 40,000 standard cubic feet when air and steam are used in desulfurization. Methods for purification of the gas stream remove better than 99.5% of $H_2S$ present in the gas.

**Treatment in a Closed, Pressurized Cyclic System**

*W.C. Schroeder; U.S. Patent 4,013,426; March 22, 1977* describes a process which comprises the removal of sulfur from coal or other sulfur-containing materials by contacting the coal in a closed pressurized cyclic system with hot hydrogen-containing recycle gases, a portion of which have been burned with oxygen to provide heat for the desulfurization reaction. No steam is used other than that generated within the system. Coal and oxygen are the only materials fed to the closed cyclic system. Desulfurized coal, sulfur and excess gases are removed as products.

Coal is fed at system pressure into a desulfurization zone of a closed cyclic system where it is contacted with the hydrogen-containing recycle gases which are at a temperature sufficiently high to raise the temperature of the coal to about 1100° to 1800°F, preferably 1500° to 1800°F. In this temperature range, a number of reactions occur. Sulfur present in the coal can be in at least three forms, i.e., as pyrites, organic sulfur and sulfates. The pyrites form iron sulfide and sulfur which react with the hydrogen to form metallic iron and $H_2S$. If steam is present some of the pyrites may be converted to FeO. Organic sulfur compounds are converted by hydrogen into C, CO and $H_2S$. If steam is present it will hydrolyze the organic sulfur compounds into $CO_2$ and $H_2S$. Since there is usually some moisture in the coal fed to the desulfurization zone some steam will necessarily be generated when the coal is contacted with the hot gases.

The equilibrium of the desulfurization reactions toward production of $H_2S$ is strongly favored by high partial pressure of hydrogen or hydrogen-steam in the gases in contact with the coal. Sulfur will be removed from FeS so long as the $H_2S$ is below about 1.8 volume percent of the hydrogen. At this concentration of hydrogen, at the temperatures involved, $H_2S$ will also be formed from the organic sulfur compounds. It is preferred to use pressures in the range of about 5 to 30 atmospheres.

It has been found that the removal of S as $H_2S$ from most coals does not require the generation of hydrogen from coal and steam. The reason for this can be made apparent by reference to the analysis of a typical bituminous coal from the north central part of the United States, which is shown below (all values are %):

| | As received | Moisture & Ash Free |
|---|---|---|
| $H_2$ | 5.3 | 6.2 |
| C | 67.0 | 78.8 |
| N | 1.4 | 1.6 |
| O | 7.9 | 9.5 |
| S | 3.4 | 3.9 |
| Moisture | 6.1 | — |
| Ash | 8.9 | — |
| TOTAL | 100.00 | 100.0 |
| Sulfur Forms | | |
|  As sulfate | 0.37 | |
|  As pyrite | 0.98 | |
|  As organic | 2.07 | |
| TOTAL | 3.42 | |

When this coal is heated to a high temperature hydrogen is released and reacts with sulfur to form $H_2S$ and with N to form $NH_3$. At temperatures above 1500°F, in the presence of the large excess of carbon, the hydrogen does not react with oxygen to form water to any considerable extent. Most of the oxygen is removed as CO with a small amount of $CO_2$. Complete reaction of the sulfur and nitrogen to $H_2S$ would require 0.586 pound of $H_2$ per 100 pounds of moisture and ash-free coal. Hydrogen in the coal is 6.2 pounds, of which about ½ to ⅔ is made available by heating the coal, which is about 5 times the amount needed for the reactions shown.

In addition to hydrogen released as elemental hydrogen from the coal, some hydrogen will be formed from the volatile matter in the coal. Temperatures required for desulfurization are high enough to decompose many hydrocarbons to carbon and hydrogen. It should also be noted that while dried coal is fed to the process this normally means coal still containing 3 to 6% water. At high temperature this water will react with carbon and hydrocarbons to produce hydrogen.  As heretofore noted, steam generated in or added to the system will also react with sulfur compounds in the coal to form $H_2S$.  Steam is therefore effective in removing sulfur from the coal so long as the products of this reaction are carried away by sufficient circulating gas to maintain the concentration of $H_2S$ in the gas below the equilibrium conditions at the existing temperature and pressure.  Steam may be added to the system if it is desired to produce more hydrogen in the desulfurizing zone but this addition is not essential to the desulfurization reaction because sufficient hydrogen and steam for this purpose are produced within the closed system.

These various hydrogen producing reactions yield considerably more hydrogen than needed in the desulfurization process and in fact for most coals provide sufficient gas (as a mixture of hydrogen with a low concentration of methane) to allow it to be burned with oxygen to heat the coal and to furnish some excess gas which can be withdrawn from the process.

The reaction of S with $H_2$ at the temperatures involved, will only go to completion when sufficient $H_2$ is present to maintain the $H_2S/H_2$ equilibrium below 1.890% $H_2S$.  This is accomplished by removing $H_2S$ from the effluent gases from the desulfurization zone and recycling a sufficient volume of the resulting $H_2$-containing gas through the coal as the $H_2S$ is formed.  It is clear, therefore, that most coals can be desulfurized without a separate step to produce $H_2$ gas, but a stream of hydrogen or hydrogen and steam must be circulated through the coal to carry away the $H_2S$ that is formed.

The selected temperature of operation in the desulfurization zone depends upon the coal, pressure of operation and retention time.  In general it will be between 1100° and 1800°F.  The process gasifies a small amount of coal varying from about 75 to 200 pounds per ton of coal fed, based on dried coal to the process.  Oxygen use is small, amounting to about 2,000 to 3,000 scf per ton of dried coal.

**Beneficiation of Low Rank Coals and Lignite**

According to the process developed by *E.L. Cole, H.V. Hess and J. Wong, Jr.; U.S. Patent 4,047,898; September 13, 1977; assigned to Texaco Inc.* there is provided a process for the beneficiation of a low rank solid fuel which comprises forming a mixture of particulate low rank solid fuel and water, heating the mixture to a temperature between about 300° and 700°F at a pressure sufficient to maintain liquid water in the reaction zone and in the presence of hydrogen for a period of time sufficient to reduce the sulfur and ash content and increase the Btu value of the fuel.

The solid fuels to which the process may be applied are low rank solid fuels II, III and IV as classified in the *1973 Annual Book of ASTM Standards, Part 19, page 57.*  The solid fuel should be reduced to a particulate form in which the particles have a maximum dimension of not greater than one inch.  Preferably, the maximum dimension is less than ½ inch and still more preferably less than ¼ inch.  The water and the particulate solid fuel are mixed in an amount to provide a mixture containing from about 0.5 to 6 parts water, preferably from 1 to 4 parts water per part fuel on a dry basis by weight.  If the process is of the batch type, the coal and water may be charged separately to the reaction zone such as an autoclave or they may be charged together as a slurry.  In such latter event, the water should be present in the slurry in an amount between about 40 and 75% by weight, preferably between 40 and 60% by weight as if the water content is less than 40%, the slurry becomes difficult to pump. Such a slurry is also used when the process is of the continuous type where the slurry is, for example, passed through an elongated tubular reaction zone.

The hydrothermal treatment in the presence of added hydrogen may be effected under either static or dynamic conditions. In one embodiment, the slurry of solid fuel in water is introduced into a pressure vessel such as an autoclave. Since the hydrothermal treatment is effected under nonoxidizing conditions, advantageously the pressure vessel is swept with inert gas prior to the introduction of the slurry. In the alternative, the slurry is introduced into the vessel which may then be swept with hydrogen or with an inert gas and then hydrogen. After removal of the oxygen-containing gases, the vessel is pressured with hydrogen and then heated under autogenous pressure to a temperature between about 300° and 700°F preferably between 400° and 650°F, the pressure being such that water in liquid state is maintained in the reaction vessel. After a period of time between about 1 minute and 2 hours the vessel is vented and the slurry removed therefrom.

*Example:* 150 grams of powdered Lake DeSmet coal was placed in a 1,740 milliliter autoclave together with 300 ml of distilled water. The autoclave was flushed and pressured with nitrogen to 450 psig followed by heating to 550°F. The autoclave was held at this temperature for 1 hour, cooled to room temperature, vented, flushed with hydrogen and pressured with hydrogen to 400 psig. The autoclave was heated to 550°F and held at this temperature for 24 hours and then vented; 1.45 ft$^3$ of gas, 314 grams of water and 103 grams of coal was recovered.

Where the hydrothermal treatment was pressured with nitrogen, the average percent desulfurization was 28.3 and the average percent increase in gross heating value was 13.4, whereas when the hydrothermal treatment and venting were followed by a hydrothermal treatment in the presence of added hydrogen, the average percent desulfurization was 35.3 and the average percent increase in gross heating value was 22.6.

**Using Aqueous Inorganic Acid Following Hydrogenation**

A process of reducing the total sulfur content of coal is disclosed by *G.C. Sinke; U.S. Patent 4,071,328; January 31, 1978; assigned to The Dow Chemical Co.* The process comprises hydrogenating the coal to remove at least a portion of the recoverable sulfur combined as pyritic sulfur. The hydrogenated coal is subsequently contacted with a sufficient amount of an aqueous inorganic acid solution to remove at least a portion of the remaining sulfur initially combined as pyritic sulfur.

In one embodiment of the process a coal selected from the group consisting of bituminous and lignite is crushed and sized to a suitable particle size. A particle size of less than about 100 U.S. Standard mesh is preferred. Preferably, the coal is then transferred to a reaction vessel capable of withstanding elevated temperatures and pressures. The reaction vessel is thereafter sealed and evacuated. A reaction vessel such as closed reaction bomb has been found satisfactory.

Hydrogen is introduced into the reaction vessel to react with at least a portion, and preferably substantially all, of the recoverable sulfur combined as pyritic sulfur in the coal. The hydrogen pressure within the reaction vessel can be any pressure above atmospheric pressure. However, the efficiency of desulfurization is increased when the hydrogen pressure is maintained at the preferred level of from about 200 to 550 pounds per square inch gauge.

The hydrogen may be mixed with inert gases such as nitrogen or argon. However, inclusion of other gases increases the time required for the reaction. The reaction vessel, during pressurization, is heated by means of any suitable heating device, such as an electrical heater, furnace or the like. Sufficient heat is applied to the reaction vessel to increase the temperature of its contents to a temperature sufficient to remove at least part, and preferably substantially all, of the recoverable sulfur combined as pyritic sulfur. The temperature should preferably be sufficient to minimize loss of coal as volatiles within the reaction vessel and prevent excessive caking of the coal. It has been found that substantial desulfurization occurs when the temperature of the contents of the reaction vessel is from about 300° to 350°C, and more preferably from about 320° to 350°C.

The preferred temperature and pressure conditions of the hydrogenation reaction are maintained for a sufficient time to allow at least a portion, and preferably substantially all, of the recoverable sulfur combined as pyritic sulfur in the coal to be removed by reaction with the hydrogen within the reaction vessel to form a product consisting essentially of iron(II) sulfide (FeS) and hydrogen sulfide ($H_2S$). The course of the reaction may be followed by monitoring the gaseous stream containing the hydrogen sulfide. When the concentration of the hydrogen sulfide drops to a predetermined level, the hydrogenation may be discontinued.

The total time for the reaction of the coal may be adjusted to compensate for the effects of temperature, pressure, particle size, and cost. When the hydrogenation is carried out at the preferred temperature, pressures and particle size disclosed above, it has been determined that a reaction time of at least about 2 hours is preferred for at least a portion of the recoverable sulfur combined as pyritic sulfur to be removed. However, a reaction time of from about 2 to 4 hours allows substantially all of the recoverable sulfur combined as pyritic sulfur to be removed.

The conditions of temperature and time under which this process may be practiced are interdependent. Generally, longer times allow lower reaction temperatures. However, certain considerations, as indicated above, indicate a preferable range for time and temperature under which substantially all of the recoverable sulfur combined as pyritic sulfur in the coal can be removed.

After the reaction vessel has been heated for a sufficient time for at least a portion, and substantially all of the recoverable sulfur combined as pyritic sulfur to be removed, the reaction vessel is allowed to gradually cool to about room temperature. Air cooling of the reaction vessel is preferred, although any other suitable means of gradual cooling such as stepwise reduction of the heating temperature may be employed.

When the reaction vessel has cooled to about room temperature, the hydrogen pressure within the reaction vessel is reduced to about atmospheric pressure.

The hydrogenated coal is contacted with a sufficient amount of an aqueous inorganic acid solution to react with any iron(II) sulfide (FeS) formed during hydrogenation and thereby remove at least a portion, and preferably substantially all, of the remaining recoverable sulfur originally combined as pyritic sulfur, and substantially entirely in the form of iron(II) sulfide (FeS). The contacting of the hydrogenated coal with an aqueous inorganic acid solution can be carried out in a number of ways including digesting the hydrogenated coal with the aqueous inorganic acid solution or by acid leaching or the like.

Strong aqueous inorganic acid solutions of HCl, $H_2SO_4$, $HNO_3$, HBr, HI, or HF may be employed in the process. However, HCl does not leave contaminated residue in the coal and is the preferred aqueous inorganic acid solution of this process. An aqueous inorganic acid concentration of from about 1 to 3 N gives satisfactory results.

The mixture of hydrogenated coal and aqueous inorganic acid solution is agitated and heated by any suitable means. The hydrogenated coal is contacted with the aqueous inorganic acid solution for a sufficient time to allow at least a portion and preferably substantially all, of the iron(II) sulfide (FeS) to react with the acid. A time from about 0.5 to 1 hour gives good results and is preferred for such a reaction. The temperature range at which the contacting takes place may vary. A temperature range of from about 40° to 80°C gives satisfactory results.

After sufficient contact time, the hydrogenated-acid contacted coal product is separated from the inorganic acid solution by, for example, filtration, although other separation methods could be employed. Following separation, the hydrogenated-acid contacted coal product is treated to remove at least a portion and preferably substantially all, of the water-soluble reaction products [such as iron(II) chloride ($FeCl_2$)] and any excess inorganic acid. Washing the coal product with a solvent such as water has been found to be a suitable means of treating the hydrogenated-acid contacted coal product.

The following example illustrates the process. Bituminous coal samples from the Allison Mine of eastern Ohio were used in the following example. The total sulfur content of the coal was 4.49% by weight as determined by the standard sulfur test of the American Society for Testing and Materials ASTM Designation: D2492-68. The coal contained 2.20% by weight pyritic sulfur, 2.21% by weight organic sulfur and 0.08% by weight sulfate.

*Example:* Pieces of coal were crushed and sized to a particle size of less than about 100 mesh. A 20.2 gram sample of the crushed coal was transferred into a 1,400 cc reaction vessel, which was then pressurized with substantially only hydrogen gas to about 525 psig.

During pressurization, the reaction vessel was heated by an electric heater to increase the temperature of its contents to about 315°C. The temperature and pressure were maintained for 2 hours. After 2 hours, the reaction vessel was allowed to air cool to about room temperature over a period of about 16 hours. The pressure was then reduced to about 1 atmosphere. The hydrogenated coal was removed from the reaction vessel and weighed. 19.65 grams of residue and iron(II) sulfide (FeS) were recovered. A 5.00 gram sample of the hydrogenated coal was transferred to a flask and contacted with an aqueous solution of 1 N HCl by digesting the hydrogenated coal with about 50 ml of the 1 N HCl solution at about 60°C for about 2 hours.  The inside walls of the flask were occasionally washed down with water.  After about 2 hours, the resulting slurry was vacuum filtered, using a Number 42 Whatman filter paper and a Buchner funnel.  The solid residue was collected on the filter paper and washed with water to remove any water-soluble reaction products.

The filtrate was collected and analyzed for $Fe^{++}$ by complexing the filtrate with 1,10-phenanthroline and subsequently collecting colorimetry readings using a Beckman UV spectrophotometer.  The percentage of pyritic iron removed from the coal was calculated by subtracting the amount of iron soluble in the aqueous HCl solution before hydrogenation from the amount of iron recovered from the filtrate after hydrogenation and acid contacting.  This quantity was divided by the amount of iron initially present in the coal as determined by ASTM D2492-68.  In this example about 0.25% by weight of the initial coal was acid-soluble iron, and about 1.92% by weight was pyritic iron.  About 87.0% by weight of the pyritic iron was removed by this process.

The hydrogenated-acid contacted solid residue was analyzed for residual sulfur by either (a) combusting the residue in a Leco Sulfur Determinator furnace Model 532 and absorbing and titrating the resulting $SO_2$ and $SO_3$ vapors or (b) by heating the residue with Eschka mixtures as described in ASTM-D-271, dissolving the mixture in water and determining sulfate ion in the solution with a Summerson-Klett colorimeter.  The percentage by weight of pyritic sulfur removed was determined by standard calculation.  In this example about 97.8% of the pyritic sulfur was removed.

The discrepancy between the value of the pyritic iron and sulfur removed is thought to be partly due to the removal of some organic sulfur and partly due to limitations in the accuracy of the testing procedures.


## OXIDATION

### Reaction with Ferric Chloride Solution

According to the process developed by *R.A. Meyers; U.S. Patent 3,768,988; October 30, 1973; assigned to TRW Inc.* finely divided coal or solid coal derivatives containing pyrite are reacted with a ferric ion solution; $FeCl_3$ is particularly suitable.  The ferric ion is reduced to ferrous ion and free sulfur is formed.  The solution is then filtered from the coal which is then washed and heat dried under low pressure.  Most of the free sulfur is volatized from the coal due to the heat drying; additional free sulfur can be removed by additional washing and heat drying and/or solvent extraction techniques.  At least 60% of the pyrite sulfur and pyrite iron is removed using this process.  If desired, the ferrous chloride can be regenerated; this permits iron oxide to be recovered as a by-product.

Typical treatment temperatures may vary from 50° to 110°C.  Reflux times are typically ½ to 2 hours and higher.  Typical coal particle sizes may vary from –200 mesh to ½ inch pieces.  Atmospheric pressure may be employed, but higher pressures can also be used.

The effective amount of the ferric ion solution employed for the extraction depends upon the amount of coal to be treated and on its pyritic sulfur content, the amount of sulfur that is desired to be extracted from the coal, the extraction times, the extraction temperatures, concentration of the ferric ion in the solution, etc.

Ferric chloride make-up solution and coal are fed into a pyrite reactor maintained at atmospheric pressure and about 212°F. Pyrite ($FeS_2$) is extracted from the coal and the slurry containing unreacted ferric chloride, ferrous chloride, sulfur, ferrous persulfide and the treated coal is fed to a coal filtration unit. Vacuum disk filters in the coal filtration unit are used to separate the bulk of the iron chloride solution from the treated coal.

In the coal washing sections, four stages of countercurrent washing with intermediate filtration steps are used to reduce the residual chloride content of the coal to less than about 100 ppm. A suitable residence time of the coal in each of the washing stages is about 15 minutes; rotary vacuum disk filters are used to separate the coal and wash the solution between washing stages.

The washed coal is then fed to a coal drying unit where rotary steam tube dryers are employed to remove the residual water from the washed coal, this operation being carried out at atmospheric pressure and about 212°F. The heated dry coal is then forwarded to a sulfur vaporization unit where free sulfur, which was produced in the extraction reaction, is vaporized at atmospheric pressure and a temperature of about 450°F or under reduced pressure (30 minutes) and at 250° to 350°F. The vaporized sulfur is removed by nitrogen gas into a sulfur condensation unit and cooled to about 225°F causing it to condense. The sulfur vapor is then passed to a recovery unit as bright sulfur. The treated coal with reduced pyrite content is then forwarded for use.

The table below shows the effect of $FeCl_3$ extraction on various coals. The table shows that 72 to 93% of the pyritic sulfur content may be removed in 2 hours by 0.5 M $FeCl_3$ solution from a wide variety of coals. Further, the process was applicable to all the coals and in the case of Indiana No. V, the extraction efficiency was excellent.

| Coal* | Total Sulfur Removed | Pyritic Sulfur Removed | Total Sulfur Before | Total Sulfur After |
|---|---|---|---|---|
| | . . . . . . . . . . . . . . . . . . . . . .% . . . . . . . . . . . . . . . . . . . . . . | | | |
| Lower Freeport | 48 | 75 | 3.87 | 2.01 |
| Lower Freeport | 64 | 72 | 3.40 | 1.23 |
| Bevier | 36 | 72 | 4.60 | 2.94 |
| Indiana No. V | 51 | 93 | 3.28 | 1.67 |
| Pittsburgh | 39 | 78 | 1.81 | 1.10 |

*All coals were –14 mesh except Bevier which was –200 mesh.

The process is extremely efficient in that at least 60% of the pyrite sulfur is extracted and the iron employed for extraction is easily recovered (about 85 to 90%) and may be reused. Furthermore, iron removal is facilitated since the iron contained in the $FeCl_3$ extraction solution and the iron in the pyrite are indistinguishable; thus, no special techniques are required to separate different metals from the wash-extraction operation if metal recycling is desired.

## Treatment with Water and Air

In the process developed by *T.J. Dillon and A. Warshaw; U.S. Patent 3,824,084; July 16, 1974; assigned to Chemical Construction Corporation* the pyritic sulfur content of coal is effectively reduced or eliminated by reaction with water and air at elevated pressure and temperature. The process is generally applicable to any type of coal, such as anthracite, bituminous, subbituminous, lignite, peat, petroleum coke, etc., and the term pyritic sulfur refers to sulfur bound in chemical combination with iron within the coal in the form of iron pyrites, which is generally designated by the formula $FeS_x$, where x may be any whole number or fraction from about 0.5 to 4.

The coal containing pyritic sulfur is initially ground to a fine particle size in the presence of water, so as to produce a coal-water slurry containing discrete particles of finely divided coal. The slurry is pumped to an autoclave where it is heated and pressurized with air. The typical reaction equations which take place in the presence of air and water at elevated temperature and pressure are as follows:

$$(1) \qquad FeS_2 + H_2O + \tfrac{7}{2}O_2 \longrightarrow FeSO_4 + H_2SO_4$$

$$(2) \qquad 2FeSO_4 + H_2SO_4 + \tfrac{1}{2}O_2 \longrightarrow Fe_2(SO_4)_3 + H_2O$$

The ferrous and ferric sulfates formed are water soluble and are separated from the coal by filtration of the treated slurry, yielding a solid low sulfur coal product and a liquid tailing containing dissolved ferrous and ferric sulfate. Any elemental sulfur contained in the original coal is melted during the elevated temperature processing and is also discarded with the liquid tailings when filtration is made at temperatures above the freezing point of the sulfur.

Since the solubility of ferric sulfate decreases with increasing temperature, some of the iron pyrites may be converted to insoluble basic ferric sulfate. When this compound is formed, flotation and filtration following autoclaving yield the low sulfur coal. The ferrous and ferric sulfates are natural depressants for any remaining iron pyrites. Both depressants are active in acid mediums, so flotation of the autoclave effluent readily separates solid basic ferric sulfate from an overflow slurry of coal particles which is filtered to yield the low sulfur coal.

The advantage of the process is that a low sulfur coal product is produced by inexpensive processing using readily available air and water. The system does not use a strong oxidizing agent to remove sulfur. It manufactures its own depressive agent in the event that a final flotation purification step is required or desired.

Referring to Figure 5.2, a flowsheet showing alternative embodiments of the process is presented. Coal stream **1** is derived from any suitable coal source or deposit from which coal may be mined by underground or strip mining methods. In this preferred embodiment of the process, the stream consists of coarse lumps of bituminous coal containing in the range of about 1 to 5% pyritic sulfur by weight which is passed into the grinding device or apparatus **2**, which is any suitable device for reducing solid particulate matter to a finely divided state, such as a ball mill, a rod mill or a hammer mill, water stream **3** is concomitantly passed into unit **2** and wet grinding of the coal particles to finely divided particles having a particle size distribution typically in the range of 10 to 100 mesh

FIGURE 5.2:  THE DILLON-WARSHAW PROCESS

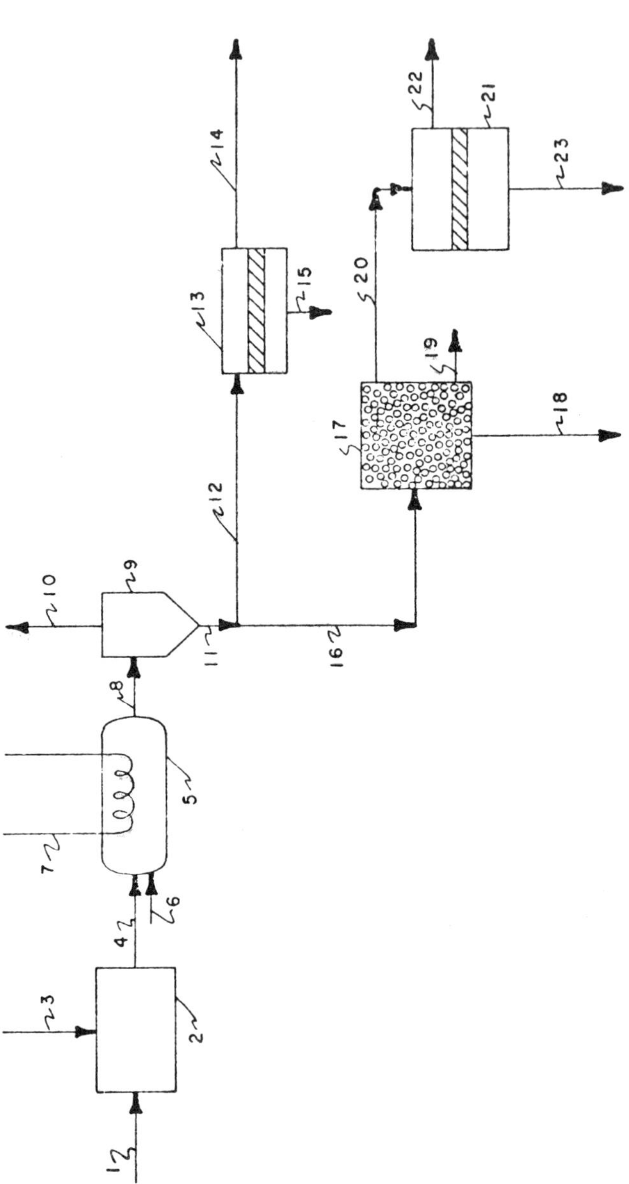

Source:  U.S. Patent 3,824,084

takes place. The mass flow rate of stream 3 is generally regulated to be about 20 to 80% of the mass flow rate of stream 1, so that an aqueous coal slurry stream 4 containing finely divided coal particles dispersed in an aqueous medium is discharged from unit 2. The processing of the coal in this unit serves to free elemental sulfur and trapped or bound iron pyrites.

Steam 4 is pumped into high pressure autoclave or reactor 5, which may be any suitable container or vessel capable of containing the process stream at high pressure for a suitable retention time, so that the desired reactions described supra may take place. Unit 5 may be provided with suitable internal baffles, agitators, or stirrers or the like to attain uniform dispersion of air or other oxygen-containing gas into the aqueous liquid slurry phase. Air stream 6 is concomitantly passed into unit 5 and sufficient air is pumped into it at elevated pressure so as to maintain an elevated pressure level typically in the range of 2 to 50 kilograms per cubic centimeter.

As the mixture of air and aqueous coal slurry proceeds through unit 5, the mixture is heated to an elevated temperature typically in the range of 90° to 200°C by the provision of heating coil 7. Steam or other suitable hot heat exchange fluid is circulated through the coil to provide heating of the mixture. The reaction mixture is typically retained within unit 5 for a time interval in the range of about 10 to 100 minutes and at least a portion of the pyritic sulfur present is oxidized in situ to ferrous sulfate and ferric sulfate, which concomitantly dissolve in the aqueous liquid phase. In most instances, a major portion of up to 90% or more of the pyritic sulfur is converted to water-soluble sulfate compounds in unit 5; however, in some instances, depending on the characteristics of the original minerals and pyritic sulfur present in the original coal feed steam 1, as well as specific operating parameters in a particular installation, some of the iron pyrites may be converted to solid basic ferric sulfate within unit 5.

In any case, a process effluent stream 8 is discharged from unit 5 which contains a solid phase in most instances consisting essentially of coal particles of diminished sulfur content, a liquid phase consisting essentially of an aqueous solution containing dissolved ferrous sulfate, ferric sulfate and molten sulfur and a gaseous phase consisting of oxygen-depleted air. Stream 8 is preferably initially passed into gas-liquid separator 9, which is a baffled or cyclonic means or device for separating the gaseous phase of stream 8 from the liquid slurry phase. The separated gaseous phase is discharged from unit 9 via stream 10 and a substantially gas-free slurry stream 11 is removed from the lower portion of unit 9.

Stream 11 may be processed by alternate procedures, depending on whether it contains a substantial proportion of solid basic ferric sulfate. In most instances the stream will be substantially devoid of solid basic ferric sulfate, and in this case it is passed via stream 12 to filter 13, from which a solid phase stream 14 consisting of a finely divided low sulfur coal product is passed to product utilization, which will generally consist of the combustion of stream 14 in a steam boiler or furnace or the like. The separated aqueous liquid phase stream 15 is also removed from unit 13.

Stream 15 is a liquid tailings consisting essentially of an aqueous solution containing dissolved ferrous sulfate and ferric sulfate. Stream 15 may be passed to suitable waste disposal, or stream 15 may be processed to recover ferrous sulfate and/or ferric sulfate for product sales or utilization by evaporative crystallization or the like.

In instances when stream **11** contains an appreciable proportion of solid basic ferric sulfate, stream **11** is processed to initially remove the solid basic ferric sulfate by flotation.  In this procedure, the dissolved ferrous sulfate and ferric sulfate act as in situ natural depressants and aid in the flotation separation process.  Stream **11** in this case flows via stream **16** to flotation vessel **17**.  A flotation-inducing stream **18** which may consist of air or water or other fluid is introduced into the lower portion of vessel **17** and rises through the fluid slurry body maintained in vessel **17**, thereby selectively moving coal particles upwards while allowing the particles of solid basic ferric sulfate to remain in the lower portion of vessel **17**.  A stream **19** consisting essentially of an aqueous slurry of solid basic ferric sulfate is removed from the lower portion of unit **17**.  Stream **19** may be passed to solids tailings disposal, or stream **19** may be further processed as desired to produce a salable sulfate product.

An aqueous coal slurry stream **20** is removed from the upper portion of unit **17**.  Stream **20** contains product solid low sulfur coal particles and an aqueous liquid phase containing dissolved ferrous sulfate and ferric sulfate.  Stream **20** is processed in a manner similar to the processing of stream **12** described above.

*Example:*  The autoclave charge was 200 grams of coal and 467 grams of water.  The coal was ground to 100 mesh and initial analysis was 2.55% total sulfur, 0.94% organic sulfur, and 28.6% volatiles.  The autoclave conditions were 175°C and 42.2 kg/cm$^2$ pressure, provided via air injection.  Reaction time was 10 minutes and agitation at 600 rpm was provided.  The autoclave contents were blown through a preheated filter at the end of the run.  The coal was then dried and analyzed for sulfur.  Total sulfur by Eschka method was 0.48%.

### Treatment with $NO_2$ and Other Gases

The process described by *A.F. Diaz and E.D. Guth; U.S. Patent 3,909,211; September 30, 1975; assigned to KVB Engineering, Inc.* provides an improved method for desulfurizing coal while producing concentrated sulfuric acid as a commercially useful by-product.  The preferred embodiment involves the following steps.

Coal is first converted into particulates, preferably no larger than approximately one-fourth inch in diameter.  The pulverized coal is placed into a reaction chamber into which is passed a combination of four gases, with the interior of the chamber being maintained at a temperature in the range of 100° to 500°F for 1 to 30 minutes for continuous reaction, or for 0.5 to 5 hours for batch reaction, at a pressure in the range from 1 to 20 atmospheres.  The process can be either on a batch or continuous basis as desired.

The gases used are preferably $O_2$ (0.5 to 20 volume %), NO (0.25 to 10 volume %), $NO_2$ (0.25 to 10 volume %) and $N_2$ the remainder.  The resulting sulfur-containing products from this step will typically be $FeSO_4$, $SO_3$ or $SO_2$ gas and various organic sulfoxides or sulfones.

$FeSO_4$ is removed by water extraction as this salt is soluble in water.  The $SO_3$ is converted into concentrated $H_2SO_4$ by being passed into a condenser containing a solution of $H_2SO_4$.  The $SO_2$ is recycled to the reactor where it is subsequently oxidized to $SO_3$.  The $SO_2$ or other sulfur-containing compounds may be converted to $SO_3$ by exposure of the $SO_2$ to oxygen within the reactor by further reacting the reactor effluent gas before contacting the gas with sulfuric acid, or the $SO_3$ gas is reacted with compounds such as calcium oxide or sodium hydroxide to form calcium sulfate or sodium sulfate instead of sulfuric acid.

Referring to Figure 5.3, there is shown a representative continuous process arrangement for carrying out the process. A batch processing arrangement is also understood to be within the scope of this process.

## FIGURE 5.3: TREATMENT WITH $NO_2$ AND OTHER GASES

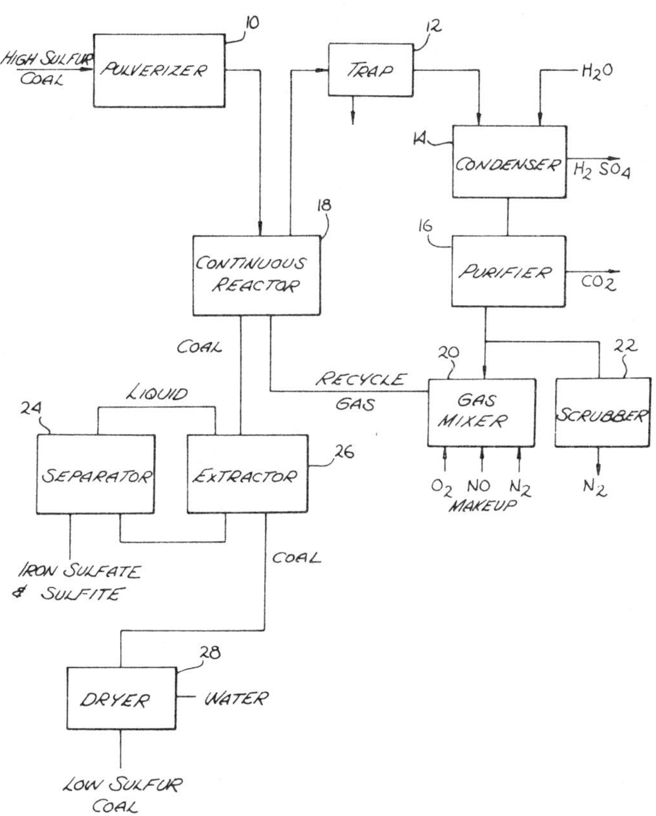


Source: U.S. Patent 3,909,211

Coal in crushed or raw form is initially fed into a pulverizer **10** which serves to convert the raw coal into particles to be processed, the size of which will range from 200 mesh to as large as one-fourth inch in diameter. The converted coal is then fed into a reactor **18** which receives the coal particles from pulverizer **10**, together with a predetermined quantity of a combination of the following four gases in the listed relative quantities: 0.5 to 20 volume % $O_2$; 0.25 to 10 volume % NO; 0.25 to 10 volume % $NO_2$; and the balance $N_2$. Upon being heated for a period of time of from 1 to 30 minutes at a pressure in the range from 1 to 20 atmospheres at a temperature of from 100° to 500°F in reactor **18**, the output from reactor **18** will typically be a combination of materials, some solid, some gaseous, including $FeSO_4$, desulfurized coal and some additional hydrocarbons containing some sulfur. The solid portion of the output from the reactor **18** is directed to an extractor **26**.

Water is added to the reactor products in the extractor and the inorganic sulfur present as sulfates or sulfites dissolves and passes to separator **24** with the liquid stream. This soluble portion includes the sulfur initially present as iron pyrites which is converted to sulfates and sulfites in the reactor. To aid in the removal of the sulfates and sulfites in the extractor, a soluble caustic such as sodium hydroxide may be added to the water in the extractor. The water may also be kept warm to facilitate solubility.

The liquid phase is cooled to precipitate the inorganic sulfates and sulfites, which are then removed by filtration. The water is then heated and recycled through the extractor. The solid phase product of the extractor is then passed to dryer **28** where it is dried. The output of the drier is coal having a substantially lower sulfur content than that entering the process.

The gaseous products from reactor **18** flow through trap **12** which removes volatile fuels and entrained coal particles which are carried in the gas stream. Suitable traps are commercially available and commonly known.

The clean gas from trap **12** which contains sulfur dioxide and sulfur trioxide given up by the coal in reactor **18** is bubbled through a sulfuric acid solution in condenser **14** dissolving sulfur trioxide in the acid solution. As previously mentioned, the $SO_3$ may be reacted to form other compounds instead of sulfuric acid (i.e., with ammonium hydroxide to form ammonium sulfate, with sodium hydroxide to form sodium sulfate, with calcium oxide to make calcium sulfate). These compounds could also be made by reacting the sulfuric acid with the appropriate basic compound.

*Example:* Using the batch reactor process a sample of Lower Kittanning coal previously pulverized and segregated as to size was treated. Coal of –14+28 mesh particle size was loaded into the reactor, the reactor assembled and heated to 200°F. For 3 hours a gas mixture consisting of air with 9% nitrogen oxide added was passed through the reactor and the bed of coal at one atmosphere pressure. The flow rate was such that about 10 times the stoichiometric quantity of oxygen required to oxidize the sulfur to sulfate forms was passed through the reactor.

The coal had an initial sulfur content of 4.3% (3.6% pyritic, 0.7% organic, a trace of sulfate). After treatment, the coal was removed from the reactor, washed with water and dried. The total sulfur content was 1.58% (i.e., 63% of the sulfur was removed). The coal was then washed with 10% aqueous sodium hydroxide followed by water and then dried. The total sulfur content was 0.47% (i.e., 87% of the sulfur was removed).

## Sulfurous Acid as Oxidizing Agent

*R.A. Meyers; U.S. Patent 3,926,575; December 16, 1975; assigned to TRW Inc.* has found that it is possible to react the pyrite contained in the coal with a solution containing an effective amount of sulfurous acid. A typical reaction proceeds substantially as follows:

Primary: Oxidation-Reduction

$$4FeS_2 \text{ (pyrite)} + 3SO_2 + 12HCl \rightarrow 4FeCl_3 + 11S\uparrow + 6H_2O$$

Secondary: Oxidation-Reduction

$$4FeCl_3 + 2FeS_2 \text{ (pyrite)} \rightarrow 6FeCl_2 + 4S\uparrow$$

Overall Reaction:

$$4FeCl_3 + 3SO_2 + 12HCl \rightarrow 6FeCl_2 + 15S\uparrow + 6H_2O$$

In addition to these major reactions, it is to be assumed that a small part of the free sulfur formed initially may be further oxidized to sulfite, sulfate, thiosulfate, etc. Formation of the secondary products can be further lessened by minimizing reaction times, acid concentration and temperature.

The solution containing mainly free sulfur, ferrous chloride and any unconsumed ferric chloride and sulfurous acid is removed from the coal by filtration. The coal is then washed and dried, preferably by heating in a vacuum; this results in most of the free sulfur being volatized. If desired, a further wash, filtration and heating will remove additional sulfur and more ferrous ion. Finally, one or more extractions with a suitable organic sulfur solvent such as benzene, kerosene, gas oil, or para-cresol at temperature of 50°C up to solvent reflux is employed to further reduce the sulfur content of the coal.

Regeneration of the unused ferric chloride and ferrous chloride solution may be accomplished by first evaporating most of the water to concentrate the solution. Cooling the concentrated solution precipitates the ferrous chloride from the ferric chloride, the latter still remaining in solution. The ferrous chloride precipitate is air oxidized to ferric chloride and iron oxide; finally, the ferric chloride is recycled or sold as a by-product and the iron oxide recovered.

Typical pyrite extraction temperatures may vary from 110° to 140°C. Reflux times are typically ½ to 2 hours and higher. Typical coal particle sizes may vary from –200 mesh to ½ inch particles. Atmospheric pressure may be employed, but higher pressures can also be used.

The effective amount of the sulfurous acid employed for extraction depends on the amount of treated coal and its pyritic sulfur content, the amount of sulfur desired to be extracted, extraction times, extraction temperatures, concentration of the sulfurous acid, etc.

Coals for which this process may be employed include anthracites, charcoal, coke, bituminous coals, lignites, etc. In addition, solvent refined coals such as hydrocracked coal and middlings are all capable of being refined by this extraction process.

In general, the procedure employed was to reflux an aqueous solution of sulfur dioxide and hydrochloric acid with pulverized coal. This converted the ferrous persulfide (pyrite) to ferric chloride and produced free sulfur. Additional pyrite is removed by interaction with ferric chloride. The resultant solution of ferrous chloride was then separated from the coal by filtering. Following a water wash, the coal was then heated to dryness under vacuum thereby vaporizing some of the free sulfur. Most of the remaining free sulfur in the coal was extracted with a suitable solvent such as benzene, kerosene, gas oil, or para-cresol. In addition, the para-cresol removes a portion of organic sulfur compounds contained in the coal. If desired, the solution containing ferrous ion can be recycled for subsequent reuse and/or oxidized to iron oxide; these products may be recovered as noted previously.

*Example:* In a typical case, aqueous sulfurous acid (10 times stoichiometric excess over pyrite content of coal), hydrochloric acid (where designated), and pulverized coal were introduced into a glass aerosol stirred bomb. The mixture was heated for the temperature and time shown; this caused a pressure rise to about 20 to 30 psig. The mixture was then cooled. The coal was then filtered from the aqueous phase and washed with hot water to remove residual acid. Some of the samples were slurried with benzene for 20 minutes and then filtered again. All samples were dried in a vacuum oven at 160°C/30 min to constant weight (ca 24 hours). Elemental sulfur was distilled out and collected on the oven window and in the pump trap and lines. The table below shows the effect on –14 mesh Indiana No. V containing pyrite, after treatment with a sulfurous acid solution.

### Removal of Sulfur from –14 Mesh Indiana No. V Coal with Sulfurous Acid*

| Temp. (°C) | HCl (M) | $H_2SO_3$ (M) | Retention Time (hr) | Benzene Post Treatment | Sulfur (%) | Ash (%) | Btu Content | Percent Removed — Total Sulfur | Percent Removed — Pyritic Sulfur | Percent Removed — Ash | Btu Change (%) |
|---|---|---|---|---|---|---|---|---|---|---|---|
| 140 | 3.6 | 0.9 | 20 | No | 3.06 | 7.4 | 13,000 | 15 | 30 | 37 | +4 |
| 100 | 3.6 | 0.9 | 20 | No | 2.76 | 7.8 | 12,600 | 23 | 46 | 33 | +1 |
| 100 | 3.6 | 0.9 | 20 | Yes | 2.69 | 7.5 | 13,000 | 26 | 52 | 36 | +4 |
| 100 | 3.6 | 0.9 | 2 | No | 2.69 | 8.2 | 11,900 | 26 | 52 | 30 | –5 |
| 100 | 0 | 0.9 | 2 | No | 2.82 | 8.7 | 12,800 | 22 | 44 | 26 | +2 |
| 100 | 0 | 0.9 | 20 | Yes | 3.19 | 8.9 | 12,950 | 12 | 25 | 24 | +4 |
| 100 | 0 | 0.9 | 20 | No | 3.53 | 8.9 | 12,800 | 2 | 4 | 25 | +2 |

*Starting Indiana No. V: 3.62% total sulfur; 0.03% sulfate; 1.79% organic sulfur; 1.80% pyrite sulfur; heat content is 12,500 Btu.

## Removal of Sulfur as Sulfate

*J.C. Agarwal, R.A. Giberti and L.J. Petrovic; U.S. Patent 3,960,513; June 1, 1976; assigned to Kennecott Copper Corporation* have developed a process for recovering from 50 to 90% of the free and combined sulfur in coal. Figure 5.4 illustrates pressure oxygen leaching of pyritic sulfur containing coal in a neutral starting solution. The process in the drawing is illustrated as coupled to an agglomerator or an existing power plant showing the various stages for effective removal of the impurities.

With reference to Figure 5.4, a coal as mined and/or washed and designated as Illinois No. 6 is introduced into a crusher **11** wherein it is crushed to at least minus 2.0 inches. Thereafter, the crushed coal is introduced into a batch or plug flow reactor **12** which may be provided with a stirrer or some other means for agitation such as fluidization by recirculating reaction liquid or oxygen, or oxygen-laden gas, e.g., air. Coal introduced in the reactor has the following sulfur assay.

| | . . . . Weight Percent Sulfur as . . . . . | | | |
|---|---|---|---|---|
| | $SO_4$ | Pyrite | Organic | Total |
| Ohio No. 6 | 0.17 | 2.10 | 0.78 | 3.05 |

# FIGURE 5.4:  COAL OXIDATION

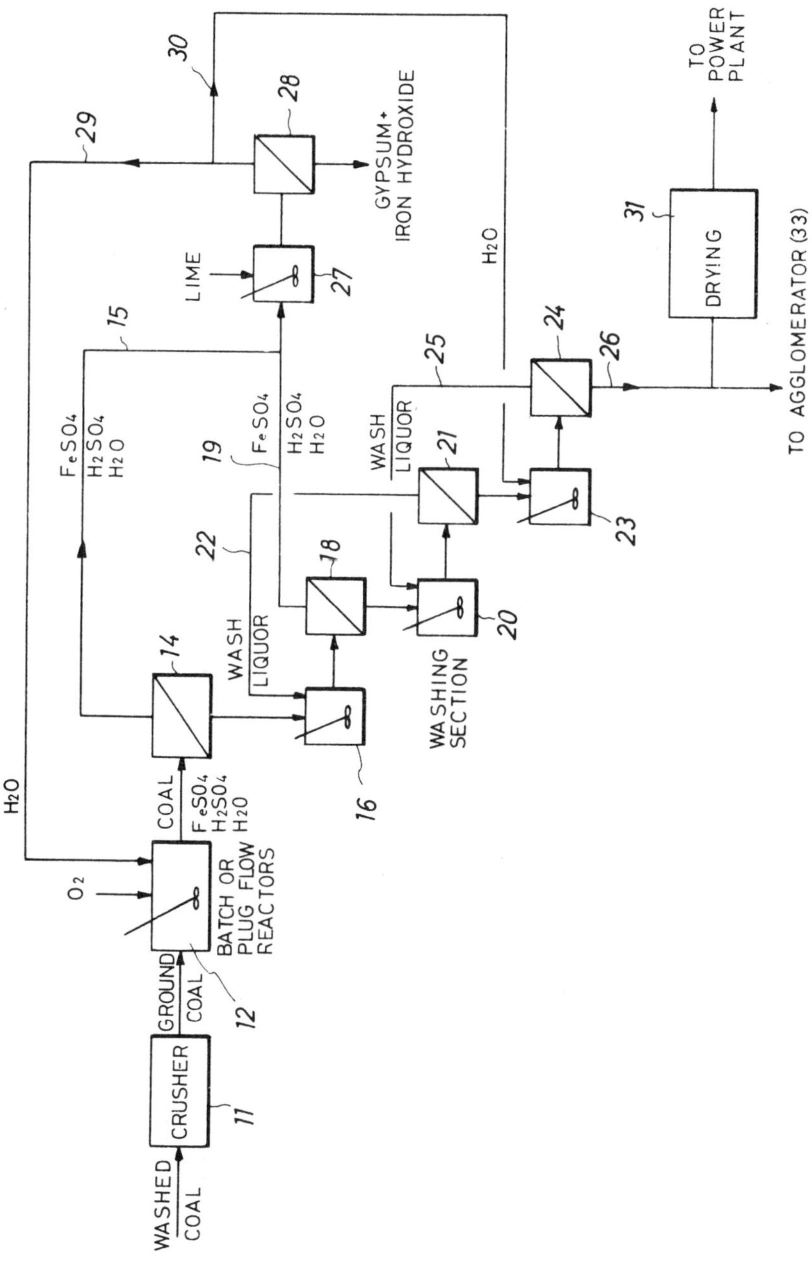

Source:　U.S. Patent 3,960,513

Generally, all types of coal containing sulfur may be subjected to the process. Generally, oxygen is used for reacting with sulfur in the coal. Instead of oxygen, oxygen-enriched air may also be used with oxygen enrichment being in a weight range from 0 to 100%, a range from 80 to 100% is preferred. In order to prevent the combustion of coal, it is reacted with oxygen in an aqueous phase and generally in a ratio of coal to aqueous phase of 1 to 60% by weight, preferably from 10 to 25%. The pressures in the reactor employed are from 0 to 1,000 psig, preferably from 30 to 350 psig. Although air can be used, power requirements may be excessive to pump air. The reactor generally is filled with the constituent parts in the following proportions, the ratios being given by weight: coal, 1 part; $O_2$, from 0.085 to 0.18; and $H_2O$, from 99 to 0.667.

From the above, the slurry density is evident. Heating coils may be provided in the reactor so that the reaction may be carried out at a temperature of 50° to 450°F, preferably from 120° to 300°F. Although the reaction of oxygen with sulfur produces a certain amount of heat, it still may be necessary to augment or remove heat and for that purpose the heating coils are used or introduced, e.g., by heat exchange between slurry, before and after reactors (not shown in the figure).

After the coal has reacted for a period from about 15 minutes to 24 hours (preferably from 1 to 4 hours), the liquid reaction phase rich in ferrous, ferric sulfate, and sulfuric acid is introduced in a liquid-solid separation device **14**, such as a thickener or rotary filter, where the liquid laden with iron sulfate, sulfuric acid, and water is removed. The solid material, that is coal, is introduced into a means for washing such as mixer-settler **16**. After washing, the coal is then introduced in a solid-liquid separation stage **18** and separated in device **14** from wash liquor. The wash liquor contains lesser amounts of iron sulfate and sulfuric acid. Generally, per ton of coal produced, from 1,000 to 10,000 gallons of wash water is used. The constitution of liquid in the flow stream designated as **15** in the figure is as follows: $FeSO_4$, 0.003 to 0.75 M; $Fe_2(SO_4)_3$, 0.003 to 0.75 M; and $H_2SO_4$, 0.003 to 0.75 M. The constitution of liquor in the flow stream **19** in the figure is as follows: $FeSO_4$, 0.0015 to 0.06 M; $Fe_2(SO_4)_3$, 0.0015 to 0.06 M; and $H_2SO_4$, 0.0015 to 0.06 M.

The coal from the separation means, such as **18**, is then introduced into another washing device **20** or it may be introduced directly into a power plant if no additional washing is required. If further washing is needed, then coal is subjected again to washing in a countercurrent fashion, sent to a separation stage **21** where the wash liquor from that stage is introduced in washing section **16** via line **22**.

Again, coal from the separation stage **21** may be introduced into an additional washing section **23**, washed such as for a period of 5 to 30 minutes which is of about the same duration as in the washing device **16** and **20**. From the final separation stage **24**, coal is ready for discharge or suitably prepared for burning in a power plant (such as by drying in a dryer **31** or by centrifuging). The wash liquor from the separation stage **24** via line **25** is introduced into wash section **20**; and thus, there is very little carryover liquor which is being discharged with coal. Generally, the moisture content of the coal discharged via line **26** from separation device **24** is about 10 to 30%.

*Example:* With reference to Figure 5.4, coal which is introduced into a ball mill **11**, is ground to –100 mesh size and fed into the reactor **12** by adding re-

cycled water thereto such that the reaction medium has 4.3% by weight coal
slurry. Under a reaction pressure of 300 psig oxygen, the liquid medium is held
at 130°C for 6.5 hours (residence time); sulfur from pyritic sources is then substantially completely converted to a soluble sulfate species such as ferrous sulfate, ferric sulfate and sulfuric acid.

A solid-liquid separation as illustrated in the figure is then carried out in a sequence as indicated. In accordance with this procedure, Illinois No. 6 coal containing 1.53% by weight pyritic sulfur was converted to a coal containing less
than 0.03% by weight pyritic sulfur with no change in the sulfate sulfur in the
material balance and approximately 10% decrease in the amount of organic sulfur being present in the coal.

The coal leaching process described above may be carried out in an acidic solution, using 0.075 molar sulfuric acid, or in a basic solution, using ammonium
hydroxide up to 3.0 molar.

**Treatment with Aqueous Solutions of Metal Oxidants**

The application of mild oxidation reactions to remove the pyrite from coal is
described in U.S. Patent 3,768,988. This process employs the ferric ion as the
oxidizing agent and is hereinafter referred to as the Meyers process. Essentially,
the Meyers process employs aqueous ferric sulfate or chloride to oxidize the
pyritic sulfur to elemental sulfur. About 60% of the pyritic sulfur content of
the coal is oxidized to the sulfate which dissolves in the aqueous leaching solution. The free sulfur is then removed from the coal matrix by solvent extraction with an organic sulfur solvent such as benzene, kerosene, p-cresol, etc., or
by steam or vacuum vaporization. The aqueous oxidizing agent is regenerated
and recycled to the oxidation step.

According to the process described by *R.M. Dessau; U.S. Patent 4,022,588;
May 10, 1977; assigned to Mobil Oil Corporation* the sulfur content of coal is
reduced by an oxidative solubilization process where the pyritic sulfur contained
in coal is reacted with an aqueous solution containing an effective amount of a
manganese, vanadium, or cerium metal oxidant. The reacted coal is then separated from the solution and the elemental sulfur which remains trapped within
the coal matrix is removed by known processes, such as those used and described
above in the Meyers process.

The manganese, vanadium and cerium metal oxidants used are known materials
conventionally employed in the art as oxidants. In general, it may be stated
that the oxidants are those metal compounds which have a metal ion existing
in an oxidation state higher than the lowest oxidation state of the metal ion.
The oxidants are further characterized as being water soluble or capable of being
solubilized in acid aqueous solution.

The preferred metal oxidants are the higher valence compounds of vanadium,
manganese and cerium such as manganese dioxide, ammonium metavanadate and
ceric ammonium nitrate.

The amount of metal oxidant utilized should be at least a stoichiometric mol
ratio of oxidant to the pyritic sulfur content of the coal. In general, a stoichiometric excess is utilized in mol ratios of 5:1 or higher. On a weight basis, the

concentration of metal oxidant ranges from about 0.5 to 5.0%. The coal, prior to reaction with the aqueous solution containing the metal oxidant, is preferably prepared by grinding so that it will have a particle size less than 10 mesh. The suitably prepared coal is then reacted in an aqueous solution which contains an effective amount of metal oxidant.

In a typical example, the coal is wet ground to a finely divided state through the use of a ball mill, rod mill, or hammer mill, etc., to a particle size of about 10 to 100 mesh. Water is then added in an amount sufficient to provide an aqueous slurry having a solids content of 1 to 20%. The aqueous solution is preferably acidified to a pH of about 1 to 5 through the use of a mineral acid such as sulfuric acid, hydrochloric acid, or the like. Thereafter the aqueous solution is reacted with the metal oxidant under reflux conditions at 100°C. The reflux time is not critical and may vary from 4 to 16 hours.

Upon completion of the reaction, the aqueous solution will contain ferrous sulfate and ferric sulfate as a result of oxidizing the pyritic sulfur in the coal. In most instances, essentially all of the pyritic sulfur is removed from the coal whereas free elemental sulfur remains in the coal. The aqueous oxidizing solution is separated from the coal and the treated coal is thereafter purified by known methods of washing or extraction to remove the free sulfur.

## SOLVENT REFINED COAL

### Use of Solvent Derived from the Coal Itself

The process developed by *W.C. Bull, L.G. Stevenson, D.L. Kloepper and T.F. Rogers; U.S. Patent 3,341,447; September 12, 1967; assigned to U.S. Secretary of the Interior and Gulf Oil Corporation* for upgrading carbonaceous materials such as coal, lignite, peat, etc. is a great economic improvement on previous solvent extraction processes because it provides for essentially complete solubilization of the coal and it is cyclic and does not require any substantial addition of solvent after start-up.

This process consists essentially of dissolving substantially all of the potentially available fuel fractions of naturally occurring carbonaceous fuels, such as coal, lignite, peat, and the like, using a solvent derived from the original fuel feedstock under critically controlled conditions of temperature, pressure and atmosphere together with a very critical holding period in processing temperature, at which the normally insoluble portions of the fuel are decomposed, generally, by depolymerization of the feed fuel, to fractions soluble in the solvent.

By observing the critical process parameters of this process, to be specified hereafter, these naturally occurring fuels are not only upgraded into low-ash, low-oxygen, low-sulfur fuels, but in addition, this solution of the feed fuel is accompanied by a generation of additional quantities of solvent from a portion of the raw feed fuel which more than makes up any amount of solvent lost by mechanical losses thereof in the system.

More specifically the process may be described as follows: A raw feed fuel, such as Kentucky No. 11 coal which has been ground by a suitable means, such as a hammer mill (preferably the coal is finely ground to approximately 80%

through 200 mesh, U.S. Standard, is fed by means of a conveyor or the like to an agitated tank where it is mixed with solvent (obtained from previous processing) at a ratio of about 1:1 to 4:1 solvent to coal. If desired, the coal-solvent slurry in the tank may be heated to any suitable temperature which will flash off any moisture which may be present in the coal.

Since the solvent used in this process is derived from the coal being dissolved, its composition may vary, depending on the analysis of the coal being used as feedstock. In general, however, the solvent employed is a highly aromatic solvent obtained from previous processing of fuel and will generally have a boiling range of about 150° to 750°C, a density of about 1.1 and a carbon to hydrogen mol ratio in the range from about 1.0 to 0.9 to about 1.0 to 0.3. Generally, any good organic solvent for coal may be used as the initial start-up solvent in the process.

A typical solvent is, for example, middle oil obtained from coal and having a boiling range of 190° to 300°C. A solvent found particularly useful as a start-up solvent is anthracene oil or creosote oil having a boiling range of about 220° to 400°C. However, the selection of a specific start-up solvent is not particularly critical since during the process dissolved fractions of the raw feed fuel form substantial quantities of additional solvent which when added to the solvent originally fed into the system provide a total amount of solvent which is greater than the original amount put in the process.

Thus, regardless of what the original solvent may have been, it will lose its identity and approach the constitution of the solvent formed by solution and depolymerization of the raw fuel fed into the process. As a result the composition of the solvent approaches that of the same general composition as the deashed product of the processed feed fuel but of lower molecular weight, with the actual composition in each case determined by the composition of the particular raw feed fuel employed. For this reason the solvent which is employed may be broadly defined as that obtained from a previous extraction of raw carbonaceous fuels, in accordance with the process.

In all events the ratio of solvent to coal in the slurry mixed in the slurry tank will be in the preferred range of 1:1 to 2.3:1. The slurry is then fed by means of any suitable method, for example, a positive displacement pump to a preheater and then into a dissolver which is suitably heated to maintain the slurry at its elevated temperature. It is essential that hydrogen be added to the slurry ahead of the preheater or dissolver at a partial pressure of at least 500 psi. The hydrogen employed is normally that recycled from previous processing together with any fresh make-up hydrogen required to provide the necessary hydrogen content. Although there is no critical upper limit to the hydrogen pressures employed, for practical reasons, the pressures employed will normally be in the range of 500 to 1,500 psi, with 1,000 psi preferably employed.

The hydrogen pressurized fuel-solvent slurry is heated in the preheaters to rapidly raise the temperature thereof to about 370° to 500°C, and preferably about 375° to about 440°C.

The tubes are sized, as to diameter and length, so as to enable the flow of slurry therethrough to be heated to operating temperatures as rapidly as is practical, which in experiments has been obtained as low as 12 to 20 seconds. In general, the preheater will be designed to provide the desired rate of heating to the slurry at a heat flux of not greater than about 10,000 Btu/hr/ft$^2$ of tube surface area. Normally in order to maintain good heat transfer, the diameter of the tubes will be sized so as to maintain the slurry in turbulent flow through the preheater.

The length of time in which the solution of the hydrogen pressurized slurry is continued in the dissolver is critical to the success of this process. Although the duration of treatment (solution) will vary for each particular carbonaceous fuel treated, it was found that the mechanics of the solution provide an accurate guide for the residence time of the slurry in the dissolver. In this regard it was found that the viscosity of the solution obtained during processing of the slurry increases with time in the dissolver followed by a decrease in viscosity as this solubilizing of the slurry is continued and then followed by a subsequent increase in the viscosity of the solution on extended holding in the dissolver.

One of the criterions used for determining the completion of the solution process, in accordance with this method, is the relative viscosity of the solution formed which is the ratio of the viscosity of the solution to the viscosity of the solvent, as fed to the process, both viscosities being measured at 210°F. Accordingly, the term Relative Viscosity (R.V.) is restricted to and defined as the viscosity at 210°F of the solution formed divided by the viscosity of the solvent at 210°F fed to the system, i.e.,

$$\text{Relative Viscosity} = \frac{\text{Viscosity of Solution at 210°F}}{\text{Viscosity of Solvent at 210°F}}$$

This R.V. can be employed for more specific clarification of the residence time for the solution in the dissolver. It may be noted that as the solubilizing of the slurry proceeds, the R.V. of the solution first rises above a value of 20 to a point at which the solution is extremely viscous and in a gel-like condition. In fact, if low solvent-to-coal ratios are used, for example, 0.5/1, the slurry would set up into a gel. After reaching the maximum R.V., well above the value of 20, it begins to decrease to a minimum after which it again rises to higher values.

It was found essential for the success of this process that the solubilization be allowed to proceed until the decrease in R.V. falls to a value of at least 10 and the resultant solution separates from undissolved residue of the coal before the R.V. again rises above 10. Normally, the decrease in R.V. will be allowed to proceed to a value less than 5 and preferably in the range of 1½ to 2.

It may be noted that, during the formation of the solution, the feed fuel depolymerizes in the presence of hydrogen to form fractions which are soluble in the solvent, and it was observed that the depolymerization of the coal during extraction is accompanied by the evolution of hydrogen sulfide, water, carbon dioxide, methane, propane, butane, and other higher hydrocarbons which will comprise part of the atmosphere in the dissolver.

It was found that the depolymerization of the feed fuel consumes hydrogen up to a weight equivalent to about 2.0% of the weight of the feed fuel (as received) and under preferred conditions, in an amount of about 0.5 to 1% of the feed fuel. Upon completion of the solubilization in the dissolver, the resultant solution is then charged to a filter for separation of the coal solution from undissolved residue (i.e., mineral matter) of the feed fuel. The filter is a conventional rotary drum pressure filter suitably adapted for pressure let down and venting of gases. The gases vented from the filter are then passed into a gas treating unit in which hydrogen sulfide and carbon dioxide are scrubbed out in any suitable manner, as for example by caustic solutions. The remaining gas may be given any further treatment desired.

The hydrogen sulfide and carbon dioxide free gas recovered are then recycled to the process by feeding to fresh slurry being fed to the preheater with all make-up hydrogen being supplied by fresh hydrogen. Analysis of the recovered gases has shown that the overall consumption of hydrogen throughout the system is quite low with the actual consumption of hydrogen being not greater than 2% (based on the feed fuel) and quite often as low as 0.5%.

The filter cake or residue is withdrawn for additional processing as desired, as for example, solvent recovery. In practice, further processing of the filter cake is desired since, in addition to mineral water, it may contain from between 40 to 50% by weight of solvent that can be recovered by pyrolytic distillation. Often this dried filter cake contains about 50% carbon and runs approximately about 7,000 Btu per pound. The further processing of the cake is particularly significant since the coal mineral matter and included solvent may comprise about 10 to 20% of the total weight of the slurry fed to the system. Since the filter cake is comprised of approximately 40 to 50% solvent, the amount of recoverable solvent comprises between about 5 to 15% of the original solvent fed into the system. As will be understood, if the filter cake is processed for the recovery of the solvent, the recovered solvent may be recycled into the system for the preparation of additional slurry for feeding into the system.

The filtrate is then pumped at a temperature of about 270° to 430°C through suitable nozzles into a simple vacuum flash evaporator for removal of the majority of the solvent. In accordance with conventional practice, the evaporator will be equipped with a demister to remove any entrained liquid in the vapor, and, also, suitable heaters will be located on the outside of the evaporator to make up for heat losses. The flashed solvent from the evaporator may be passed to a still or recycled directly to the system.

The remaining product from the evaporator which is liquid at this point, may be used as such, or it may be pumped onto a continuous rotating steel belt where it is cooled and solidified. This product, which solidifies upon cooling, is very brittle and breaks into flakes as it finally falls from the belt. Other methods may also be used to recover the product as for example spray cooling or prilling.

*Example:* Kentucky No. 11 coal was dissolved in a solvent recovered from previous extraction runs (in accordance with this process) under the following conditions: 1,000 psig total gas pressure; 80% hydrogen in gas; 410°C temperature; and 3/1 ratio of solvent/coal.

The coal/solvent slurry was heated within 12 to 20 seconds to the reaction temperature of approximately 410°C and dissolution continued until the R.V. of the solution rose and then dropped to 3.43. Thereafter, the solution was filtered at this R.V. from the undissolved residue of the coal, followed by treatment of the filtrate to separate recycle solvent from the product. The properties of the product of the example are tabulated with the original analysis of the Kentucky No. 11 coal, for purposes of comparison.

|                      | Process Product | Kentucky No. 11 Coal (maf) |
|----------------------|-----------------|----------------------------|
| Ash, %               | 0.48            | 7.33                       |
| Carbon, %            | 88.16           | 78.45                      |
| Hydrogen, %          | 5.23            | 5.20                       |
| Nitrogen, %          | 1.51            | 1.19                       |
| Sulfur, %            | 1.17            | 3.74                       |
| Oxygen, %            | 3.42            | 11.40                      |
| Volatile matter, %   | 33.6            | 42.88                      |
| Btu per pound        | 15,768          | 13,978                     |
| Density, g/cc        | 1.24            | 1.33                       |
| Melting point, °C    | 220             | —                          |

## Formation of Indurated (Hardened) Pellets

In the process described by *E. Gorin; U.S. Patent 3,748,254; July 24, 1973; assigned to Consolidation Coal Company* coal is partially converted by solvent extraction to a mixture of extract, solvent and undissolved carbonaceous residue. The mixture is separated into a low solids-containing fraction and a high solids-containing fraction. The composition of the latter is adjusted so that its admixture of solids and liquid binder (i.e., extract and solvent) is such as to make it pelletizable. Pellets are formed from the pelletizable composition, preferably in a rotary drum and indurated either concurrently with formation or subsequently thereto. The indurated pellets serve either as solid fuel or as a source of carbon in carbon-steam reactions.

Any coal may be used in the process, nonlimiting examples of which are lignite, bituminous coal and subbituminous coal. The feed coal, in a finely divided state and free of substantially all extraneous water, is subjected to solvent extraction in a suitable solvent extraction zone at an elevated temperature. The solvent extraction process may be any of the processes commonly used by those skilled in the art, for example, continuous, batch, countercurrent or staged extraction; and conducted at a temperature in the range of 300° to 500°C, a pressure in the range of 1 to 6,500 psig, a residence time in the range of 1 to 120 minutes, a solvent-to-coal ratio of 1/1 to 4/1 and, if desired, in the presence of a catalyst and/or up to 50 scf of hydrogen per pound of maf (moisture- and ash-free) coal.

Polycyclic, aromatic hydrocarbons which are liquid at the temperature and pressure of extraction are generally recognized to be suitable solvents for the coal in the extraction step. At least a portion of the aromatics may be partially or completely hydrogenated, whereby some hydrogen transfer from solvent to coal may occur to assist in the breakdown of the large coal molecules. Mixtures of the hydrocarbons are generally used and these may be derived from subsequent steps in the process of this method. Other types of coal solvent, such as oxygenated aromatic compounds, may be added for special reasons, for example, to improve the solvent power, but the resulting mixture should be predominantly of the type mentioned.

The coal and the solvent are maintained in intimate contact at the elevated temperature until up to about 80 weight percent of the maf feed coal has been converted, i.e., depolymerized, hydrogenated, etc., to a product soluble in the solvent at the temperature employed. The product, for want of a better term, is called extract even though more transpires in the conversion than simply dissolving the coal. Generally, in order to attain depths of extraction above 50 weight percent, hydrogen must be added to the coal during extraction. The hydrogen may be added by means of a hydrogen-transfer solvent of the type mentioned above, or simply as hydrogen gas.

The objective of the partial separation step is to provide (a) a low solids-containing product consisting essentially of extract and solvent, with such solids as may unavoidably be present, and (b) a high solids-containing product consisting essentially of extract, solvent and solids, with the solids in the predominant amount. The partial separation may be accomplished by sedimentation, use of hydrocyclones, filtration, centrifugation, or by a combination of two or more of these unit operations. If complete, or substantially complete separation is effected, as might be the case if filtration is used, then extract and perhaps some high-boiling or nondistillable liquid, must be added to the solids to serve as binder, as will be more fully described later. In any case, separation is effected at elevated temperatures at or close to the temperature maintained in the extraction step.

Cooling of the extraction product may result in selective precipitation of some of the higher molecular weight portions of the extract. At times, this may be done deliberately to improve the ease of separation of the solids from the extract and to improve the quality of the extract produced. The above type of selective precipitation process may be further intensified, if desired, by addition of a saturated, i.e., paraffinic or naphthenic, solvent. The precipitated extract serves also as part of the binder required in the subsequent pelletizing step. The low solids-containing product is separately recovered for use per se as a fuel, or as an intermediate in the production of liquid fuels by hydrogenation.

The adjustment of the relative proportions of solids, solvent and extract in the high solids-containing product is the key step of this process. The primary objective of this step is to provide a high solids-containing product which is pelletizable in a pelletizing zone. The product, in order to be pelletizable, must contain sufficient binder to permit formation of the pellets. The extract itself is a viscous nondistillable material containing little or no material distillable at 400°C and atmospheric pressure and having a softening point well above room temperature. It serves as the principal source of binder. However, the high molecular weight portion of the solvent may also serve as part of the binder.

In addition, binder may be added extraneously to supplement the extract and solvent. For instance, high boiling residual oils may be used, or even coking coal, in limited amounts. When the term extract is used herein, it is intended to include extract itself and any extraneously added binder.

The composition of the high solids-containing product is adjusted with two objectives in mind: the first to provide a composition which is flowable and the second to provide a composition which is pelletizable in the pelletizing zone. If the conditions maintained in the pelletizing zone are such as to cause volatilization of the solvent, then the feed to the pelletizing zone may be richer in solvent. But if the conditions in the pelletizing zone do not cause volatilization of solvent, then the feed must be of the necessary composition for pelletization. The pelletizable composition, that is the composition excluding any solvent that may be vaporized in the pelletizing zone, is as follows:  55 to 75 weight percent extraction residue solids; and 25 to 45 weight percent extract and solvent, where the extract is at least 30 weight percent of the extract-solvent mix, and is preferably greater than 50 weight percent.  The pellets will, of course, have substantially the same composition unless the pelletization is conducted under carbonizing conditions which result in production of some gas and tar.

Obtaining the desired proportions of solids, solvent and extract in the feed to the pelletizing zone generally requires adjustment of the composition of the high solids-containing product after it leaves the separation zone.  However, part of the adjustment may be achieved in the separation zone itself if it is necessary to increase the amount of extract in the high solids-containing product.  The necessary additional extract may be obtained by the addition of a precipitating solvent to the separation zone.  Generally, the problem of adjustment is twofold, one of removal of extract, and the other of removal of solvent.  The removal of solvent may be readily accomplished by distillation.  The removal of extract to the extent desired may be effected by washing with solvent.

Adjustment of the relative proportions of the three ingredients may also be effected by adding recycled off-size pellets made in the pelletizing zone or from other sources.  Generally, such pellets will have the same composition as the rest of the feed to the pelletizing zone except for losses in the pelletizing zone due to distillation.  If pelletizing is conducted under carbonizing conditions, then the recycled pellets would constitute part of the solids fed to the pelletizing zone, and as such, would be considered as part of the extraction residue solids for the purpose of the abovementioned formulation of solids and binder.

The primary objective of the pelletizing zone is to form pellets out of the flowable mass received from the separation zone after suitable adjustment of the composition as described above.  The pellets may be made by any one of the many known pelletizing processes, for example, briquetting, extrusion, or agglomeration, under carbonizing or noncarbonizing conditions at a temperature above the softening point of the binder.  The process selected should be one which is adapted to be used at an elevated temperature at which the extract is fluid.  The size of the pellets is generally between 14 mesh Tyler Standard screen and two inches, suitable for use as a fuel or in subsequent treatments.

To be useful as fuel or as reactant, the pellets must be sufficiently hard to permit handling without breakage or attrition.  If the pellets are formed in the pelletizing zone under carbonizing conditions, that is at a temperature above 750°F, the resulting pellets will be sufficiently hard to permit handling for almost any use.  However, if no carbonization occurs in the pelletizing zone, then congelation of the pellets subsequent to the pelletizing step is required.  The extract will solidify upon cooling below the softening point of the solution of extract and solvent which is generally above 300°F.  The resulting congealed pellets are of sufficient hardness to be handled for most uses as a fuel or reactant.

### Dissolving Coal in Mixture of CO and Steam

In accordance with the process described by *W.C. Bull and B.K. Schmid; U.S. Patent 3,808,119; April 30, 1974; assigned to The Pittsburgh and Midway Coal Mining Company and the U.S. Secretary of the Interior* a low-sulfur fuel is recovered from coal in combination with valuable by-products.  This is accomplished by dissolving all or a substantial portion of the potentially available fuel fraction of a naturally occurring carbonaceous fuel such as coal in a suitable solvent and in the presence of both carbon monoxide and steam or carbon monoxide, steam and hydrogen, then separating the undissolved portion of the carbonaceous fuel from the solution, and thereafter recovering a low-ash, low-oxygen, low-sulfur, solid, carbonaceous fuel from the solvent. The liquid and gas products will be obtained by flashing and/or fractionation of the liquid media from the solvation step. It is important to control the operating parameters such as temperature, pressure and atmosphere and to effect the solvation within a relatively narrow range of holding times. In a preferred embodiment, the particle size of the carbonaceous material treated will range between about 0.006 and 0.008 inch in diameter.

In general, any of the solvents known in the prior art to be useful for the purpose of dissolving the available fuel portion of a carbonaceous material may be used in the method of this process. Suitable solvents include the highly hydrogenated aromatic materials, generally, boiling within a range of about 200° to 900°F such as anthracene oil or creosote oil. A particularly preferred solvent, however, is one obtained by the extraction of the carbonaceous fuel itself. In this regard, it should be noted that a liquid by-product is obtained as a result of dissolving the carbonaceous feed material and a portion of this liquid material is suitable for use as a solvent therein.

Moreover, a sufficient quantity of such a solvent will be produced during the extraction to satisfy the process needs therefor In either a batch or continuous operation. As will be readily apparent, the composition of the solvent thus produced will vary with the particular carbonaceous feed material but that portion of the liquid by-product having an initial boiling point within the range of about 100° to 700°F and a final boiling point within the range of about 700° to 1100°F, a density of about 1.1 and a carbon to hydrogen ratio in the range of about 1.0:0.9 to about 1.0:0.3 will be satisfactory for use in the method of this process.

During the solvation step, it is essential that sufficient solvent be employed to effectively dissolve the available fuel portion of the carbonaceous material being treated without forming a high viscosity slurry which would be difficult, if not impossible, to process upon dissolution of the dissolved fuel fraction therefrom. It is, therefore, essential that the ratio of hydrocarbon solvent to the carbonaceous material being treated (on a dry basis) be at least 0.5:1 and ratios of hydrocarbon solvent to dry carbonaceous material within the range of about 0.5:1 to about 5:1 will be operable.

In general, the fuel fraction of the carbonaceous material being treated by this method will be dissolved at a temperature sufficiently high to facilitate the solvation but not so high as to cause excessive decomposition of the fuel fraction, which is sought to be recovered, or the solvent employed in the extraction or solvation step. Temperatures within the range of about 700° to 950°F have been found suitable for use.

Since the solvation of the available fuel fraction of the carbonaceous material is accomplished at an elevated temperature and in the presence of hydrocarbon materials having a boiling point below these temperatures, it is essential that the extraction or solvation step be accomplished at an elevated pressure. Moreover,

since elevated pressure enhances the solvation of the available fuel fraction of
the carbonaceous material, it is most desirable to effect the solvation at elevated
pressures. In general, pressures within the range of about 500 to 5,000 psig will
be effective.

The essence of the process resides in the discovery of unexpected advantages
which are realized when part or all of the available fuel fraction of the carbonaceous material is dissolved in an atmosphere containing both carbon monoxide
and steam. Generally, these advantages will be realized when carbon monoxide
is present in an amount ranging between about 1.5 and 40 scf of CO per pound
of dry coal in combination with steam within the range of about 0.2 to 1.5
pounds per pound of dry coal. It is not, however, essential that carbon monoxide and steam be the only gaseous components present in the gas phase during
the extraction or dissolving step and, in fact, it has been found beneficial in
many cases to also have hydrogen available in the gas phase.

In this regard, and when hydrogen is employed, the ratio of hydrogen plus carbon monoxide to dry carbonaceous feed material will range between about 1.5
and 40 scf per pound of the carbonaceous material and the ratio of steam and/or
water to dry carbonaceous material will range from about 0.2:1 to 1.5:1 on a
weight basis. Moreover, when hydrogen is employed, the ratio of hydrogen to
carbon monoxide will range between about 0.1:1 to 10.0:1. In general, the
length of time during which the solvent and carbonaceous fuel will be contacted
at the process temperature will vary between about 3 and 180 minutes.

During the extraction or solvation step, materials other than those which are
soluble in the solvent are either formed or liberated. Such materials include hydrogen sulfide, carbon dioxide, methane, propane, butane and other higher hydrocarbons and these materials will comprise part of the atmosphere in the dissolver. Generally, however, their presence therein will not adversely affect the
extraction or solvation of the carbonaceous material. Care should, however, be
exercised so as to prevent a buildup of these materials to the extent that the
partial pressures of carbon monoxide and steam (and hydrogen, when used) are
reduced to inoperable values. In this regard, it should be noted that these materials may be separated from any recycled gas by conventional means.

*Example:* In this example, a Kentucky No. 11 bituminous coal was ground such
that 100 weight percent thereof passed through a 100 mesh (U.S. Standard)
screen and then slurried with a mixture of water and a highly aromatic solvent.
The solvent to dry coal ratio in the slurry was 2 to 1. The ratio of water to dry
coal was 0.25 to 1. The slurry was heated in an atmosphere comprising 50 mol
percent carbon monoxide and 50 mol percent hydrogen at an initial pressure of
1,500 psig and held at these conditions for 30 minutes (at 425°C, the autogenous
produced pressure was 3,800 psig). During the 30 minute residence time, 94.1%
of the available fuel fraction of the carbonaceous feed material was dissolved in
the highly aromatic solvent or otherwise converted to a lower molecular weight
liquid or gas material. After the 30 minutes residence time, the undissolved portion of the carbonaceous feed material was separated from the solution by filtration and the gaseous materials flashed so as to yield a solution containing an upgraded carbonaceous fuel.

The upgraded carbonaceous fuel was then recovered by vacuum distillation of
the solvent and other liquid products. The upgraded carbonaceous fuel was

recovered in a yield of 48.7 weight percent based on initial coal feed. A liquid product boiling within the range of 100° and 800°F was obtained in a yield of 25 weight percent based on initial coal feed and a gas product consisting of $C_{1-4}$ hydrocarbons was recovered in a yield of 1.4 weight percent based on the initial coal feed. The hydrogen-to-carbon ratio in the upgraded carbonaceous fuel was 0.73:1 while that in the liquid product was 0.86:1. The separation of the undissolved portion of the carbonaceous feed material from the solution was accomplished in a laboratory filtration apparatus in one hour.

## Pelletized Coal plus High Btu Gas for Steelmaking

The primary purpose of the process developed by *J.T. Clancey, E. Gorin, E.H. Reichl and C.H. Rice; U.S. Patent 4,008,054; February 15, 1977; assigned to Consolidation Coal Company* is to provide a process for converting any coal, by itself, whether it be caking or noncaking, low sulfur or high sulfur, low ash or high ash, low volatile or high volatile, to a desired spectrum of clean gaseous, liquid and solid fuels. The solid fuel, for purposes of a steel plant, is coke, or formcoke. The term formcoke defines coke obtained by the calcination of pre-formed or preshaped carbonaceous solids and is used to distinguish it from coke obtained as broken pieces of all sizes and shapes from conventional coke ovens. The advantages of the process include the following:

(1) A single coal, whether it be caking or noncaking, high sulfur or low sulfur, high volatile or low volatile, high ash or low ash, may be used to satisfy the energy requirements of an ore reduction plant.

(2) The process may be readily operated continuously to yield a spectrum of fuels whose distribution and respective compositions may be regulated in response to conditions maintained in the coal liquefaction zone and the separation zone.

(3) The process permits autogenous maintenance of solvent balance and hydrogen requirements.

(4) The process is especially valuable in its application to high sulfur and high ash coals. A variety of ash-free fuels which are substantially free of inorganic sulfur and much reduced in organic sulfur may be readily obtained.

Figure 5.5 is a schematic flow sheet of the process in its broadest aspects. Referring to the figure, coal and liquefaction solvent are introduced into a coal liquefaction zone **10**. The coal is introduced through line **11**. The liquefaction solvent is introduced through lines **12** and **13**.

Any coal may be used in the process, nonlimiting examples of which are lignite, bituminous coal and subbituminous coal. The coal may be noncaking, weakly caking or caking. It may be low in sulfur or high in sulfur, or in between; it may be low in ash or high in ash, or in between; or it may be low volatile or high volatile, or in between. The process is of especial value in the case of non-caking or weakly caking coals and high sulfur coals since normally such coals have only limited utility.

A suitable liquefaction solvent is a mixture of polycyclic, aromatic hydrocarbons which is liquid under the conditions of temperature and pressure maintained during coal liquefaction. A suitable boiling range for such a solvent, for example,

FIGURE 5.5: PELLETS PLUS HIGH BTU GAS FOR STEELMAKING

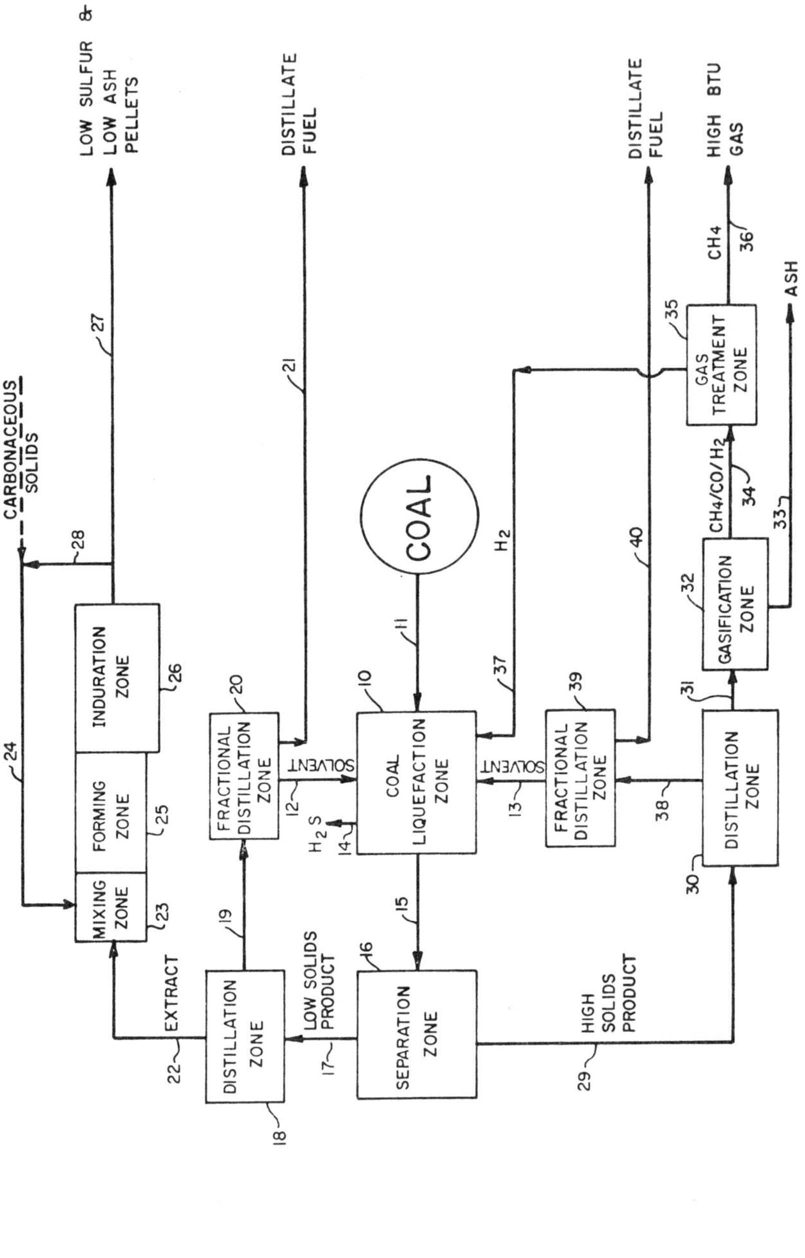

Source: U.S. Patent 4,008,054

is within the range 230° to 475°C. The solvent may be conveniently derived as
a distillate fraction from one or more of the unit operations of this process.
The selected coal, in a finely divided state, is subjected to simultaneous solvent
extraction and hydrotreatment in the liquefaction zone **10** under hydrodesulfuriz-
ing conditions. The extraction operation may be any of those used by those
skilled in the art, for example, continuous, batch, countercurrent or staged extraction. Hydrogen for the hydrotreatment may be supplied as gaseous hydrogen,
or by means of a hydrogen-donor solvent. If gaseous hydrogen is used, the solvent may be any suitable polycyclic aromatic hydrocarbon, or mixtures of polycyclic aromatic hydrocarbons which are liquid at the temperature and pressure
of extraction. If a hydrogen-donor solvent is required, at least a portion of the
polycyclic aromatic hydrocarbons is partially hydrogenated. Such a solvent is
rehydrogenated to maintain its effectiveness as a hydrogen donor.

The conditions maintained in the extraction zone are those which are effective
in desulfurizing the extract, that is, in reducing the organic sulfur content of the
extract. Those conditions are generally known and are typically as follows:

|  | Gaseous<br>Hydrogen | H Donor |
|---|---|---|
| Temperature, °C | 400–450 | 400–450 |
| Pressure, kg/cm$^2$ | 70–200 | 5–50 |
| H addition, wt % maf coal | 0.70–2.5 | 0.70–2.5 |
| Solvent to coal (wt ratio) | 1.5–3.0 | 1.5–3.0 |

The product consists of an effluent gas (containing $H_2S$) which is rejected
through a line **14** and an effluent slurry product (containing solvent, extract and
undissolved hydrocarbonaceous solids and ash) which is conducted by a line **15**
to a separation zone **16**.

Separation of the effluent slurry from the liquefaction zone into at least two
parts is effected in the separation zone **16** at elevated temperature, generally
close to that of the liquefaction zone. The first part is a low solids-containing
coal extract and solvent. The second part is a high solids-containing slurry product containing coal extract, solvent and the major portion of the undissolved ash-
containing hydrocarbonaceous solids. The separation may be accomplished by
settling, hydroclones, centrifugation, filtration or by a combination of two or
more of these unit operations.

Cooling of the extraction slurry before separation may result in selective precipitation of some of the higher molecular weight portions of the extract. At times,
such cooling may be done deliberately to facilitate separation of the solids from
the extract and to improve the quality of the extract. Selective precipitation
of this type may be further intensified, if desired, by addition of a precipitating solvent, for example a paraffinic or naphthenic hydrocarbon.

The low-solids liquid product issuing from the separation zone **16** is conducted
without intentional cooling by a line **17** to a distillation zone **18**. The primary
purpose of the distillation zone is to recover the major portion of the solvent
along with that part of the extract which boils below or in the boiling range of
the solvent. Some distillate may be left in the extract in order to lower the
naturally high viscosity of the extract. The temperature of the molten extract
(which, in a solvent-free state, is a solid at temperatures below 100°C) should be

kept below that at which any coking of the extract might occur.  The distillation zone may be any suitable distillation equipment for flashing off most of the distillables, leaving a bottoms product consisting essentially of nondistillable extract, some solvent and whatever solids are in the feed to the distillation zone.  The vaporous products are conducted through a conduit **19** to a fractional distillation zone **20** where they are fractionally distilled to yield a distillate fraction suitable for use as a distillate fuel and a higher boiling distillate fraction which is solvent.  The distillate fuel fraction is withdrawn through conduit **21** and the solvent is recycled through conduit **12** to the liquefaction zone **10**.

The coal extract is then conducted through a conduit **22** to a suitable mixing zone **23** where low ash and low sulfur carbonaceous solids are added in sufficient quantity to form a high solids-containing product which is pelletizable when hot, in forming zone **25**.  The amount of carbonaceous solids required for this purpose is a function of the binding properties of the extract and of the solvent, as well as the relative proportions of the extract and retained solvent.  Furthermore, binder may be added extraneously to supplement the extract and solvent. If the conditions maintained in the subsequent forming zone **25** are such as to cause volatilization of the solvent, then the pelletizable mixture may, of course, be richer in solvent.  But if the conditions in the forming zone do not cause volatilization of solvent, then the pelletizable mixture must be of the necessary composition for pelletization.

In general, the pelletizable composition, excluding any solvent that may be vaporized in the forming zone, is typically as follows:  35 to 75 weight percent carbonaceous solids; and 25 to 65 weight percent extract and solvent, where the extract is generally at least 75 weight percent of the extract-solvent mix and is preferably greater than 90 weight percent.  The pellets will have substantially the same composition unless the pelletization is conducted under carbonizing conditions which result in production of some gas and tar.

The carbonaceous solids admixed with the extract in the mixing zone **23** are supplied through a conduit **24**.  They are preferably derived from the process itself. However, in its broadest aspects, the process may resort to extraneous sources of carbonaceous solids, however derived.

The pelletizable composition, suitably admixed in the mixing zone, is transferred to the forming zone **25**.  The primary objective of the forming zone is to form pellets out of the pelletizable mass received from the mixing zone.  The pellets may be made by any one of the many known pelletizing processes, for example, briquetting, extrusion or agglomeration, under carbonizing or noncarbonizing conditions at a temperature above the softening point of the binder. The process selected should be one which is adapted to be used at an elevated temperature at which the extract is fluid.

If the pellets are formed in the forming zone **25** under carbonizing conditions, that is at a temperature above 400°C, the resulting pellets will be sufficiently hard to permit handling.  However, if no carbonization occurs in the forming zone, induration of the pellets subsequent to the forming step is required.  In such event, the pellets formed in the forming zone **25** are advanced to an induration zone **26**.  Hardening of the pellets in the induration zone **26** may be effected by heating the pellets under carbonizing conditions, that is at a temperature above 400°C or by simply cooling them below the softening point of the solution of extract and solvent which is generally above 150°C. Carbonization of the pellets, whether effected in the course of pellet formation in the forming zone or subsequently in a separate zone, yields a vaporous product which contains a fraction corresponding to the solvent used for the coal extraction which may be recovered for use in the liquefaction zone **10**. In the preferred embodiment mixing, forming and induration need not necessarily be performed separately, but in any suitable or appropriate combination of steps.

Hardened pellets are withdrawn from the induration zone **26** through a conduit **27**. Depending upon the forming process selected, the pellets may contain off-size pellets, that is pellets which are not suitable for the intended use, either because they are too small or too large. In such event, the off-size pellets, after suitable size adjustment, may be recycled for use as part or all of the carbonaceous solids admixed with extract in the mixing zone **23**. Such recycle is shown by the lines **24** and **28**. Those pellets which are within the desired range may be used as in certain operations. For example, they may serve as part of a coke oven feed, or as solid reductant in certain ore reduction furnaces. Or they may be subjected to a temperature between 800° and 950°C to yield blast furnace coke.

At least a portion of the high solids product issuing from the separation zone **16** is transferred through a conduit **29** to a distillation zone **30** which is maintained at a temperature sufficiently high to distill off solvent, leaving a hydrocarbonaceous ash-containing solid. If extract is present, as may well be the case unless exhaustive solvent washing has been resorted to, the temperature maintained in the distillation zone **30** should be sufficiently high to carbonize the extract to coke or char and vaporous products. A suitable temperature is in the range of 425° to 760°C.

The ash-containing solids are withdrawn from the zone **30** and conducted by a conduit **31** to any conventional type steam-carbon gasification zone **32** operated under conditions to yield $CH_4$, CO and $H_2$. The solid gasification residue rich in inorganic sulfur compounds is discharged from the system through a conduit **33**. The gaseous product is conducted by a conduit **34** to a conventional-type gas treatment zone **35** for the recovery of hydrogen. The other principal product, $CH_4$, is recovered through a conduit **36** as high Btu gas. The hydrogen is conducted by a conduit **37** to the liquefaction zone **10**. As pointed out earlier, a portion of the required coal liquefaction solvent is introduced into the liquefaction zone by a conduit **13**. The solvent is derived from the distillation zone **30**. The vapors issuing from the latter zone are conducted by a conduit **38** to a fractional distillation zone **39** which is a conventional type.

The fraction having the desired boiling range for the solvent is recovered for that purpose, while the balance may be withdrawn through conduit **40** to serve as distillate fuel. Instead of returning solvent directly to the liquefaction zone, the solvent may first be hydrogenated by the hydrogen from conduit **37** to thereby yield a hydrogen-donor solvent which may then be used as the source of hydrogen in the liquefaction zone.

**Solidifying Solvent Refined Coal (SRC)**

In the process developed by *R.H. Wolk; U.S. Patent 4,021,328; May 3, 1977; assigned to Electric Power Research Institute, Inc.* the nondistillable residue from solvent refined coal, after extraction and distillation, is sprayed into a boiling pool of a nonsolvent hydrocarbon liquid, whose boiling point is at least 50°F below the melting point of the residue.  The solidified product may then be separated by mechanical means and further cooled by countercurrently contacting with a precooled nonsolvent hydrocarbon liquid which is employed to remove the sensible heat of the solvent refined coal liquid.  The product is thus further cooled from the boiling liquid bath temperature.  The resulting product is found to be hard, nonporous, nontacky and resistant to disintegration and powder formation.

In accordance with the process, a solvent refined coal product is obtained after removal of the volatiles which boil at or below a temperature at atmospheric pressure of about 800°F, more usually 650°F.  The resulting hot residue, generally at a temperature above about 500°F is sprayed by means of a nozzle at moderate pressures into an immiscible liquid, whereby the residue solidifies to solid, nonpermeable, sturdy particles.  The immiscible liquid will boil at least in part at or below the solidification temperature of the SRC residue, which is generally in the range of about 300° to 375°F.  The resulting particles, which can be of a variety of shapes depending upon the manner of spraying, may then be separated from the liquid and further cooled by contacting the particles with precooled immiscible liquid.  The particles are then ready for use as a fuel.

Solvent refined coal processes are well known in the art and need only be discussed generally here.  A wide variety of coals may be processed, such coals being illustrated by lignite, bituminous coal and subbituminous coal.  The feed coal, in a finally divided state and substantially free of water, is extracted with a suitable solvent in an extraction zone at an elevated temperature and pressure.

For the most part, the extractants are polycyclic aromatic hydrocarbons, which are maintained in the liquid state at the temperature and pressure of extraction. Particularly desirable solvents are aromatic solvents, which are partially hydrogenated, such as tetralin and dihydronaphthalene.  The solvent employed is generally a mixture of hydrocarbons, although oxygenated aromatic compounds and other heterocyclic compounds may also be included.

The coal employed usually contains substantial quantities of ash.  This ash is normally removed by a filtration along with unconverted coal after the extraction step and prior to the vacuum distillation step.  The solvent refined coal product is essentially free of metal values.  The final product of this process is then composed substantially of combustibles and economies are achieved in that only materials having heating values are processed and transported.

The extraction may be carried out in a continuous, batch, stage or countercurrent process.  The temperatures employed will normally be in the range of about 570° to 930°F.  The pressures will normally be from about 1 to 6,500 psig with a residence time of about 1 minute to 2 hours.  The solvent to coal ratio may be varied widely, generally being in the range of 14:1.  The process may be carried out noncatalytically or catalytically, with iron, cobalt and nickel being the usual catalysts.  Generally, hydrogen will be employed, with up to 50 standard cubic feet of hydrogen per pound of coal being employed.  Usually, at least about 75 weight percent of the coal is extracted, generally not more than about 95 weight percent of coal.

Initial separation of the liquid may be achieved mechanically by sedimentation, filtration, centrifugation or combination thereof, so that a substantially solids-free fraction is obtained, with a residue having a high solids content. The volatiles may then be further removed by vacuum distillation, generally at pressures in the range of 0.05 to 0.5 atmosphere and at temperatures in the range of about 500° to 750°F, more usually about 525° to 610°F.

The resulting hot residue will generally be at a temperature at which the distillation is carried out or slightly below. Preferably, significant cooling is avoided, usually the temperature being maintained within at least about 120°F, more usually at least about 75°F of the temperature in the still which will be in the range of about 500° to 750°F.

The viscous liquid is then pressurized to a pressure from about 25 to 150 psig, more usually 35 to 75 psig and sprayed by means of a nozzle into an immiscible liquid. Depending upon the nature of the nozzle, the resulting particles can be formed in a wide variety of shapes and sizes. Also, by varing the pressure, the size and nature of the product can also be varied. The nozzle will generally be from about ¼ to 1 inch inner diameter.

For the most part, the immiscible liquid will be a hydrocarbon liquid which is highly paraffinic and/or naphthenic, preferably 50% or greater paraffinic. The liquids may be obtained as a product from the extraction of the coal, or as the distillate of the light liquid phase, followed by solvent extraction and/or hydro-treating. The ratio of liquid to SRC will generally be in the range of 0.1 to 0.3 part by weight per part of liquid. This liquid will generally have a boiling range in the temperature range of 225° to 550°F, usually 250° to 450°F, with the mid-boiling point being in the range of 275° to 375°F.

The API gravity will be in the range of 35 to 75, generally 50 to 60. The initial boiling temperature will generally be from about 0° to 50°F below the melting point of the SRC residue. The temperature of the bath will be maintained substantially constant by the refluxing of the liquid into which the residue is sprayed.

By controlling the rate of addition of the residue to the immiscible liquid, the volume of the immiscible liquid as compared to the amount of residue added and the initial boiling point of the immiscible liquid, porosity can be minimized, so that the particles are capable of being handled and shipped without powdering.

The particles may then be collected by any convenient means such as filtration, sedimentation, centrifugation, or the like. The temperature of the particles will still be substantially greater than ambient temperature and may be further cooled by treatment with an immiscible liquid which has been precooled to below 140°F, preferably below about 105°F, and more preferred at about ambient temperature. The process can be carried out in a countercurrent manner, where the fluid flows by the particles. By appropriate choice of the hydrocarbon liquid, the heat of the particles can be absorbed as vaporization. By condensation of the vapor, the cool liquid is continuously returned to the contact zone.

The particles may then be transferred away from the cooling zone and freed of any residual hydrocarbon liquid by a gas stream or other means and are then ready for use as a fuel. The particles have good mechanical integrity, are relatively impermeable and can be shipped and mechanically handled without powdering. As illustrative of this process, typically an SRC unit will process 10,000 tons per day of coal, e.g., Illinois No. 6 coal, operating at a total pressure of 1,700 psig, a hydrogen partial pressure at the reactor outlet of 1,200 psig, a temperature of 845°F, a coal space velocity of 25 pounds of coal/hr/ft³ of reactor volume and a solvent to coal feed ratio of 2 parts of solvent to 1 part of coal, and about 40,000 scf of hydrogen is circulated through the reactor per ton of coal feed.

After the slurry leaves the reactor, the gas and slurry phases are cooled, separated and the slurry phase filtered to remove solids, e.g., ash and unconverted coal. The liquid is then distilled first at atmospheric pressure, then in vacuo to recover the solvent to slurry the feed coal. Fifty-five hundred tons per day of vacuum tower bottoms are the solvent refined coal product.

A centrifuge pump is used to remove the vacuum bottoms from the tower at 600°F. This pump develops a head of 50 psig and feeds the coal to a multiplicity of ¼ inch diameter nozzles submerged in a 55° API heavy naphtha pool in a large vessel operating at atmospheric pressure. The naphtha is maintained at its initial boiling point of 275°F and heat is recovered by condensing the naphtha vapors with cold water.

The average residence time of the SRC in the naphtha pool will vary from 5 to 60 minutes and the average concentration is about 0.15 part of SRC per part by weight of naphtha. The residence time will vary with the size of the particle formed, with the size of the particle varying from about 0.25 to 1 inch and the length varying from 1 to 6 inches.

### Conversion to Liquid and Solid Fuel

*H. Owen, P.B. Venuto and T.-Y. Yan; U.S. Patent 4,077,866; March 7, 1978; assigned to Mobil Oil Corporation* describe a solvent refining process for conversion of coal to liquid and solid fuels which comprises, in its broadest aspects, the following essential steps: (1) subjecting a finely divided coal to treatment with a solvent and an inorganic, solid sulfur scavenger without added hydrogen at a temperature sufficient to dissolve substantially all the organic material in the coal; and (2) partially separating the total extraction effluent in a slurry settler operation to produce a clarified extract/solvent overflow and a solids-containing underflow.

The clarified extract/solvent overflow from the slurry settling step may be further processed by any of the methods practiced in the art of solvent refining of coal, including distillation, coking, cracking of separated coal extract fractions, etc. Similarly, the solids-containing underflow may be further treated by methods taught in the art. However, in a preferred example, the underflow from the slurry settler is subjected to a subsequent solid-fluid separation by such means as filtration, centrifugation, multicloning, etc. The solid sulfur scavenger may then be efficiently recovered from the solids-containing stream produced in this subsequent separation step by regenerative gasification of the solids-containing stream to remove hydrocarbon values and sulfur, followed by separation of the scavenger from the remaining solid residue by means suited to the properties of the particular scavenger employed. The dissolution solvent may be any of the hydrogen-donor solvents which are extensively discussed in the prior art. However, the use of highly refractory aromatic petroleum solvents is preferred, particularly sulfur- and solids-containing aromatic solvents derived from the catalytic cracking of petroleum. The unique advantages of the process are particularly valuable in such an aspect because the solid, sulfur scavenger employed in the dissolution step will result in desulfurization of both the coal and sulfur-containing solvent.

Moreover, the slurry settling step efficiently removes solids contained in the total extraction effluent which are derived from both the coal and the solids-containing solvent. Catalyst fines and other solid impurities are also efficiently removed from the solvent in the process and therefore, prior separation steps to improve solvent quality in terms of sulfur and insoluble solids content are surplusage. Thus, substantial savings in terms of both equipment and energy requirements for the overall processing of coal and petroleum to clean fuel products result.

Furthermore, viewing this aspect as part of an integrated process for the treatment of coal and petroleum, an efficient method of augmenting the supply of heavy fuels from petroleum refining operations is achieved since coal is incorporated into the refining operation with minimum additional process requirements. The product of the incorporation is a low sulfur, low ash, heavy fuel suitable either for outside marketing to the public or, in the event of fuel emergencies or legislative pressures, for use in supplying the refiner's internal energy needs.

Any solid carbonaceous material may be employed as coal in the method of this process including natural coals such as high- and low-volatile bituminous, lignite, brown coal, etc., or solvent-refined coal or related modified coal. The coal may be high ash, high metals, high sulfur, and have poor caking characteristics, and still be quite suitable for this process scheme. The process scheme is particularly useful in removing the last, most difficult-to-remove sulfur in order to meet combustion specifications in natural coals or solvent-refined coals.

The sulfur scavenger employed may be any material with an affinity for sulfur in organic moieties derived from coal or in petroleum fractions. It is essential that the sulfur scavenger be precipitable, or capable of separation from a liquid medium by sedimentation, settling, multicloning, centrifugation, filtration or other density-related fluid/solid separation techniques. Any inorganic, organic or organometallic element or compound capable of conversion to an insoluble sulfide or sulfur complex and/or capable of removing or displacing organic sulfur as a volatile sulfur compound may be used in this process. However, iron and iron oxides (particularly $Fe_2O_3$) are preferred because of potential low cost and ease of availability. For example, the iron could be easily obtained from scrap iron, tramp iron, etc.

Included within the scope of the term coal liquefaction solvent as used herein are all solvents employed in the coal liquefaction step of solvent refining processes known in the art of coal conversion. Polycyclic, aromatic hydrocarbons which are liquid at the temperature and pressure of the extraction are generally recognized to be suitable solvents for the coal in the liquefaction step. Preferred solvents suitable are thermally stable, highly polycyclic aromatic mixtures which result from one or more petroleum refining operations. A highly preferred solvent is an FCC main column bottoms fraction which is obtained from the catalytic cracking of gas oil in the presence of a solid porous catalyst. This bottoms fraction is recovered as a slurry containing a suspension of catalyst fines. The slurry oil is directly suitable for use as a liquefaction solvent in the process, or it can be subjected to further treatment to yield a clarified slurry oil. The further treatment can involve introducing the hot slurry oil into a slurry settler unit in which it is contacted with cold heavy cycle oil to facilitate settling of catalyst fines out of the slurry oil. The overhead liquid effluent from the slurry settler unit is the clarified slurry oil.

*Example:* Pulverized high volatile bituminous A coal (60 grams) was mixed with FCC main column bottoms (90 grams, 10° API) and powdered iron oxide, $Fe_2O_3$ (6 grams having an average particle size of 75 microns). The mixture was heated at autogenous pressures in a closed autoclave for 1 hour at 750°F with mechanical stirring and without any added hydrogen.

After cooling, the gasiform product was vented, collected and analyzed. A uniform, dark-colored, highly viscous fluid was recovered from the reactor. Total recovery was over 95%. After workup, which included a series of selective extractions, precipitations, and filtrations, the following approximate component breakdown was obtained:

| Components | Weight Percent* |
|---|---|
| Ash | 4.9 |
| Pyridine-insoluble | 4.0 |
| Benzene-insoluble | 31.6 |
| Benzene-soluble/hexane-insoluble | 7.5 |
| Gas | 1.1 |
| Water | 0.9 |
| Hexane-soluble | 50.0 |

*No loss basis

For purposes of comparison, a control run was also performed under the same conditions as described for the iron oxide run. As the following sulfur analysis indicates, a substantial further sulfur reduction occured in the iron oxide run.

| | . . . Sulfur Analysis, wt % . . . | | Decrease in |
| Product | Control Run | Iron Oxide Run | Sulfur over Control, % |
|---|---|---|---|
| Benzene-insoluble | 0.73 | 0.53 | 27.4 |
| Benzene-insoluble/hexane-insoluble | 0.98 | 0.65 | 33.7 |

Not only was there sulfur reduction in the liquid-solid coal-derived products shown above, but there was also complete removal of sulfur (i.e., no hydrogen sulfide), as was shown in a subsequent analysis.

The gases coproduced from these coal dissolution methods show gross heats of combustion of ~1,120 to 1,250 Btu/scf, and thus constitute useful fuel gas streams, with the gas from the iron oxide run constituting a sulfur-free fuel gas. The foregoing indicates the effectiveness of sulfur removal attained by employing a solid sulfur scavenger, particularly an iron sulfur scavenger, in the coal dissolution step of solvent refining processes for converting coal to clean fuel products.

## AQUEOUS ALKALI TREATMENT

### Treatment with Steam and an Alkali Metal Compound

A process for desulfurizing coal, coke or petroleum is described by *R.M. Murphy and H.C. Messman; U.S. Patent 3,387,941; June 11, 1968; assigned to The Carbon Company*. An intimate mixture of the carbonaceous material, such as coal, is treated with steam and an alkali metal hydroxide, oxide, carbide, carbonate, or hydride at a temperature of about 500° to 850°C at which the hydroxide of the alkali metal is a liquid, and the sulfur impurities volatilzed during the treatment are removed.

By treating carbonaceous materials in this way, the sulfur contents can be reduced by at least 35% when the carbonaceous material being treated initially has sulfur contents of between 1% by weight and 2% by weight on a volatile-free basis; and in the cases of carbonaceous material initially having sulfur contents of more than 2% by weight, the sulfur contents of the carbonaceous material so treated can be reduced by at least 50% by weight on a volatile-free basis. A typical alkali metal reagent which can be used in the process consists of sodium hydroxide (NaOH) which in its commercial form is commonly known as caustic soda or lye.

When solid carbonaceous materials are to be desulfurized, it is preferable that the maximum dimension of any particle be no greater than 0.05 inch and that the material to be used be in particles of a size range that is no larger than 16 x 0 U.S. Standard Sieve. Further, it is usually preferable to provide caustic soda which is equivalent to at least approximately twice the weight of the sulfur to be removed from the solid carbonaceous materials. In order to achieve optimum desulfurization in a minimum length of time, it is preferable to use caustic soda in excess of this proportion when the initial sulfur content of the carbonaceous material being treated is less than 3% on a volatile-free basis.

Dry caustic soda may be mixed with the carbonaceous material to be desulfurized before heating. It is preferable to place the requisite amount of caustic soda in solution with water, and to thoroughly mix the solution with the solid carbonaceous material to be desulfurized at ambient temperatures. This mixture should then be heated to a temperature of at least 500°C, which temperature should be maintained for at least 10 minutes in such a manner that the hot mixture is periodically exposed to an atmosphere containing at least 2% moisture or steam. In most instances it has been found preferable to maintain the temperature range between 550° and 850°C for periods between 10 and 180 minutes.

The intermittent introduction of steam into the hot mixture in the reactor vessel meets the chemical reaction requirement; but some movement of gas through the material is desirable to sweep out the volatile sulfides which are released by the reaction. Since native volatile matter, which is incidentally driven out of the solid carbonaceous material, may serve as part of this sweep gas, it is preferable to supplement this sweep gas with additional steam as may be required. During the reaction period, at least two-thirds of the sulfur being removed will be released in gaseous form. Most of the residual sulfur remaining in the mixture in sulfide or hydrosulfide form may be removed by washing with water.

*Example:* An air-dried Ohio steam coal sample analyzed 4.75% sulfur, 39.32% volatile, 10.98% ash and 1.93% moisture. This sample of coal was first crushed to 16 x 0 mesh. It was then devolatilized and activated in a bed fluidized with steam at a temperature between 600° and 650°C for 20 minutes. Following this treatment the sample analyzed 6.2% volatile, 2.3% moisture, and it had been activated to the extent of approximately 2% carbon loss.

1,000 grams of the devolatilized and preactivated sample, after cooling, was then thoroughly mixed with 100 grams caustic soda mixed in 100 cubic centimeters water. The resulting mixture was then placed in a fluid bed reactor and heated to 500°C without fluidization. Heating continued and, beginning at 500°C the bed was alternately fluidized with steam for 2 minutes and let rest for 5 minutes over a period of 91 minutes up to a final temperature of 650°C. The resulting product was then rapidly cooled and was washed in water before drying. After drying the sample analyzed 1.4% sulfur.

## Followed by Acidification

*L. Reggel, R. Raymond and B.D. Blaustein; U.S. Patent 3,993,455; November 23, 1976; assigned to the U.S. Secretary of the Interior* has found that a substantial reduction of the amount of pyrites in coal may be achieved by treatment of the coal with aqueous alkali at slightly elevated temperature, followed by acidification with a dilute strong acid.

The feed material may be any coal containing a substantial proportion of pyritic sulfur and/or other mineral matter. This will usually be a bituminous coal having a substantial pyritic sulfur and ash content. The coal is preferably employed in a finely divided form, e.g., –200 mesh, but coal up to ¼ inch in particle size may be used. Suitable particle size reduction may be achieved by conventional techniques such as grinding, pulverizing, etc.

The preferred aqueous alkali consists of an aqueous solution of sodium hydroxide, although solutions of other alkalis, such as potassium hydroxide, may also be used. In the case of NaOH, the concentration may vary over a range of 50 to 420 grams per liter. Optimum concentration of the alkali will, however, depend on the composition of the particular coal, i.e., the pyrite and mineral matter content, state of subdivision of the coal, amount of aqueous alkali, temperature, etc., and is best determined experimentally.

The ratio of the amount of coal to the amount of alkali solution may also vary over a considerable range, with about 40 to 200 grams of coal per liter of alkali solution generally being suitable. A dispersion of the coal in the solution is prepared and maintained during the course of the reaction by conventional means, such as stirring.

Temperature of the aqueous alkali treatment is suitably from about 175° to 350°C, preferably about 225°C, in a closed vessel having an inert atmosphere. Optimum reaction time will also vary with the above variables, and may vary from about 15 minutes to 6 hours. Generally, however, a reaction time of 2 hours is preferred in this step, in which it is believed that the pyritic sulfur is dissolved. The reaction may be carried out in any conventional reaction vessel, such as stirred or rocking autoclaves.

Following the reaction of the coal with aqueous alkali, the reaction mixture is cooled to approximately room temperature and acidified with a dilute solution of a strong acid. The amount and concentration of the acid should be sufficient to adjust the pH of the reaction mixture to 2 or less. The preferred acid is sulfuric acid, in a concentration of about 6 normal. However, other acids, such as hydrochloric acid or $SO_2$ may also be used in concentrations sufficient to provide similar pH values. Weak acids, such as carbonic, are not effective.

Agitation of the reaction mixture is performed during and after addition of the acid, for a time sufficient to permit reaction of the acid with ingredients in the mixture. It is believed that these ingredients include some of the constituents originally present in the acid-insoluble portion of the mineral matter of the coal, which constituents have now been transformed into an acid-soluble form as a result of treatment with the alkali. Usually a reaction time of 30 minutes to 6 hours is sufficient.

Following reaction with the acid, the reaction mixture is filtered and centrifuged and washed with water to recover the product coal of low mineral matter and pyrite content. Such a coal is highly desirable for combustion in generation of electricity, e.g., in steam plants, gas turbines or MHD (magnetohydrodynamic) generators. It would also extend the life of catalysts used for catalytic hydrodesulfurization of coal.

*Example:* 30 grams of –200 mesh Illinois No. 6 high volatile B bituminous coal was treated with a solution of 24 grams of sodium hydroxide in 240 milliliters of water for 2 hours at 225°C in a stirred autoclave. The reaction mixture was cooled to room temperature and acidified with 125 milliliters of 6 N sulfuric acid. The product coal was isolated by centrifugation and decantation of the supernatant aqueous layer. This was followed by repeatedly stirring the coal with water, aspiration of the aqueous layer, and centrifugation, until the aqueous layer gave only a very faint test for sulfate ion with barium chloride solution.

The product coal was then dried and analyzed for ash and pyritic sulfur contents, and for heating value. Ash and pyritic sulfur contents were found to be 0.5% and 0.2%, respectively, and the coal had a heating value of 14,000 Btu per pound, on a moisture-free basis. By comparison, the starting coal contained 9.8% ash and 1.0% pyritic sulfur, and had a heating value of 12,400 Btu per pound, on a moisture-free basis.

## Addition of Oxygen

Carbonaceous fuels may be desulfurized by treatment with hot, molten caustic materials, such as sodium hydroxide, potassium hydroxide, and the like. Aqueous solutions of caustic materials may also be employed to desulfurize carbonaceous fuels but the degree of desulfurization is not nearly as great as when a hot melt of the caustic materials is employed.

*J.R. Longanbach; U.S. Patent 4,054,420; October 18, 1977; assigned to Occidental Petroleum Corporation* has discovered that greatly improved results in desulfurizing carbonaceous fuels are achieved with an aqueous caustic solution if an oxygen-containing gas is introduced into the solution. A carbonaceous fuel, such as coal or coal char, is admixed with an aqueous solution of caustic. The carbonaceous fuel is preferably prepared in small particles prior to admixing with the aqueous caustic solution to expose a substantial amount of the fuel's surface to the solution. The materials used to form the caustic bath are preferably sodium hydroxide, potassium hydroxide, but any other caustic substance, for example, calcium oxide or sodium carbonate, or any combination thereof, should work as well.

The aqueous caustic solution may comprise 1 to 70% by weight of caustic material. In the preferred embodiment, the aqueous caustic solution comprises a range of 10 to 50% by weight of caustic, and preferably is 25% by weight of caustic. The ratio of caustic to carbonaceous fuels should be maintained at a level which will effect maximum desulfurization. This ratio is at least 10 parts of caustic to 1 part of sulfur in the carbonaceous fuel, and preferably is at least 20 parts of caustic per 1 part of sulfur. This ratio may be increased but it will not result in any appreciable increase in desulfurization relative to the increased ratio of caustic to fuel.

When the carbonaceous fuel is introduced into the aqueous caustic solution, it may be mixed in by means of a motor-driven stirrer or other equivalent means which will provide a thorough mixing of the fuel with the aqueous caustic solution for intimate contact therebetween.

Oxygen-containing gas is introduced into the aqueous caustic system by suitable means and in an amount to provide at least 2 grams of oxygen for every 1% by weight of sulfur present in the initial fuel. This aqueous-caustic-oxygen system is maintained at a temperature of 200° to 1600°F for a period of time of 1 to 300 minutes to effect desulfurization. The fuel is thereafter washed with water to remove the caustic materials and the water-soluble sulfate salts formed as a result of the aforesaid treatment.

*Example:* 50 grams of –50 mesh coal characterized as run-of-the-mine (ROM), Hamilton high volatile bituminous coal (HVCB), was mixed with 100 grams of sodium hydroxide, 100 grams of potassium hydroxide and 200 milliliters of water in a Hastelow C autoclave liner until the caustic was dissolved in the water. The liner was sealed in an autoclave, heated to 660°F, and vented to a pressure of 800 psig. Oxygen gas, under a pressure of 800 psig, was introduced into the mixture at a flow rate of 400 milliliters per minute.

The temperature of the system and oxygen pressure and flow rate were maintained as above for a period of 10 minutes. Thereafter, the autoclave was sealed and allowed to cool overnight. The following morning the autoclave and liner were opened. The coal was separated from the aqueous caustic solution, washed in water to remove the caustic, filtered and dried. The dry weight recovery was 48.85% of the original weight of the coal. The recovered coal was analyzed and the sulfur content was found to be 4.01% of the original content of the sulfur. An analysis of the coal before and after treatment is set forth in the table below.

| | . . . . . . Before . . . . . . | | . . . . . . . After. . . . . . . | |
| | Percent | Percent Dry Basis* | Percent | Percent Dry Basis |
| --- | --- | --- | --- | --- |
| Water | 7.16 | 0 | 1.14 | 0 |
| Ash | 8.30 | 8.94 | 5.32 | 5.38 |
| Carbon | 67.50 | 72.70 | 81.85 | 82.79 |

(continued)

|  | . . . . . . Before . . . . . | |  . . . . . . .After. . . . . . | |
|---|---|---|---|---|
|  | Percent | Percent<br>Dry Basis* | Percent | Percent<br>Dry Basis |
| Hydrogen | 5.35 | 4.90 | 5.17 | 5.10 |
| Nitrogen | 1.58 | 1.70 | 1.70 | 1.72 |
| Sulfur | 2.63 | 2.83 | 0.23 | 0.23 |
| Oxygen | — | 8.93 | — | 4.78 |
| Total weight, g | 50 | 46.42 | 22.94 | 22.68 |

*The water content was excluded from the total analysis which was recalculated.

## Leaching Out Sulfur with Alkali plus Heat

The process for extracting a substantial proportion of the pyritic, organic and sulfate sulfur compounds and ash present in coal or coke described by *E.P. Stambaugh and G.F. Sachsel; U.S. Patent 4,055,400; October 25, 1977; assigned to Battelle Memorial Institute* comprises mixing the fuel with an aqueous alkaline solution containing a sodium, calcium, or ammonium carbonate, hydroxide, sulfide, or hydrosulfide, or a plurality thereof, heating the resulting mixture to at least 125°C at a pressure of at least 25 psig to leach out the sulfur compounds and ash, separating the easily removable leached out materials from the remainder of the fuel, and washing the remainder of the fuel.

The concentration of alkali in the solution typically is about 1 to 35% by weight and the mixture typically is heated for at least 5 minutes at 350°C, about 1 hour at 300°C, about 2 hours at 250°C, about 4 hours at 200°C, about 10 hours at 125°C or for at least a time approximately proportionately between the foregoing times at an intermediate temperature. The pressure may be provided at least in part by oxygen, hydrogen, nitrogen, oxygen plus nitrogen, or hydrogen plus nitrogen.

The method may be carried out either in a batchwise fashion or in a substantially continuous operation. Where the extraction is to be substantially continuous, the method typically comprises the steps of continuously introducing the fuel at a preselected rate into the aqueous alkaline solution to form a slurry, moving the slurry through a region maintained at the elevated temperature and pressure to leach out the sulfur compounds and ash, moving the slurry outside the leaching region and separating the easily removable leached materials from the remainder of the fuel by filtering, moving the remainder of the fuel away from the separated leached out materials and washing the remainder.

The resulting solution containing the leached out materials typically is regenerated by removing the leached out materials therefrom and is recycled as the liquid aqueous medium in the continuous process.

The approach which has met with success in this process for removing impurities from coal is known as the hydrothermal process. Pressure leaching is employed here for the removal of impurities from coal. The coal, after washing or directly from the mine, i.e., unwashed, is crushed to the desired particle size and leached in an autoclave to extract the sulfur and ash values. After removal of the leach liquor, i.e., the filtrate containing the impurities (leached out materials), the leached coal is washed to remove the absorbed liquor and dried. The filtrate is regenerated and recycled to the process. In addition to the removal of the impurities from coal, the reactivity of the coal may be greatly increased due to the fact that the coal structure is opened by (1) the removal of the pyritic sulfur from between the layers of coal, (2) disruption of the coal structure during the removal of the organic sulfur, and (3) swelling of the coal as a result of the elevated temperature and penetration of the coal structure by the hot water, steam, and the leachant.

The temperature at which the extraction is carried out typically is within the range of 125° to 350°C and preferably within the range of 125° to 250°C. Reaction time may also vary over a wide range and it is obvious that the exact time period depends upon such factors as temperature, particle size, and alkali concentration. In general, the reaction is carried out for 5 minutes to 10 hours or more. The use of ground or crushed coal is preferred in order to facilitate the reaction.

A continuous, closed-loop process can be used for the commercial production of clean coal. By this process, the slurry of the coal and the leachant is pumped continuously through a heated reactor to selectively extract the impurities. The product after separation from the product slurry is processed as described above, and the leachant then is regenerated for recycle to the system.

Using, for example, caustic soda as the leachant, the aqueous medium contains sulfur compounds such as $Na_2S$, $NaHS$, possibly $Na_2S_x$, and sodium derivatives of the organic sulfides and soluble ash values. Regeneration of the leachant for recycle and recovery of the sulfur values as elemental sulfur is achieved by first treating the solution with steam to remove the organic sulfur compounds. Sulfur is recovered from these by hydrogen reduction to produce $H_2S$, which is readily converted to elemental sulfur by the Claus reaction. The organic compounds may be of commercial value.

Treatment of the remaining solution with $CO_2$ releases the remaining sulfur as $H_2S$, converts the sodium values to $NaHCO_3$, and precipitates extracted ash values such as alumina. The $H_2S$ is combined with that from the reduction of the organic sulfides and is converted to elemental sulfur by the Claus reaction. Treatment of the aqueous carbonate solution with lime converts the $NaHCO_3$ to $NaOH$ for recycle and to $CaCO_3$. The $CaCO_3$ is thermally decomposed to $CaO$ and $CO_2$ which is then recycled to the system. By this scheme, recycle of all chemicals is achieved, coal is converted to low sulfur, low ash fuel, and sulfur is recovered in the elemental form. In accordance with this process it has been found that low sulfur, low ash coal can be prepared by leaching coal in aqueous alkaline media in an autoclave.

*Example 1:* Approximately 50% of the organic sulfur, 100% of the pyritic sulfur, and approximately 70% of the sulfate sulfur were leached from a washed Ohio coal using sodium hydroxide as the leachant. This work involved heating a ground coal, 70% –200 mesh in caustic solutions at 200° to 250°C for 2 hours, after which the slurries were cooled to approximately 90°C and filtered.

The resulting cake (leached coal) was washed by displacement washing and dried for analysis. The final coal product contained 0.97% total sulfur (0.04% sulfate sulfur and 0.93% organic sulfur). The original (washed) coal contained 5.18% total sulfur (3.28% pyritic sulfur, 1.78% organic sulfur, and 0.12% sulfate sulfur).

*Example 2:* Desulfurization of coal using an aqueous medium was also performed in a continuous reactor whereby a slurry of coal and aqueous medium was pumped in one end of the reactor and the slurry containing the leached coal was removed at the other end. The feed slurry was made up of 150 g of coal, ground to −325 mesh, and 750 ml of an aqueous sodium hydroxide solution containing 150 g/liter of sodium hydroxide. Reactor operating conditions were: 250°C temperature, 600 psig and 14 minutes residence time. Under these conditions the sulfur level was reduced from its original value of 5.18 to 2.47%, giving an overall extraction efficiency of 52.3%.

## OTHER PROCESSES

### Removal of Pyrites by Pretreatment

The process described by *L.P. Crecelius; U.S. Patent 2,369,024; February 6, 1945; assigned to The Coal Processing Company* is one for the treatment of coal which contains relatively large amounts of iron sulfur compounds, such as iron pyrites, for the purpose of improving the burning of the coal, obviating the effect of deleterious flue gases and tube slagging, and improving the ash characteristics. These objects are accomplished by preparing a coal spray which has a composition such as the following:

|                         | Percent by Weight |
| ----------------------- | ----------------- |
| Calcium chloride        | 92                |
| Chromic oxide           | 3                 |
| Manganese dioxide       | 3                 |
| Dextrin                 | 1                 |
| Potassium acid fluoride | 1                 |

The composition, mixed with water, is sprayed onto the coal at the mine, or some time before it is burned, at a rate which is approximately 5% of the sulfur content of the coal which is to be burned.

The presence of chlorine (from the deliquescent halogen compound $CaCl_2$) effects a decomposition of the iron sulfides forming $F_2O_3$ and $SO_2$. The chromic oxide is a catalyst for the process.

The colloidal characteristic of the dextrin maintains the metallic oxides in suspension in the liquid and materially assists in securing a proper distribution of the metallic oxides when the mixture of calcium chloride, manganese dioxide and chromic oxide is sprayed upon the coal. Likewise, the maintaining of the metallic oxides in suspension obviates the clogging of pipes and nozzles which may be a part of the apparatus used in spraying the coal. To some extent, the dextrin functions to retain the other chemical substances on the coal. The potassium acid fluoride is used for the purpose of making more effective the elimination of the tube slagging.

Owing to the high temperature of the furnaces (in excess of 2140°F) the ferrous sulfide particles which are carried by the coal in many cases accumulate on the surfaces of the tubes of the boiler. In some instances seemingly a silica film forms upon these iron sulfide particles (the silica coming from silica-containing impurities in the coal), which film is not readily attacked by the chlorine. But where potassium acid fluoride has been used in connection with the mixture with which the coal is treated, this potassium acid fluoride liberates fluorine. The quantity of fluorine liberated is small because the percentage by weight of the potassium acid fluoride which is used is low in the composition. However, the fluorine thus liberated will attack any silica or siliceous film upon the iron sulfide particles that may have adhered to the tube, in effect eradicating the film and exposing the iron sulfide to the action not only of fluorine but mainly to the action of the chlorine, which will be decomposed to the stable oxide or iron and sulfur dioxide.

## Water plus Heat in a Nonoxidizing Atmosphere

It was found by *R.W. Rieve; U.S. Patent 3,660,054; May 2, 1972; assigned to Atlantic Richfield Company* that if coal is mixed with water and heated in a nonoxidizing atmosphere at a temperature near the critical temperature of water, but not above the critical temperature of water, the sulfur and/or ash content of that coal is reduced.

According to this process a coal, which can be one or a mixture of two or more of anthracite, bituminous, semibituminous, subbituminous, lignite, peat, and the like, is subdivided, for example, so that substantially all of the coal particles are no larger than 1 inch in their largest cross-sectional dimension (effective diameter), and mixed with water.

The mixing can be carried out in any desired manner so long as an intimate mixture of coal particles and water is achieved and can be conducted at subambient, ambient, or superambient conditions of temperature and pressure. Preferably, the coal and water are simply mixed together at ambient conditions of temperature and pressure. The mixture is preferably composed of coal and water in the weight ratio range of coal/water of from about 2 to about 0.1.

The mixture is then heated in a nonoxidizing atmosphere, e.g., steam, inert gas, reducing gas, and the like, at a temperature of from 600°F to about the critical temperature of water, i.e., 706°F. Although a broad temperature range of 600° to 706°F can be employed, a preferred range is 650° to 706°F. It is important that the temperature of the water during the reaction be near but not above the critical temperature so that the water will be more monomeric than polymeric and more acidic than normal. It is also important that the water not be in the vaporous state to any substantial extent.

The pressure should be maintained sufficiently elevated so that substantially all of the water will be maintained in a liquid state at the temperature prevailing. The pressure therefore can be the autogenous pressure of water vapor at the temperature of reaction and will generally range from 1,545 to 3,226 psia.

By following the above conditions substantial amounts of sulfur and/or ash can be removed from the coal. The reaction should therefore be carried out for a time sufficient to achieve these results. This time will generally be at least 1 minute, preferably at least 5 minutes, e.g., from 5 minutes to 60 minutes. The coal, after completion of the reaction time, can be recovered with or separate from the water present in the reaction. The recovered coal can be utilized, either with or without drying, as a solid fuel or as a feed stock for a coal gasification and liquifaction process.

*Example:* Pittsburgh No. 8 seam coal subdivided so that substantially all passes a 100 mesh sieve and is retained on a 200 mesh sieve (Tyler), was mixed with water in a weight ratio of coal/water of 0.25 by simple stirring at ambient conditions of temperature, pressure and under one atmosphere of nitrogen to provide an intimate mixture of the coal particles and water. The mixture was then heated in an autoclave at 650°F under an autogenous pressure of 2,350 psig for 1 hour after which the autoclave was vented to the atmosphere and most of the water thereby expelled. The standard coal analysis for the coal before and after treatment was as follows:

**Analysis of MAF Coal, wt %**

|                 | Raw Coal | Treated Coal |
|-----------------|----------|--------------|
| Carbon          | 76.16    | 80.77        |
| Hydrogen        | 5.21     | 4.76         |
| Oxygen          | 5.96     | 5.49         |
| Nitrogen        | 1.55     | 1.61         |
| Sulfur          | 2.74     | 1.27         |
| Ash             | 8.38     | 6.10         |
| Volatile matter | 39.06    | 33.66        |
| Fixed carbon    | 52.56    | 60.24        |

It can be seen from the above data that the sulfur content was reduced by 53.6% and that the ash was also reduced by 27.3%. With such a substantial reduction in both sulfur and ash content, the treated coal was a much more acceptable burning fuel and liquid fuel feedstock since it would introduce substantially less sulfur and ash into the combustion products or other liquid and gas products obtained therefrom.

## Treatment of Agglomerative Coal to Produce Char and Fuel Gas

*A. Sass, C. Finney, H. McCarthy and P. Kaufman; U.S. Patent 3,736,233; May 29, 1973; assigned to Occidental Petroleum Corporation* have described a continuous process for the pyrolysis of agglomerative bituminous coal to recover the volatiles therefrom. It comprises forming a high velocity stream composed of particulate agglomerative coal, particulate char, and a substantially inert carrier gas in a pyrolysis zone, such that the char and coal particles are intimately admixed and entrained within the gaseous portion of the stream.

The solids content of this stream in the zone ranges from 0.1 to 10.0% by volume based upon the total volume of the stream in the pyrolysis zone. The stream at its initiation is made up of from 5 to 25 parts by weight of particulate coal having a particle size of less than 65 microns and an initial temperature of less than 300°F, based upon the total weight of the coal and char being utilized in forming the stream; and from 95 to 75 parts by weight of particulate char having a particle size of less than 65 microns and an initial temperature ranging from 1150° to 1600°F, based upon the total weight of the coal and char being utilized in forming the stream.

The residence time of the stream in the pyrolysis zone ranges from 0.01 to 3 seconds, in which the particulate coal is heated to a temperature ranging from 900° to 1200°F. The products produced in the pyrolysis zone are removed, the char particles are separated from the carrier gas and hydrocarbons produced upon pyrolysis of the coal.

Exemplary of gases suitable for use as carrier gases in this process are nitrogen, argon, methane, hydrogen, carbon monoxide and any other gas which will not deleteriously react with or oxidize the organic portion of the matter in the system.

This process is designed for the use of agglomerative particulate bituminous coal, well known for its tendency to plasticize and become sticky at relatively low temperatures, i.e., 750° to 850°F. The particulate char is added to the particulate coal in the process both to prevent agglomeration and to provide at least a portion of the heat required for pyrolysis. The system is essentially designed to heat the agglomerative coal particles to a temperature ranging from 900° to 1100°F to remove the maximum amount of volatiles therefrom.

The effluent from the pyrolysis zone is composed of char, volatilized hydrocarbons, product gas, and inert carrier gas. The char solids can be readily separated therefrom by any conventional solids/gas separator, such as a cyclone separator and the like. The volatilized hydrocarbons and carrier gas can be separated and recovered by conventional separation and recovery means.

*Example 1:* This example is given to illustrate the process. Commercial nitrogen was utilized as the carrier gas to pass the agglomerative bituminous coal and char through the pyrolysis zone. The zone consisted of a pipe 10 feet long having an inside diameter of 1 inch which was wrapped in electrically powered heating units to provide the heat for pyrolysis, 2.3 pounds per hour of mixture of char and coal particles having a maximum particle size of 25 microns and a weight ratio of 3 parts char solids to 1 part coal solids.

The solids were heated to a temperature of 1025°F in the pyrolysis zone. The flow rate of nitrogen through the zone during pyrolysis was maintained at 14 standard cubic feet per minute. The system was operated for a period of 1 hour and both the product and the system were monitored to check for any deleterious effects arising due to agglomeration of the coal in the system. Minimal effects were noted but were so small as to permit the process to be carried out indefinitely without requiring steps to be taken to clean the pyrolysis zone. In fact, the gas pressure differential drop in the pyrolysis zone was less than 0.2 inch of water during the course of the experiment.

The yield of liquid hydrocarbons was equal to 37.9% of the initial dry weight of the coal. By comparison, the Fischer assay test for the coal yields only 15.7% by weight of liquid hydrocarbons.

*Example 2:* This example is given to show the effectiveness of magnetic treatment of char to remove sulfur. In this example samples of the unpyrolyzed coal and pyrolyzed char from Example 1 having a maximum particle size of 25 microns were run through a roll type high intensity magnetic separator. The magnetic separator was a high intensity induced roll magnetic separator run under test operating conditions of 3.0 amperes, with a splitter setting of 0, a drum speed setting of 80, a separation gap between rotor and magnet of ⅛ inch and a particulate coal or char feed rate of 2.6 grams per minute.

The initial sulfur content of the unpyrolyzed particulate coal was 2.55% and the sulfur content of this particulate coal after it had been passed through the magnetic separator was 2.56% (the initial sulfur content of the pyrolyzed char produced in Example 1 was 2.57%) and after the char particles had been passed through the magnetic separator the desulfurized char had a sulfur content of 2.01%.

## Cascade Reactor

There is a series of coal desulfurization processes beginning with the process of contacting coal by hot gases (which may be multifarious products of partial pyrolysis of coal and other components) by which the coal is heated to temperatures at which pyrolysis reactions occur at operationally significant speeds. There is also the process of contacting coal (or partially pyrolyzed coal or totally pyrolyzed coal) by the same multifarious gaseous products, or modified products, or alternate fluids but all containing hydrogen such that reaction between the gaseous hydrogen and sulfur in the coal or coal residue occurs with the production of hydrogen sulfide.

A related subsequent process involves the contacting of the resulting hydrogen sulfide and more relevant solid absorbent (such as calcium oxide) in such manner that the hydrogen sulfide is partially or totally removed from the gaseous stream, to a specified degree and/or as may be possible according to the details of the reactor design, with the calcium oxide forming calcium sulfide to some permissible or appropriate percentage of conversion.

The resulting calcium sulfide (that may simultaneously contain unchanged calcium oxide and/or other solid component that is relevantly inert in this context) is then contacted by hot gases (in a manner as determined by the operation of the cascade reactor) such that the calcium carbonate is heated to a temperature at which a thermal decomposition reactor will occur at operationally significant speeds.

Such reactions as indicated above may be permitted to occur in the same reactor or in a sequence of reactors but the desired net result is the removal of significant amounts of sulfur from the coal, to the hydrogen-containing gases, then to the calcium oxides. Commercial grade sulfur is then extracted from the calcium oxide to upgrade its purity sufficiently to permit recycling.

*R.H. Essenhigh; U.S. Patent 3,801,469; April 2, 1974; assigned to Scientific Research Instruments Corporation* has developed an efficient mechanism for carrying out coal desulfurization which utilizes a dual channel cascade type of reactor. Here, a dual channel baffled cascade apparatus is provided whereby coal containing sulfur is passed down one side of the apparatus in a cascade fashion from one baffle to the other while a solid material, such as calcium oxide, is passed down the other side of the apparatus to pass in a parallel but separate channel between similar successive baffles. While both of the cascade channels are exposed to one another for rapid gas transfers therebetween, provisions are made for insuring that the solid particulates flowing in each of the baffled channels do not mix with each other.

Thereafter, a hot gas (containing hydrogen) passes first through the cascading coal channel and then through the cascading oxidized particulate channel and back again, etc., with the residence times for the solid particulate flow rates being above a certain level and the flow rate of the gas from one channel to the other being above a certain level.

In this manner, the hydrogen in the gas passing through the coal reacts with the sulfur therein to form hydrogen sulfide which is then carried by the flowing gas stream to the opposite but separate channel of oxidized particulate calcium oxide which reacts with the gaseous hydrogen sulfide to form calcium sulfide thus purifying the hydrogen in the gaseous carrier for reuse in the next higher level of the apparatus.  The same gas carrier is continually recycled as it passes upward in the apparatus from one channel to the other and from one level to the other. First it picks up sulfur from the coal channel and then transfers it to the other separate but parallel channel before being reused in a similar cycle at the next higher level.  Later conventional processes extract commercial grade sulfur from the oxidized particulate material flowing in the second channel of the apparatus. An exemplary dual channel cascade reactor is shown in Figure 5.6.

## FIGURE 5.6:  DUAL CHANNEL CASCADE COAL DESULFURIZER

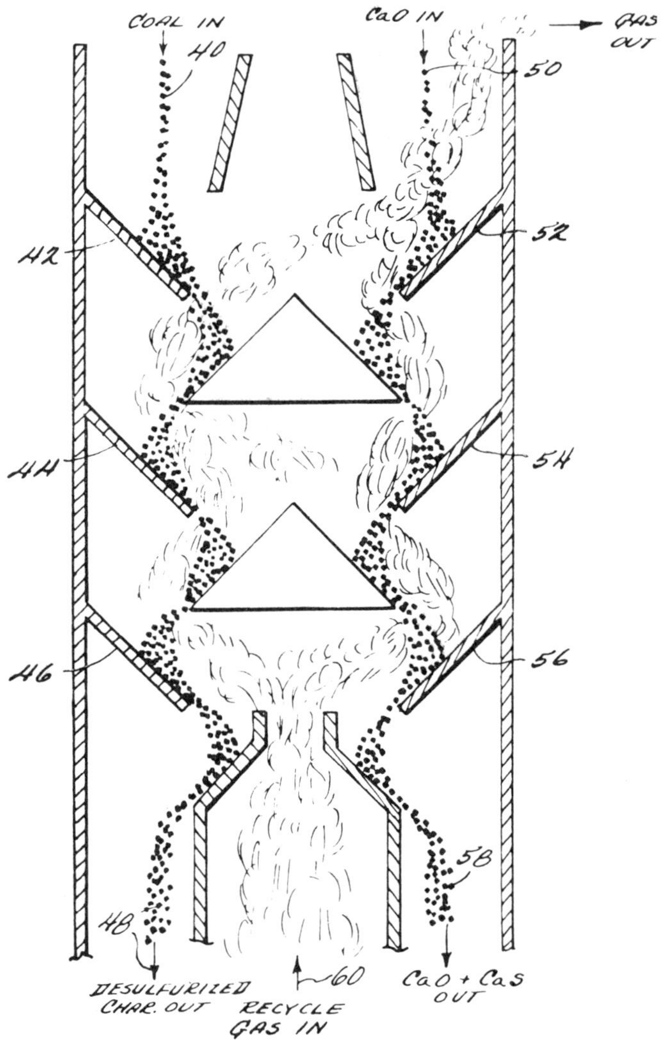


Source:  U.S. Patent 3,801,469

As shown in the figure, a particulate coal stream is input on the left side at **40** and it cascades from baffle plate **42**, to plate **44**, to plate **46** until it exits at the left bottom side as desulfurized char at **48**.

A particulate absorbent, such as CaO, is input at the right hand top at **50** and it cascades from baffle plate **52**, to plate **54**, to plate **56**, until it exits at the right bottom side as CaO and CaS at **58**. Of course, the reason the coal loses sulfur and the CaO picks up sulfur on its downward trip is that simultaneously a hot hydrogen-bearing gas input at **60** has been passing upwardly and back and forth between both channels of the reactor as shown. The hydrogen in the gas is the sulfur carrier and the residence time of the particulate materials is maintained at higher than predetermined levels and the transit time of the gas between channels is maintained at lower than predetermined levels to insure that sufficiently complete reactions occur, and that the hydrogen sulfide is not reabsorbed by the char itself.

In general, the dual cascade reactor is such that coal and calcined dolomite can be introduced into the top in parallel flows, each separately mixed intimately with an upflowing reaction gas so that there is a rapid and complete evolution of sulfur in a gaseous state from the coal and an efficient absorption onto the calcined dolomite without physical mixing of the two parallel particle streams.

In order that the overall rate process can be commercially feasible, the residence times of the coal and calcium oxide particles must be long enough to complete the desulfurization reactions and the consequent absorption of hydrogen sulfide and calcium oxide. Also, the oscillations of gas flow between the coal channel and the channel of calcined dolomite must have appropriate gas residence times to capture the hydrogen sulfide in the calcium oxide without the recapture in the reverse reaction by the char.

The following exemplary requirements should prove satisfactory: (a) 1 second per vertical foot residence time for the solid particles, and (b) 90% gas transfer from one channel of the reactor to the other within 1 second.

The coal entering the top left of the reactor is first dried and lightly preoxidized to reduce agglomerization tendencies. The coal then enters the pyrolyzer stage where it is treated by convection of the recycled pyrolysis gas between the hot (1000°C) calcined dolomite and the coal. The heat required in this reactor is provided by preheating the incoming recycled gas against the exiting char and by the hot incoming calcium oxide. The heat required for calcination of the dolomite may be provided by combustion of char fines carried over with the dolomite and by combustion of a fraction of the fuel gas produced from the pyrolyzer.

### Two-Stage Froth Flotation

Sulfur content of bituminous coals ranges from less than 1% to as much as 6% or more. Pyritic sulfur, i.e., sulfur in the form of pyrite or marcasite, essentially iron sulfide, generally makes up 40 to 80% of the total sulfur present, with the balance being chiefly organic sulfur compounds. The pyritic sulfur is present in macroscopic and microscopic forms, the macroscopic form occurring chiefly as veins, lenses or beds, nodules, or pyritized plant tissue. The microscopic pyrite occurs as finely disseminated globules, veinlets, or euhedral crystals as small as 1 or 2 microns.

Obviously, the sulfur content of some coals could be significantly reduced if most of the pyrite were physically removed during coal preparation. However, because coal must be crushed to a fine size in order to liberate pyrite, such a separation is extremely difficult with conventional coal preparation techniques. Specific gravity separation methods are ineffective because of the extreme fineness of some of the crushed material, and conventional froth flotation suffers from the tendency for the fine size pyrite to float with the coal.

*K.J. Miller; U.S. Patent 3,807,557; April 30, 1974; assigned to the U.S. Secretary of the Interior* has found that these problems can be largely overcome by means of a process in which the coal, in a finely divided form in an aqueous pulp, is initially subjected to froth flotation to float the coal and remove most of the coarse, or essentially free, pyrite as underflow, along with the other nonfloatable refuse, such as clay and shale. The froth product from this first step of the process is then repulped with fresh water and froth flotation, using a coal flotation depressant, and a pyrite flotation collector is employed to float a substantial portion of the remaining fine size pyrite, while removing the coal product as underflow.

The feed material in this process may be any coal containing a substantial proportion of pyritic sulfur. This will usually be a bituminous coal having a pyritic sulfur content as discussed above. It is employed in a finely divided form, i.e., in a particle size of less than 35 mesh (500 microns). Suitable particle size reduction may be readily achieved by conventional techniques, such as grinding, pulverizing, etc.

The initial stage of the process consists of conventional froth flotation of the finely divided coal from an aqueous pulp consisting of 8 to 15% coal. This pulp is prepared by conventional means comprising addition of the coal to water in a flotation cell, or prior to addition to the cell, followed by thorough dispersion, as by mechanical stirring means. The pH of the pulp is not critical but will usually be 4.5 to 8.5.

A conventional frother may also be added to the dispersion in an amount of 0 to 0.001% by weight. It has, however, been found that optimum results are obtained when the amount of frother employed is held to near the minimum amount required to float the coal. Suitable frothers include those commonly used in the froth flotation of coal and other minerals, e.g., pine oil and aliphatic alcohols, such as methyl isobutyl carbinol (MIBC) and 2-ethylisohexanol. Flotation of the coal is then accomplished by aeration at a flow rate of 0.3 to 1.2 cubic feet of air per minute per gallon of slurry, generally for a period of 1 to 3 minutes. Subsequently or simultaneously the resulting froth product is separated by conventional means, such as froth scrapers or paddles.

In the second stage of the process, this froth product is repulped in the flotation cell with fresh water to form a pulp having 6 to 14% solids content. The pH of this pulp should be from 4.5 to 8.0, since values substantially above or below this range may have an adverse effect on the pyritic sulfur content of the second stage froth concentrate refuse. This pulp is then subjected to froth flotation, employing a coal flotation depressant and a pyrite flotation collector in order to float a substantial proportion of the fine-size pyrite, with the desired coal product remaining as underflow. Preferably the second-stage pulp is initially conditioned with the coal flotation depressant by maintaining 0.002 to 0.007% by weight of the depressant dispersed in the pulp for a period of 5 seconds to 1 minute. The depressant consists of an organic colloid. This material may consist of a carbohydrate such as dextrin or modified, i.e., the so-called soluble starches such as modified corn or potato starch.

The organic colloid depressant may also consist of a proteinaceous material such as glue, gelatin, albumin, casein or whey. It may also consist of complex polyhydroxy carboxylic acids and glucosides of high molecular weight such as quebracho extract, tannin or saponin.

About 0.001 to 0.005 weight percent of pyrite flotation collector is then added and the mixture again conditioned for a period of 5 seconds to 1 minute. The collector consists of a xanthate, of the formula ROC–SM, where R is an alkyl radical of at least 3 carbon atoms and M is a metal such as potassium or sodium. Generally, the effectiveness of the compound increases with the number of carbon atoms in R, potassium amyl xanthate being a preferred compound. However, other xanthates such as sodium isobutyl xanthate and sodium isopropyl xanthate are also effective.

A frother such as those disclosed above is then added in similar amount, and the pulp is aerated in the same manner as in the first flotation stage. The froth is again removed, as in the first stage, and the coal product recovered from the remaining slurry by means of conventional procedures such as vacuum filtration or centrifugation.

The process will be more specifically illustrated by the following examples. In these examples, flotation tests were conducted in a standard laboratory rotor-stator flotation cell of the subaeration type. The impeller speed was set at 1,800 revolutions per minute, and the aeration rate was 0.33 cubic feet of air per minute. For the first stage of each test, 200 grams or 400 grams of –35 mesh bituminous coal was mixed in the cell with 2,300 milliliters of tap water, and the slurry was conditioned for 15 minutes to insure thorough wetting of the material. The 8 or 15% solids slurries had a pH ranging from 5 to 8.

The first stage of the process was a standard flotation procedure with minimum frother during which much of the coarse, or essentially free, pyrite was removed as underflow with the other nonfloatable refuse. The froth product from the first stage was then repulped in the flotation cell with 2,300 milliliters of fresh water, and the slurry was treated with a coal flotation depressant. After a brief conditioning time, about 1 minute, the pyrite flotation collector was added. After another brief conditioning period, the frother was added, the air applied, and the froth product collected for 1 to 3 minutes.

Feed compositions, reagents and the results are given, in tabular form in the following examples. Tailings 1 in these examples is the underflow refuse from the first stage, tailing 2 the froth concentrate refuse in the seond stage, and clean coal is the underflow product from the second stage. Hercules RT 1712 is a trade name for pine oil, Aero Zanthate 350 is a trade name for potassium amyl xanthate, and Aero Depressant 633 is a trade name for a carbohydrate colloid flotation depressant. The feed material in the examples consisted of coal from the Lower Freeport coal bed in Cambria County, Pennsylvania.

Examples 1 through 3 show the effect of increased amounts of xanthate in the second stage. The depressant dosage also had to be increased to avoid floating the coal, but this did not adversely affect pyrite flotation. The clean coal products in Examples 1 through 3, which amounted to 50 to 60% of the feed, contained only 0.29 to 0.56% pyritic sulfur, whereas the tailings 2 product contained 3.83 to 7.71%. That is, although tailings 2 represented only 3.4 to 10.7% of the feed product, it contained 45 to 75% of the pyrite available in the second stage. These results clearly demonstrate the selectivity of the process. Examples 1 through 3 were run at reduced pH with HCl, but this step was not essential with this coal.

*Example 1: Analyses, Percent —*

| Product | Weight | Ash | Total Sulfur | Pyritic Sulfur |
|---|---|---|---|---|
| Feed | 100.0 | 31.7 | 2.22 | 1.79 |
| Tailings 1 | 36.8 | 67.5 | 3.31 | 3.21 |
| Tailings 2 | 7.7 | 11.5 | 4.50 | 3.83 |
| Clean coal | 55.5 | 10.8 | 1.18 | 0.56 |

| Reagents | Pounds per Ton of Coal |
|---|---|
| Hydrochloric acid | 0.9 |
| Aero Depressant 633 | 0.5 |
| Aero Xanthate 350 | 0.2 |
| Hercules RT1712 | 0.1 |
| Final pH = 6.2 | |

*Example 2: Analyses, Percent —*

| Product | Weight | Ash | Total Sulfur | Pyritic Sulfur |
|---|---|---|---|---|
| Feed | 100.0 | 25.2 | 2.12 | 1.77 |
| Tailings 1 | 34.1 | 57.2 | 3.52 | 3.52 |
| Tailings 2 | 5.8 | 12.1 | 5.63 | 5.23 |
| Clean coal | 60.1 | 8.4 | 0.99 | 0.45 |

| Reagents | Pounds per Ton of Coal |
|---|---|
| Hydrochloric acid | 0.9 |
| Aero Depressant 633 | 0.5 |
| Aero Xanthate 350 | 0.3 |
| Hercules RT1712 | 0.1 |
| Final pH = 5.7 | |

*Example 3: Analyses, Percent —*

| Product | Weight | Ash | Total Sulfur | Pyritic Sulfur |
|---|---|---|---|---|
| Feed | 100.0 | 32.1 | 2.28 | 1.85 |
| Tailings 1 | 37.6 | 68.1 | 3.40 | 3.34 |
| Tailings 2 | 10.7 | 11.6 | 4.71 | 4.13 |
| Clean coal | 51.7 | 10.1 | 0.96 | 0.29 |

| Reagents | Pounds per Ton of Coal |
|---|---|
| Hydrochloric acid | 0.9 |
| Aero Depressant 633 | 0.7 |
| Aero Xanthate 350 | 0.5 |
| Hercules RT1712 | 0.1 |
| Final pH = 6.2 | |

## Using Spent Iron Electroplating Solutions

One method known to the art for the removal of iron pyrites is the Meyers process wherein the iron pyrites are leached from coal utilizing aqueous solutions of ferric ions. In the Meyers process, coal containing iron pyrites is treated with an aqueous ferric chloride or sulfate solution at approximately 100°C to convert the pyritic sulfur content to elemental sulfur and $Fe^{2+}$. The aqueous solution is separated from the coal and the coal is washed to remove residual ferrous and ferric salt. The elemental sulfur which is dispersed in the coal matrix is then removed by vacuum distillation or extraction with a solvent such as toluene or kerosene. The resulting coal has a substantially reduced pyrite content and may be used as low sulfur fuel.

It is also well known to the art to electroplate iron articles such as sheets, foil, strips, tubes and the like wherein the iron is plated from an electroplating bath onto a suitably shaped cathode. Typically, the electroplating is accomplished with a highly concentrated $FeCl_2$ solution.

In the process developed by *R.E. Colwell; U.S. Patent 3,864,223; February 4, 1975; assigned to Continental Can Company, Inc.,* removal of iron sulfide from coal is effected by the use of spent electrolyte solution from iron plating baths wherein the spent electrolyte and coal particles containing iron sulfides are charged to a reaction chamber having a plurality of openings in its sidewalls, the size of the openings being smaller than the size of the coal particles.

The chamber is rotated with sufficient rotational speed to generate a centrifugal force to cause the coal particles to be propelled outwardly against the walls of the chamber and be fixed thereon to form a continuous bed of particles. As the chamber is rotated, the spent electrolyte is propelled outward into contact with the coal particle bed fixed on the walls of the chamber and is caused to move radially therethrough under the action of centrifugal force, the spent iron electrolyte during its passage through the particle layer entering into chemical reaction with the iron sulfides contained in the coal particles whereby the iron sulfides are converted to $FeCl_2$ and sulfur.

The coal, after contact with the electrolyte, is advanced through the reaction chamber by an oscillating pusher plate or other means, is washed to remove electrolyte, and is then discharged from the chamber.

## Production of Desulfurized Char

A method for carbonizing and desulfurizing carbonaceous materials by heating the carbonaceous material admixed with iron in a reducing atmosphere between 1200° and 1700°F and, subsequently, subjecting the resulting char to an oxidizing atmosphere at a temperature between 800° and 1100°F has been developed by *J.G. Selmeczi and J. Vlnaty; U.S. Patent 3,886,048; May 27, 1975; assigned to Dravo Corporation.*

Referring to Figure 5.7, finely divided coal which is 65% total carbon, 40% volatile material and 5% total sulfur is introduced into a fluid-state reactor **10**, such as a conventional fluidized bed, along with air heated to 600°F by a gas preheater **12** for decaking. After decaking, the coal is admixed with finely divided iron in an amount of 30% by weight, in the form of iron or iron oxide or their mix which may originate from coal pyrites, and introduced into a second fluid state reactor **14** for about 30 minutes.  The coal is carbonized and partially desulfurized in the reactor.  The FeS$_2$ is converted to more stable FeS and also to H$_2$S which may be either in free form or partially adsorbed on the carbon surface.  These reactions are explained in more detail below.

## FIGURE 5.7:  PRODUCTION OF 0.6–0.7% SULFUR CHAR

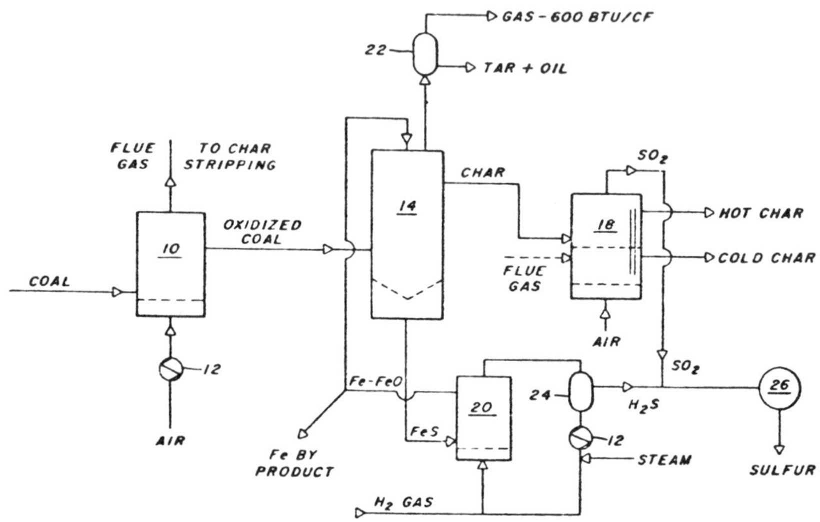


Source:   U.S. Patent 3,886,048

The sulfided iron oxide is gravitationally separated from the resulting char in the second fluid state reactor or alternatively the iron oxide may be magnetically separated from the char in a separate step.  The resulting char is then introduced into a two-stage third fluid state reactor **18** where it is oxidized at about 1000°F in the first stage and discharged hot to be burned in a boiler or cooled in the second stage by the introduction of air.  The resulting char has a carbon content of 75%, volatile material of 1% and a total sulfur content of 0.6% by weight.  It also has a thermovalue of 11,000 to 12,000 Btu/lb.

After separation from the char, the sulfided iron may be introduced into an iron regenerator **20** where it is regenerated with steam-saturated hydrogen-containing gas and recycled to be admixed with further coal or removed as by-product iron.  The off gases from the second fluid state reactor **14** are passed through a tar-oil separator **22** to produce by-product tar-oil and further off-gases.

The hydrogen-containing gas from iron regenerator **20** is separated from hydrogen sulfide and sulfur in separator **24**.  The hydrogen sulfide is mixed in a sulfur production furnace, such as conventional Claus furnace **26**, with the sulfur dioxide from the third fluid state reactor to produce by-product sulfur.

After separation in the tar-oil separator, the usable fuel gases comprise approximately 50% hydrogen, 35% methane and 15% other gases, which could be further enriched in a methanator to produce synthetic natural gas.

The initial introduction of 2,000 lb of coal results in 1,000 to 1,200 lb of fuel char, 100 lb of tar-oil, 80 lb of sulfur, 14,000 scf of fuel gas and 100 lb of by-product iron. The total percentage of iron used may be between 3 and 30% by weight, and the reducing time may be between 5 and 60 minutes. These ranges are determined by the initial sulfur content of the coal and are considerably more limited for coal of any particular sulfur content. The temperature range in the reducing stage must be greater than 1200°F to achieve the desired reaction, but less than 1700°F to prevent agglomeration of the iron.

For proper operation of the fluidized bed, the iron and coal particles must be of sufficient fineness to pass through a 100-mesh screen. This fineness is also necessary to insure intimate contact of the iron and coal particles, and consequently greater reactivity.

The oxidation with air of various types of chars and a 5% sulfur steam coal in a temperature range of 600° to 1200°F was tested with different degrees of success of sulfur removal. The best desulfurization results were achieved on chars which were prepared at 1600°F in various gaseous environments and in the presence of sulfur acceptors. The optimum temperature range for selective oxidation of sulfur from chars was found to be 800° to 1000°F.

An important reaction is the effect of metallic iron on the sulfur behavior during carbonization. An addition of 30% by weight iron powder to the coal blockaded the transformation of the main part of the sulfur into the locked-in type of organo-complexes. Consequently, the mass of the remaining sulfur was easily removed by selective oxidation. The final sulfur content in the oxidized char was reduced to 0.5 to 0.7% sulfur. These results indicate that iron has a greater effect than that of a simple sulfur acceptor such as dolomite.

## Treatment with Metal Oxides and Metal Chlorides

*R.N. Sanders; U.S. Patent 3,909,213; September 30, 1975; assigned to Ethyl Corporation* describes a process for the desulfurization of coal which contains pyritic sulfur and sulfur-containing organic compounds comprising: digesting, in a digestion zone, the coal and a Group I-A or Group II-A metal oxide salt or mixtures thereof in the presence of a fused metal chloride salt medium, which medium is capable of dissolving sulfur-containing organic compounds present in the coal and which has a melting point below the digestion temperature, until at least part of the dissolved sulfur-containing organic compounds and pyritic sulfur has been converted to Group I-A or Group II-A metal sulfides; feeding anhydrous hydrogen chloride to the digestion zone subsequent to the digestion period to convert the formed Group I-A or Group II-A metal sulfides to Group I-A or Group II-A metal chlorides respectively and hydrogen sulfide; and removing from the digestion zone an at least partially molten mixture containing desulfurized coal, the formed Group I-A or Group II-A metal chloride salts and the metal chloride salt medium.

Digestion temperatures are within the range of from 400° to 600°C. Pressures for the digestion are generally atmospheric and are not critical to the process.

The fused metal chloride salt medium can be either a single metal chloride salt or a mixture of such salts. As the high solubility in alcohol of the metal chloride salts making up the medium will be highly desirable for subsequent refining of the desulfurized coal, salts, whether they be used singularly or in mixtures, such as manganese chloride, ferric chloride, zinc chloride, stannous chloride, and bismuth trichloride are preferred. Of these preferred salts, the most preferred salt is zinc chloride due to its availability and low cost. Preferred mixtures of salts are zinc chloride-ferric chloride and zinc chloride-calcium chloride.

The Group I-A and Group II-A metal oxide salts utilized in the process are most preferably anhydrous. The absence of water in these salts is important from a safety standpoint as popping and spattering of the hot digestion mixture will be prevented. Examples of Group I-A or Group II-A metal oxide salts which may be used are sodium oxide, magnesium oxide, calcium oxide, potassium oxide, cesium oxides, etc. Mixtures of these salts may, of course, be used. The preferred metal oxides are calcium oxide and sodium oxide as they are extremely economical reactants. Of these preferred salts, calcium oxide is most preferred.

The digestion temperature should be within a range of 400° to 600°C. The digestion period should be that period of time which is sufficient to achieve at least substantially complete reaction of the organic sulfur compounds in solution and of the pyritic sulfur. Generally speaking, times ranging from ¼ to 5 hours are sufficient with the time being dependent upon the amount of organic sulfur compounds and pyritic sulfur present and the type of coal being treated. From 2 to 3 hours have been found to be sufficient for coal which contains up to 1.0% organic sulfur compounds and 0.2% pyritic sulfur.

Quantitatively, the amount of Group I-A or Group II-A metal oxide salt utilized will be that amount which provides at least a stoichiometric mol ratio of these salts to the sulfur present in the pyritic sulfur and in the dissolved organic sulfur compounds. In other words, when utilizing Group I-A metal oxide salts a mol ratio of this salt to the sulfur should be at least 2:1. When utilizing Group II-A metal oxide salts a mol ratio of 1:1 is sufficient. Amounts less than stoichiometric may be utilized; however such amounts do not result in high reaction efficiency.

Amounts in excess of stoichiometric can also be used. In fact, mol ratios up to 4:1 for Group I-A metal oxide salts and up to 2:1 for Group II-A metal oxide salts are preferred as reaction kinetics are improved by the presence of the excess of these salts. For calcium oxide, a highly preferred salt, it has been found that a mol ratio of from 1.2:1 to 1.5:1 is most preferred.

The amount of metal chloride salt used as the medium should be that amount which will result in a relatively liquid digestion mixture. Generally speaking, such a mixture can be achieved by utilizing the salt medium in an amount so that the weight ratio of medium to coal and metal oxide is within the range of from 6:1 to 2:1.

The coal, prior to digestion, should be prepared by grinding so that it will at least pass through a ½-inch mesh. This grinding of the coal is to hasten the dissolving of the sulfur-containing compounds in the fused metal chloride salt medium. The anhydrous hydrogen chloride fed to the digestion zone subsequent to the digestion period can be achieved by bubbling the gaseous acid through the digestion mixture, i.e., a mixture containing fused metal chloride salt, coal and Group I-A or Group II-A metal sulfides and remaining oxide salt. The amount of anhydrous hydrogen chloride used should be at least stoichiometric with relation to the amount of Group I-A or Group II-A metal sulfides formed by the digestion. Amounts in excess of stoichiometric are, however, preferred. Amounts ranging up to 100 times stoichiometric quantities are most preferred. By passing the anhydrous hydrogen chloride through the digestion mixture the Group I-A and Group II-A metal sulfides are converted to Group I-A or Group II-A metal chlorides, respectively, with the evolution of gaseous $H_2S$ from the digestion zone. The $H_2S$ can be either disposed of or can be utilized in processes having an $H_2S$ demand.

After the anhydrous hydrogen chloride has been fed to the digestion zone the at least partially molten material remaining in the zone is removed therefrom. The remaining material includes for the most part desulfurized coal, the metal chloride salt used as the medium and the formed Group I-A or Group II-A metal chlorides. Small amounts of Group I-A or Group II-A metal oxides and sulfides may also be present. If it is desired, further refining of this material to cleanse the desulfurized coal of the greater portions of these various salts and reaction products may be effected.

One refining technique comprises: cooling the removed at least partially molten material so as to solidify the molten components; crushing the solidified material; extracting from the crushed material, with alcohol, at least part of the metal chloride salt used as the medium and the formed Group I-A or Group II-A metal chlorides; and separating the desulfurized coal from the alcohol-salt solution. Optionally, the desulfurized coal may be even further refined by contacting the desulfurized coal with a dilute aqueous acid to extract any remaining salts or metal sulfides and separating the coal from the dilute aqueous acid-salt-sulfide solution. The recovered desulfurized coal can then be dried and pelletized to a form for consumption.

*Example:* A metallurgical coal containing from 1.0 to 1.2% total sulfur was crushed to a size between 1 and 4 inches. 6 g of the coal, along with 30 g of zinc chloride and 3 g of calcined lime, were heated together in the digestion zone to a temperature of 450°C for 2 hours. After this period anhydrous hydrogen chloride was bubbled through the glass dip tube for ½ hour. The resultant mass was then dumped from the digestion zone and cooled.

The cooled material was ground to about −16 mesh and extracted with isopropanol in a beaker for ¾ hour. The resultant mass was separated from the isopropanol-salt solution and contacted with 5% aqueous hydrogen chloride solution for 1 hour. The remaining solids were filtered from the acid solution and dried by heating. It was found by x-ray fluorescence that 62.1% of the total sulfur had been removed from the coal.

**Treatment with Undecomposed Iron Carbonyl**

A process is described by *J.K. Kindig and R.L. Turner; U.S. Patent 3,938,966; February 17, 1976; assigned to Hazen Research, Inc.* for improving coal wherein the raw coal is reacted with substantially undecomposed iron carbonyl which alters the apparent magnetic susceptibility of certain impurity components contained in the raw coal thereby permitting their removal by low intensity magnetic separators. The process is especially effective for removing pyrite from coal, while at the same time reducing ash and increasing the calorific value. The apparent magnetic susceptibility of pyrite as well as other associated impurities in coal is increased to the point where selective magnetic separation of these impurities from the coal particles is feasible. The increase is effected by contacting coal containing pyrite or other impurities liberated from the coal with an iron carbonyl-like iron pentacarbonyl under conditions at which ordinary pyrolytic decomposition of the iron carbonyl into metallic iron and carbon monoxide is not appreciable. With pyrite a chemical reaction between the iron carbonyl and the pyrite particles occurs to form a replacement shell, on the surface of the pyrite particles, of a material having a magnetic susceptibility significantly greater than that of untreated pyrite. The carbonyl treated coal product is then passed through a magnetic separator for removal of the pyrite and impurity particles.

It is desirable to have the coal comminuted finely enough to give substantial liberation of impurities from coal particles. The carbonyl is introduced as a vapor into a reaction chamber containing the coal. These carbonyl vapors can be carried into the chamber by a gas, inert to the reaction, by first passing the gas over or through a vessel holding liquid iron carbonyl.

*Example:* The process was applied to a bituminous coal from central Pennsylvania. The coal was charged into a kiln which was then rotated. To introduce the iron pentacarbonyl into the reaction zone, an inert gas was passed through liquid iron pentacarbonyl at room temperature contained in a vessel outside the kiln with the gas carrying carbonyl vapors then being introduced into the reaction zone of the kiln.

The reaction zone was held between 185° and 195°C for 1 hour, following which the kiln was purged of carbonyl vapors by the inert gas and the reaction zone cooled to room temperature.

Three products were made by magnetic separation using magnets of different field strengths, a "magnetic fraction," "weakly magnetic fraction," and "nonmagnetic fraction," with the magnetic fraction obtained from wet processing. Two magnets were used in the separation—a laboratory Davis tube tester and a small, hand horseshoe Alnico magnet. These three products were then analyzed for forms of sulfur, ash, and calorific value (Btu).

The process removed almost 70% of the pyritic sulfur. Not all pyritic sulfur was liberated at the size treated in this example so the 68% reduction may in fact represent all the liberated pyrite. The process also reduced the ash from 22.1 to 13.2%.

**Washing with Certain Polar Compounds**

*T.-Y. Chia; U.S. Patent 3,988,120; October 26, 1976* describes a method of desulfurizing high sulfur, low grade coal by immersing coal powders in an aqueous solution of a polar compound containing nonpolar groups bonded with polar groups and washing and drying the immersed mass for desulfurization, thereby providing low ash and high calorie coal. Many nonpolar and polar groups may be used in the process, and are listed in the patent; however the following polar compounds are preferred, particularly those having less than 25 carbon atoms:

alkyl sulfates, alkyl sulfonates, alkylaryl sulfonates, $\alpha$-sulfonated fatty acids, $\alpha$-sulfonated fatty esters, dialkyl succinates, fatty acid benzimidazolines, succinic amide alkylester sulfonates, polyoxyethylene alkyl ethers, polyoxyethylene alkylphenol ethers, polyoxyethylene polyol fatty esters, polyoxyethylene alkylamines, alkylpyridinium halides, alkyl phosphates and alkylaryl phosphates.

Raw coal may consist of any of peat, lignite, brown coal, bituminous coal and anthracite. Though freely selectable, the particle size of raw coal powders is preferred to be between 50 and 200 mesh. Larger particles are also desulfurized, but with a lesser degree of effectiveness because of the larger mass relative to the surface area. Particles finer than 200 mesh are effectively treated. However, the additional cost of the comminution necessary to produce minus 200 mesh powder may not be economically feasible.

This coal desulfurizing process comprises the following steps: Powders of the raw coal are immersed in an aqueous solution of any of the above-listed polar compounds having a prescribed concentration of from 0.01 to 1.0% for a sufficient time to substantially lower the sulfur content. The aqueous solution preferably contains between 0.01 and 0.06%, with 0.03 to 0.06% most preferred. The temperature of the aqueous solution is preferably ambient temperature, with up to 40°C preferred. The immersion time is preferably between 90 and 120 minutes. A longer immersion (which is more costly) does not appear to improve the product and a shorter immersion may not be sufficient to sufficiently lower the sulfur content. The coal powders should have a weight ratio of 1:3-5 to the aqueous solution of the polar compound. The coal powders taken out of the solution are washed to eliminate impurities, followed by drying. If the concentration and temperature of the aqueous solution of the polar compound are not considered to bear a great economical importance, the concentration and temperature may be higher.

*Examples 1 through 16:* 2 kg lots of 200 mesh powders of brown coal were immersed in aqueous solutions each weighing 8 kg and containing the eight kinds of polar compounds given in Table 1 below. The immersion was continued 100 minutes with occasional stirring. In the table the concentrations are given in percent by weight of aqueous solutions and the temperatures in °C.

## TABLE 1

| Ex. No. | Polar Compounds | Concentration (% by wt) | Temperature (°C) |
|---|---|---|---|
| 1 | Lauryl sulfate | 0.14 | 20 |
| 2 | Lauryl sulfate | 0.04 | 40 |
| 3 | Oleyl sulfate | 0.13 | 20 |
| 4 | Oleyl sulfate | 0.02 | 35 |
| 5 | Sodium oleylmethyl aminoethane sulfonate | 0.11 | 20 |
| 6 | Sodium oleylmethyl aminoethane sulfonate | 0.01 | 35 |
| 7 | Sodium dibutylnaphthalene sulfonate | 0.10 | 20 |
| 8 | Sodium dibutylnaphthalene sulfonate | 0.05 | 35 |
| 9 | Polyoxyethylene dodecylphenol ether | 0.11 | 20 |
| 10 | Polyoxyethylene dodecylphenol ether | 0.03 | 35 |
| 11 | Sorbitan monolaurate polyglycol ether | 0.10 | 20 |
| 12 | Sorbitan monolaurate polyglycol ether | 0.05 | 35 |

(continued)

**TABLE 1: (continued)**

| Ex. No. | Polar Compounds | Concentration (% by wt) | Temperature (°C) |
|---|---|---|---|
| 13 | Oleyl sarcoside | 0.15 | 20 |
| 14 | Oleyl sarcoside | 0.04 | 35 |
| 15 | Sodium di(n-octyl) sulfosuccinate | 0.05 | 20 |
| 16 | Sodium di(n-octyl) sulfosuccinate | 0.02 | 35 |

After the treatment the brown coal powders were removed, washed 10 times with 10 liters of running water and dried 12 hours at 150°C, providing desulfurized coal. Table 2 below indicates the sulfur content, ash content and calories of coal powders before and after treatment by the eight polar compounds. Analysis was made on the dry basis.

**TABLE 2**

| Sample | Sulfur Content (% by weight) | Ash Content (% by weight) | Calories (Btu/lb) |
|---|---|---|---|
| Untreated coal | | | |
| 1 | 11.63 | 41.88 | 7,688 |
| 2 | 11.56 | 40.24 | 7,851 |
| Desulfurized coal | | | |
| 1 | 5.22 | 34.59 | 9,182 |
| 2 | 5.39 | 34.93 | 9,156 |
| 3 | 5.34 | 35.11 | 9,208 |
| 4 | 5.63 | 35.61 | 9,105 |
| 5 | 5.28 | 34.63 | 9,185 |
| 6 | 5.23 | 34.71 | 9,134 |
| 7 | 5.39 | 34.77 | 9,213 |
| 8 | 5.44 | 35.91 | 9,159 |
| 9 | 5.30 | 35.47 | 9,202 |
| 10 | 5.34 | 35.44 | 9,202 |
| 11 | 5.29 | 36.02 | 9,020 |
| 12 | 5.28 | 34.88 | 9,178 |
| 13 | 5.31 | 35.17 | 9,184 |
| 14 | 5.28 | 35.15 | 9,210 |
| 15 | 5.47 | 36.67 | 9,039 |
| 16 | 5.40 | 35.20 | 9,158 |

## Use of Ferromagnetic Particles

*J. Iannicelli; U.S. Patent 3,999,958; December 28, 1976* has developed a process to extract sulfur from coal by treatment with ferromagnetic particles, such as steel wool and shavings, at elevated temperatures. These ferromagnetic particles are particularly chemically reactive at their surface and susceptible to corrosion, while retaining magnetic characteristics of the interior portions of their mass.

The unreacted interior portions of the individual magnetic particles provide for removal from the coal particles, by magnetic separators, of the composite of sulfur-bearing surface and magnetically susceptible interior mass and of weakly magnetic particles by the unreacted ferromagnetic particles. The large surface to mass ratio of steel wool fibrous particles is used to separate such elements from the weakly magnetic equiaxed magnetic pyrrhotite particles created by reaction of the steel wool surface and pyrite in the coal to provide for economic recovery and recycling of such particles.

The raw sulfur-bearing coal is automatically and continuously crushed and fed
to a ball mill for grinding to provide for reduction of such coal to particles of
a size of 100% -200 mesh, in order to thereby liberate magnetic particles and
to liberate sulfur-bearing particles insofar as feasible and so expose magnetic particles and sulfur-bearing components to the surface of such ground coal particles.

A dispenser automatically and continuously dispenses fibrous scavenger particles
(steel wool fibers of mechanically strained low carbon steel 0.1 to 0.12% carbon
with fiber thickness of 0.005 to 0.025 mm and width of 0.03 to 0.10 mm and
length in excess of 2 cm) into the mixer with the ground coal, in sufficient
amount to carry all the sulfur to be removed from such coal on the scavenger
particles on or near a minor portion of the surface of such scavenger particles.
The mixer operates at ambient temperature to thoroughly mix the scavenger
particles and coal particles.

The mixture of coal and scavenger is then fed by a conventional screw conveyor
to the top of an air-tight heated chamber where the temperature of the particulate contents of the tank is brought to 350°C.

A pump cycles chemically inert gas to agitate the particulate mass therein and
so improve contact of separate particles of such mass with each other and with
the heater coil and also to prevent oxidation of the particulate mass in the heater
by exclusion of air. The temperature of the particulate mass in the heater is
controlled automatically.

A cyclone or centrifugal separator separates fines carried with agitating gas from
the agitating gas stream. Such fines are passed through a gas-solids separator and
mixed with material fed to the heater and the gas is exhausted.

In a particular preferred embodiment, the steel wool scavenger particles react in
the heater for 90 minutes at 350°C with the organic sulfur in the coal and cause
release of such sulfur from its attachment to the coal particles and transfer of
such sulfur to the surface of the steel wool scavenger particles and firmly bind
such sulfur to the surface of such steel wool fibers as well as causing reduction
in the amount of sulfur held as pyrite (which is relatively poorly ferromagnetically
susceptible) and concurrent production of more strongly ferromagnetically susceptible pyrrhotite.

While the heating of the admixture of ground crude coal and finely divided steel
wool scavenger particles converts the organic sulfur in the coal to a form of sulfur adherent to the surface of those finely divided steel wool alloy particles,
such heating does not consume the interior portion of such magnetic steel scavenger particles. The interior portions of the scavenger particles remain as magnetically susceptible carriers for such sulfur and those carriers are removed by a
magnetic separator from the remainder of the treated mass of particulate coal
and its impurities and recycled.

Experiments demonstrating critical features of the above process are as follows:
The object of these experiments is to demonstrate the extraction of sulfur (principally organic sulfur) from coal by heating coal with the sulfur scavenger or acceptor
in a closed system. After reaction, the reaction system is quenched, the mixture
of reacted coal is removed, and the sulfur scavenger or acceptor is separated by
the magnetic separator. The coal is a bituminous coal designated Kentucky No.
14, having a sulfur content of 3.9% and a particle size of 100% -200 mesh.

*Example 1:* 50 g of 200 mesh coal was mixed with 0.5 g ferric oxide and then charged into a capped steel pipe ½ x 3 inches long. The pipe was sealed with a second cap and heated in a lead bath for 90 minutes at 350°C. After cooling, the pipe was uncapped and the contents dispersed in 500 ml of demineralized water containing 5 drops of tincture of green soap (potassium oleate in isopropanol).

The above slurry was passed through a column containing 430 stainless steel shavings and magnetized to a field of 5 to 10 kilogauss. Coal was flushed from the column with demineralized water and the magnetic ferric oxide was collected by deenergizing the magnet and rinsing out the ferric oxide. Testing of the ferric oxide with the Hach spot test for sulfide revealed a small amount of sulfide present, proving the ferric oxide had abstracted sulfur from coal during this autoclaving, but only in small amount.

*Example 2:* Example 1 was repeated except that 0.5 g of electrolytic iron powder was used in place of 0.5 g of ferric oxide. Analysis of the iron powder revealed by the Hach spot test showed the presence of sulfur using Hach sulfur reagents. This clearly proves that the iron powder has abstracted sulfur from coal.

*Example 3:* Example 1 was repeated except that 0.5 g steel wool (fine grade) was mixed with coal. After reaction, the steel wool was manually removed and a portion of it tested for sulfur. A very strong sulfur reaction was obtained.

**Reaction with Metal Sulfides or Polysulfides**

A coal desulfurization process described by *R. Swanson; U.S. Patent 4,018,572; April 19, 1977* comprises contacting the coal with a sulfur compound which is a nonvolatile sulfur unsaturated sulfide or polysulfide of an alkali metal or with a hydrogen polysulfide. The unsaturated sulfide or polysulfide of the alkali metal will dissolve elemental sulfur, will remove sulfur from many organic compounds, dissolve certain water-insoluble metal sulfides and form higher sulfur content alkali metal polysulfides.

The form considered to be sulfur saturated for sodium is the tetrasulfide, for potassium it is the pentasulfide, and for cesium and rubidium the hexasulfide. Although additional sulfur will dissolve in these saturated forms, the above forms represent what is generally considered the limit of sulfur which combines with each of these alkali metals with formation of definite compounds.

When these sulfur saturated compounds are made with the sulfur derived from the fossil fuels, they can be thermally decomposed into lower sulfur content polysulfides of the alkali metals with elemental sulfur separating from this melt. The temperatures required to decompose the various polysulfides are as follows from the saturated forms: potassium tetrasulfide, over 300°C; potassium trisulfide, cannot be produced directly thermally; potassium disulfide, over 850°C which decomposes the tetrasulfide; the monosulfides cannot be produced thermally; sodium trisulfide does not exist; and sodium disulfide, 550°C.

The existence of sodium pentasulfide is questionable and its production in the process is not preferred. Therefore, the tetrasulfide is considered the sulfur saturated sodium form. At least two forms of sodium tetrasulfide exist and one of these forms appears capable of forming a pentasulfide of definite chemical composition.

In melt condition, the potassium polysulfides are preferred to the sodium polysulfides. In aqueous solutions, the ability of potassium to gather sulfur is somewhat greater than that of sodium and the recycling of the polysulfides is much easier with potassium. With sodium, the sulfur is distilled from the melt at 550°C, whereas the sulfur is molten at just over 300°C with the potassium pentasulfide decomposition. In aqueous solution, there is little difference between sodium and potassium in their ability to remove sulfur from the fossil fuels.

## Treatment with Alcohol Having One to Four Carbons

*L.J. Keller; U.S. Patent 4,030,893; June 21, 1977; assigned to The Keller Corporation* describes a process for preparing a low-sulfur, low-ash fuel from coal while simultaneously recovering valuable chemical products and taking advantage of a coaction between comminution and dissolution and it is characterized as the steps of comminuting the coal in the presence of an alcohol containing 1 to 4 carbon atoms.

Herein the term "alcohols" is employed to signify the alcohols including 1 to 4 carbon atoms inclusive, including methanol, ranging in purity from the substantially pure state to the crude alcohol that is produced by the gasification of coal followed by a "methanol," or alcohol, synthesis operation. These alcohols are frequently referred to in the art as "methyl fuel." The methyl fuel may be produced at a site closely adjacent to the mined coal or it may be transported into the area in which the liquid-solid suspensoid or coal-alcohol slurry that is a feed stream for this process is prepared.

The coal particles are prepared to have suitable sizes. Specifically, they are all of a –8 mesh Tyler Standard screen size, with the majority of the particles being of –100 mesh size. If desired, the majority of the particles may be of –200 mesh size. In any event the coal particles have a settling velocity of less than 2½ centimeters per second in water. The coal particles are worked in the presence of the methyl fuel, including the methanol, so as to dissolve the water and other alcohol-soluble impurities from the coal and activate and wet the surfaces of the coal particles. This step is apparently necessary to form the sheer thinning slurry.

The method of preparing this low-sulfur, low-ash fuel from coal comprises the steps of:

    (a)    comminuting the coal in the presence of an alcohol containing 1 to 4 carbon atoms to a size wherein a major part of the coal is of –100 mesh size;

    (b)    forming an admixture of the resulting comminuted coal and the alcohol, the mixture containing at least 40% by weight of coal;

    (c)    separating the admixture into three streams, the first underflow stream comprising primarily a first slurry of high density solids, a first middling stream comprising primarily a first slurry of the coal particles, and a first overflow of stream comprising primarily alcohol, a liquor of constituents dissolved from the comminuted coal in this alcohol and colloidal constituents suspended in the alcohol and liquor;

    (d)    separating the first underflow of the first slurry of high density solids into a second underflow comprising a second viscous slurry of sulfur and solids that would form ash residue if burned with the coal, and a second overflow comprising a dilute slurry of coal in alcohol and the liquor of constituents dissolved from the comminuted coal in the alcohol;

(e) separating the first middling stream into a third underflow comprising a second slurry of high density solids, a second middling stream comprising primarily a second slurry of coal with only a minor amount of high density solids and sulfur therein, and a third overflow comprising primarily the alcohol, a liquor comprising constituents dissolved in alcohol and colloidal constituents suspended in the alcohol and the liquor;

(f) separating the first overflow into a fourth underflow and a fourth overflow;

(g) refining the fourth overflow into respective desired constituents and alcohol; and

(h) forming a low sulfur, low ash fuel from the second middling stream.

The first and second middling streams may be sent to a thickener, such as a centrifuge, to form a coal-alcohol slurry. The slurry may then be pressurized and heated and sent to a flash tower where it is flashed to atmospheric pressure.

The heat content will effect a flashing of the organic gases and the alcohol. The alcohol and other condensible constituents are condensed via suitable heat exchange and the constituents are employed as previously described.

As a consequence of the flashing off of the alcohol and the organic gases, puffed coal particles are formed as a residue, or bottoms, in the flash tower. The puffed coal particles are highly susceptible to oxidation and can be readily burned in any process to give a hot flame with very low pollution emissions, since the sulfur and other pollutants have been removed. Moreover, the ash-forming minerals and the like have been removed from the coal particles before they are formed into the puffed coal particles.

**Autoclaving Treatment for Lignite**

*E. Koppelman; U.S. Patent 4,052,168; October 4, 1977* describes a beneficiation process in which lignitic-type coals in a substantially as-mined condition containing from 20 up to 40% moisture are charged into an autoclave and heated to an elevated temperature of at least 750°F and a pressure of at least 1,000 psi for a controlled period of time to effect a controlled thermal restructuring of the chemical structure thereof and to effect a conversion of the moisture and a portion of the volatile organic constituents therein into a gaseous phase.

At the conclusion of the autoclaving step, the lignitic-type coal is cooled, preferably in contact with the gaseous phase so as to effect a deposition of the condensible organic constituent on the surfaces thereof to provide for a further stabilization of the upgraded coal product, rendering it nonhydroscopic and more resistant to weathering and oxidation during shipment and storage. The noncondensible gaseous phase is recovered and can be advantageously employed as a fuel in the process for heating the autoclave or for commercial sale.

The upgraded coal product produced is generally of a hard black glossy appearance, having an internal structure which visibly has been transformed from the original lignitic-type coal charge and which is possessed of increased heating values of a magnitude generally ranging from 12,000 up to 13,500 Btu per pound.

In contrast, consolidated lignitic coal on an as-mined basis has a heating value of about 7,000 Btu per pound, while on a moisture-free basis, has a heating value ranging from about 10,300 up to 11,900 Btu per pound.

*Example 1:*　A lignitic coal derived from a mine in Zap, North Dakota, having an average moisture content of 30% by weight and being of a slate color, is screened to provide a particulated charge of a particle size less than ½ inch. A measured amount comprising 6.64 g is placed in a stainless steel pressure vessel having an internal chamber 3 inches long and of a circular cross-sectional configuration of ⅛-inch diameter and a wall thickness of ¼ inch. The ends are capped with screw-type couplings to seal the charge within the chamber. The pressure vessel or autoclave is placed in a furnace chamber heated to 1000°F and after a 5-minute preheating period is maintained at temperature for a residence time of 30 minutes.

At the completion of the autoclaving operation, the pressure vessel is removed and cooled under tap water to room temperature, whereafter an end cap is removed to release the residual pressure and the charge is removed and subjected to a moderate drying operation to remove surface water by air drying.

The upgraded lignite product weighs 4.98 g, evidencing a loss of 25% and has an average heating value of 12,549 Btu per pound. The upgraded carbonaceous product is of a dark color and is of a glossy appearance.

*Example 2:*　The test procedure as described in Example 1 is repeated employing a lignitic charge material containing 30% moisture, which after a 5-minute preheat period is maintained at 1000°F for a period of 30 minutes. A total of 5.64 g of charge material is employed and at the completion of the autoclaving step, the pressure vessel is removed and permitted to air cool to a temperature of 350°F, whereafter an end cap is removed to release the residual gas.

An upgraded carbonaceous product comprising 3.33 g is recovered, representing a loss of about 40% and has a measured heating value of 12,978 Btu per pound.

*Example 3:*　The test procedure as described in Example 2 is repeated employing a charge of 6.71 g Colstrip subbituminous coal, which is heated for a total duration of 30 minutes at 1000°F after being brought up to temperature in 5 minutes. At the conclusion of the autoclaving step, the vessel is permitted to cool to 300°F, whereupon the cap is unscrewed and the residual pressure released.

A product comprising 3.99 g is recovered, representing a loss of 40.9% by weight of charge. The upgraded coal product has a heating value of 12,927 Btu per pound. This product is compared to a control sample of untreated Colstrip subbituminous coal in both as-received and moisture-free form, and the comparative data, as set out in the table, illustrate the increase in heating value and the decrease in sulfur and oxygen content caused by the autoclaving treatment.

Although the data in the table illustrate the improved qualities of the treated coal over the original coal, they do not emphasize the reduction in sulfur that has occurred. This can be shown by allowing a sulfur balance through the treatment. 100 pounds of as-received subbituminous coal contains 1.30 pounds of sulfur. This is converted to 59.1 pounds of upgraded product containing 1.38

weight percent sulfur, or 0.82 pound of sulfur. This shows that 0.48 pound, or 37 weight percent, of the sulfur in 100 pounds of as-received coal is removed in the treatment.

### Comparative Data—Subbituminous Coal from Colstrip, Montana

| | Control Sample | | Treated Sample | |
|---|---|---|---|---|
| | As Received | Moisture Free | As Received | Moisture Free |
| Moisture (wt %) | 16.8 | 0 | 1.9 | 0 |
| Heating Value (BTU/lb) | 9639 | 11585 | 12927 | 13177 |
| Ultimate Analysis | | | | |
| C | 56.3 | 67.7 | 77.2 | 78.7 |
| H | 2.90 | 3.49 | 2.95 | 3.01 |
| S | 1.30 | 1.56 | 1.38 | 1.41 |
| N | 0.78 | 0.94 | 1.06 | 1.08 |
| O | 28.9 | 14.5 | 5.3 | 3.5 |
| Ash | 9.78 | 11.8 | 12.1 | 12.3 |

## Magnetic Separation of Pyrites

The objects of the process devised by *T.-Y. Yan; U.S. Patent 4,052,170; October 4, 1977; assigned to Mobil Oil Corporation* are:

(1)　To provide a process for magnetically separating paramagnetic impurities from pulverized coal in which velocity passing a high-gradient magnetic separator is related to the paramagnetism of the impurities;

(2)　to provide a process for selectively drying and heating a pulverized coal while distilling oils therefrom; and

(3)　to provide a process for sequential stagewise heating of dried pulverized coal while selectively oxidizing and converting pyrite to highly paramagnetic compounds.

The process consists of:

(A)　pulverizing coal to about 200 mesh while passing a stream of either hot air or heated inert gas therethrough at sufficient velocity to entrain and fluidize the pulverized coal;

(B)　passing the hot air and entrained coal through a high-gradient magnetic separator means to remove pyritic impurities and form beneficiated coal and either feeding the beneficiated coal and the hot air to the combustion zone of a furnace, or, alternatively, drying with inert gas and separating the dried pulverized coal from the inert gas within a fluidized stage;

(C)　condensing oily distillate from the inert gas and recirculating the gas, after heating, to the coal being pulverized;

(D)　successively entraining the dried pulverized coal with a stream of oxygen-containing gas in sequential fluidized stages having successively higher temperatures, while passing the oxygen-containing gas countercurrently thereto, to a final temperature of 480° to 600°C;

(E)   passing the heated pulverized coal through a high-gradient magnetic separator means and magnetically removing iron-containing
compounds therefrom to produce a beneficiated coal.

A preferred configuration of the magnetic separator is a steel canister fitted with
steel screens.  Preferably, the steel screens are in parallel across the interior of
the canister, are spaced about 1 cm apart, and are 20 to 60 mesh.  Each canister
is preferably equipped with, or attachable to while in its discharging state, a shaking device that rapidly removes magnetically attracted particles.  The high-gradient magnetic separator has a field strength of at least 10,000 gauss.

This process is primarily one for producing rapidly burning low pyrite fuel in
situ for a large-scale electrical and steam generation, so that a complete description of the process is given in the patent, and illustrations are included of the
apparatus for desulfurizing the pulverized coal, removing the iron-containing compounds from it, and feeding the beneficiated coal to the combustion zone of a
furnace.

## Flotation with Surfactant

The process developed by *R.M. Dessau; U.S. Patent 4,076,505; February 28,
1978; assigned to Mobil Oil Corporation* for reducing the pyritic sulfur content
of coal provides under sink-floak conditions (1) contacting finely comminuted
coal with a heavy organic liquid medium having dissolved therein a dispersant
quantity of ionic surfactant which is a hydrocarbyl-succinimide derivative of an
aminoaryl sulfonic acid salt; and (2) separating the coal float phase from the
heavy organic liquid medium, and recovering coal with a reduced sulfur and ash
content.

The term "coal" includes solid carbonaceous fossil minerals, such as anthracite
coal, bituminous coal, subbituminous coal, lignite, peat, and the like.  It is an
important feature of the process that the treated coal product recovered from
the process has both a reduced sulfur content and a reduced ash content.  The
sulfur and ash contents of various coals suitable for use are as follows:

### High Volatile B

|          | Percent |
|----------|---------|
| Sulfur   | 6.81    |
| Nitrogen | 0.88    |
| Oxygen   | 4.97    |
| Carbon   | 56.90   |
| Hydrogen | 3.11    |
| Ash      | 21.01   |

### Subbituminous A

|          |       |
|----------|-------|
| Sulfur   | 2.48  |
| Nitrogen | 0.77  |
| Oxygen   | 9.16  |
| Carbon   | 54.01 |
| Hydrogen | 3.92  |
| Ash      | 18.91 |

Ball mills or other types of conventional apparatus may be employed for pulverizing raw coal in the preparation of the finely comminuted feed coal for the process. The crushing and grinding of the coal can be accomplished either in a dry state or in the presence of a liquid, such as the heavy organic liquid medium being employed in the practice of the process.

The efficiency of pyritic sulfur and ash removal is directly dependent upon the degree to which the raw coal has been pulverized. The efficiency of the pyritic sulfur and ash removal tends to increase as the average size of the coal particles decreases. Under sink-float conditions it is necessary for iron pyrites and ash to be in the form of discrete particles in order for the mechanism of gravity concentration based on differential specific gravity to be operative. Preferably, the average particle size of the pulverized raw coal being desulfurized and deashed by this process is in the range between 10 and 200 mesh.

A heavy organic liquid is employed as the parting medium in the process for gravity concentration and separation of finely comminuted coal from the sulfur and ash contaminants. Preferred parting liquid media are halogenated hydrocarbons, and mixtures thereof, which contain 1 to 2 carbon atoms and 2 to 6 halogen atoms. Illustrative of suitable halogenated hydrocarbons are methylene bromide, methylene chlorobromide, methylene iodide, dibromochloromethane, tribromofluoromethane, 1,2-dibromotetrafluoroethane, and the like.

For the purposes of the process, it is required that the heavy organic liquid parting medium have a specific gravity which is intermediate between that of the coal particles and the pyrite and ash particles. The gravity concentration of coal from impurities is effective when the density of the liquid parting medium is greater than that of the coal, and lesser than that of the pyrite and ash constituents of the coal. Illustrative of comparative specific gravities (20°C) are the following: tetrabromoethane, 2.96; methylene iodide, 3.26; pentachloroethane, 1.67; coal (average), 1.10; pyrite ($FeS_2$), 4.90; coal ash, 2.50.

The particular heavy organic liquid medium employed can have a specific gravity in the range between 1.1 and 3.3 for effective gravity concentration of pyrite from coal. For purpose of ash concentration and separation from coal, a liquid medium with a specific gravity in the range between 1.1 and 2.5 is preferred. The specific gravity of the liquid medium can be precisely controlled by custom blending of approximate organic liquids, such as methylene bromide/methylene chlorobromide, methylene bromide/perchloroethylene, and the like. A liquid medium of equal proportions of hexane and carbon tetrachloride has a specific gravity of 1.2.

An important aspect of the process for coal desulfurization is the provision of an ionic surfactant which is dissolved in the heavy organic parting medium in a dispersant quantity. The ionic surfactant has an exceptional affinity for the surface area of finely comminuted coal. This ionic surfactant is a hydrocarbyl-succinimide derivative of an aminoaryl sulfonic acid salt. In a preferred embodiment the ionic surfactant corresponds to the formula on the next page where R is an acyclic hydrocarbyl group containing between 5 and 1,000 carbon atoms; Ar is a polyvalent aromatic radical; X is a metal or $HNR'_3$ cation; and R' is hydrogen or a hydrocarbyl substituent; m is an integer having a value of 1 or 2; n is an integer having a value of 1 to 4; and p is an integer having a value of 1 or a value up to the valence of X.

$$\left[\left(\begin{array}{c} R-CH-C\overset{O}{\diagup} \\ \qquad\qquad N-Ar-(SO_3-)_n-X \\ CH_2-C\diagdown_O \end{array}\right)_m\right]_p$$

The hydrocarbyl-succinimide derivatives of aminoaryl sulfonic acid salts described previously can be prepared by reacting a hydrocarbyl-succinic acid anhydride or ester with a sulfanilate type salt, wherein the two reactants have structures corresponding in combination to the desired hydrocarbyl-succinimide derivative of aminoaryl sulfonic acid salt.

In the structural formula represented above, R is a hydrocarbyl group, such as straight chain or branched chain alkyl or alkenyl substituent, e.g., pentyl, hexenyl, isooctyl, decyl, tetradecyl, tetradecenyl, eicosyl, and the like. R can also be an oligomer of a polymerizable olefin, such as propylene, butylene, isobutylene, octene, decene, and the like.

In the structural formula, Ar is an aromatic radical derivative of compounds, such as benzene, nitrobenzene, phenol, toluene, xylene, diphenyl, naphthalene, naphthol, anthracene, phenanthrene, fluorene, acenaphthylene, and the like.

In the structural formula, X is a cation, such as ammonium, methylammonium, trimethylammonium, tributylammonium, tetraethylenepentaammonium, hexamethylenediammonium, lithium, sodium, potassium, magnesium, calcium, zinc, barium, strontium, and the like.

The ionic surfactants corresponding to the structural formula above can range in molecular weight from simple monomolecular structures to complicated high molecular weight gel structures. The ionic surfactants can contain one or a multiple of each of the hydrocarbyl-succinamido, aromatic and sulfonic moieties in the ionic surfactant molecular structures.

It is also contemplated to employ an ionic surfactant which corresponds to the structural formula represented above in which the hydrocarbyl-succinimide moiety is in the form of an acyclic amido/carboxylate hydrolysis product:

$$\left[\left(\begin{array}{c} (H)-CH-C\overset{O}{\diagup} \\ R\diagdown\qquad\qquad N-Ar-(SO_3-)_n-X \\ (H)-CH-C-O-Y \\ \qquad\overset{}{\underset{O}{\|}} \end{array}\right)_m\right]_p$$

where R, Ar, X, m, n and p are as previously defined, and Y is hydrogen or metal or other form of cation.

In a batch-type float-sink operation, in a closed system, finely comminuted coal is slurried with liquid parting medium which contains a selected sulfanilate salt ionic surfactant dissolved therein. When the mass of solid particles has become thoroughly wetted, the liquid-solids slurry admixture is allowed to stand until the main body of the liquid medium has substantially clarified. Ultrasonic vibration can be employed to accelerate the settling of pyrite and ash solids.

Purified coal is recovered as a coal float phase, and pyrite/ash is recovered as a sink phase. The density of the liquid parting medium is intermediate between the coal particles and the pyrite/ash particles. The relative weight proportion of coal to liquid parting medium is not critical, and is dictated by practical considerations. The quantity of liquid parting medium can vary from 1 to 100 in weight ratio to the comminuted coal being treated. A weight ratio between 1.5 and 10 to 1 of liquid medium to coal is suitable for typical processing systems.

A closed processing system is employed in order to prevent volatilization of the organic liquid medium. A halogenated hydrocarbon liquid medium is a relatively expensive commodity. Also, the presence of gaseous halogenated hydrocarbons in the atmosphere can constitute a serious pollution hazard.

Since this coal desulfurization process relies on physical interaction for selective separation of pyrite and ash impurities, i.e., gravity concentration, it is simple and convenient to operate the liquid-solids system over a broad range of temperature and pressure conditions. In a preferred embodiment, the coal desulfurization process is operated on a continuous basis. This is readily accomplished by a conventional arrangement of liquid reservoir, and endless belt conveyors for introduction of untreated pulverized coal into the liquid parting medium and for removal of the coal float phase and the pyrite/ash sink phase.

For the practice of the process on a continuous basis, pulverized coal is delivered to one end of the liquid reservoir on an endless belt conveyor. The delivery end of the belt is submerged in the liquid parting medium so that the coal solids are properly wetted by the liquid. The main body of coal solids floats and spreads on the liquid surface, and is urged toward the opposite end of the reservoir as it is displaced by the continuous convey or incoming raw coal.

A second conveyor belt at the opposite end of the reservoir continuously removes the coal float phase from the liquid surface and delivers it to the next processing zone of the operation. The pyrite and ash solids which settle to the bottom of the reservoir are continuously or intermittently removed by a partially submerged conveyor system.

The purified coal solids which are recovered from the float phase are heavily wetted with parting medium liquid and ionic surfactant. Separation of the adsorbed liquid from the coal can be accomplished by delivering the wetted coal solids onto a fine mesh screen, where filtering action removes a major quantity of retained liquid from the coal solids. The coal is then washed with a solvent, such as methanol, to complete the removal of heavy organic liquid and the ionic surfactant dissolved therein.

The removal of heavy organic liquid may be accomplished by washing the coal solids with water, or by treating the coal solids with steam at atmospheric or elevated pressure. The water and heavy organic liquid are immiscible, so that the organic liquid is readily recoverable for recycle in the continuous process. The ionic surfactant substantially remains dissolved in the organic liquid.

It is another advantage of the process that the heavy organic liquid parting medium is an excellent solvent for organic compounds contained in high sulfur types of coal. Hence, during the sink-float phase, the liquid parting medium leaches resins and sulfur-containing organic compounds from the pulverized coal being treated. For this reason it is advantageous to provide a means in the process for separation of solvent from solute content to prevent impurity buildup in the liquid parting medium.

As previously described, the raw coal feed can be pulverized dry or in a liquid medium. Another convenient and economical method of coal preparation is to crush and pulverize the coal in an aqueous medium. Pulverized raw coal which has an adsorbed water content of not more than 2 weight percent can be desulfurized advantageously in accordance with the process. The presence of adsorbed water in the coal solids tends to reduce the amount of heavy organic liquid which is subsequently adsorbed.

The purified coal product of the process has a substantially reduced content of inorganic and organic impurities. It is suitable for burning as fuel, or for other applications, such as the production of high purity coke for electrode manufacture. The following examples are further illustrative of the process. The reactants and other specific ingredients are presented as being typical, and various modifications can be derived in view of the foregoing disclosure within the scope of the process.

*Example 1:* Into a 2,000 ml flask equipped with a thermometer, stirrer, gas inlet tube and condenser was added 140 g (0.51 mol) of triethylammonium sulfanilate (prepared from triethylamine and sulfanilic acid) and 1,000 g (0.51 mol) of polybutenyl-succinic anhydride (diluted with 29% unreacted polybutene), the polybutenyl group being obtained by reacting maleic anhydride and polybutene of 1,300 molecular weight. The reaction mixture was heated under a nitrogen atmosphere at 165°C for 2 hours. Approximately 9 g of water were removed. The yield of product, 1,130 g, was 100% of theoretical based on the following structure:

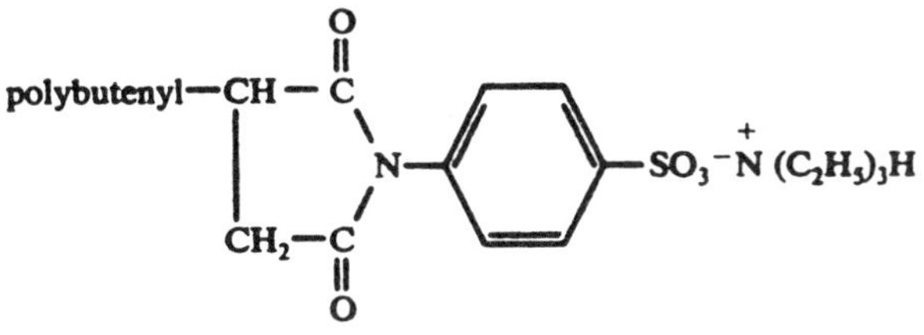

$$\text{polybutenyl—CH} \overset{\displaystyle \overset{O}{\|}}{\underset{\displaystyle \underset{\underset{\displaystyle O}{\|}}{CH_2—C}}{\overset{C}{\diagdown}}}{\diagup} N \diagdown \!\!\!\!\!-\!\!\!\!\!\diagup \text{(C}_6\text{H}_4)\!-\!SO_3^{-}\,\overset{+}{N}\,(C_2H_5)_3H$$

In the same manner, 0.51 mol of a bistrimethylammonium salt of 1-amino-2,5-benzenedisulfonic acid was reacted with 0.51 mol of polybutenyl-succinic anhydride, and a 1,182 g yield of hydrocarbyl-succinimide derivative of disulfonic acid salt was obtained.

*Example 2:* Finely comminuted medium volatile bituminous coal was desulfurized in accordance with the sink-float separation process in comparison with prior art methods. The bituminous coal had a pyritic sulfur content of 2.93% (total sulfur, 3.88%), and an ash content of 13.8%.

The heavy organic liquid parting medium consisted of bromotrichloromethane, with and without a quantity of an ionic surfactant dissolved therein, respectively, as indicated below. In each case, 20 g of bituminous coal was thoroughly slurried with 200 g of organic liquid parting medium. The admixture was allowed to stand until the main body of the liquid medium had substantially clarified by formation of a coal float phase and a pyrite/ash sink phase. The coal float phase was recovered, and the sulfur and ash content of the purified coal was determined. A reduction of sulfur content to less than 1.0 weight percent can be realized by increasing the sulfanilate salt concentration until the desired sulfur level has been achieved.

In the following table, TRS-1080 (Witco) is a petroleum sulfonate sodium salt; and the sulfanilate salt corresponding to the structure:

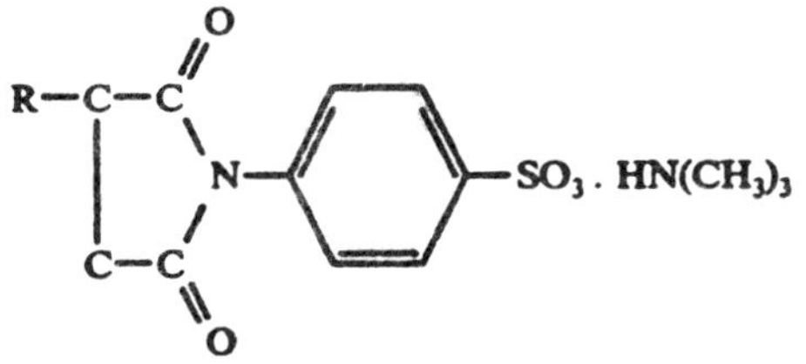

$$R{-}C{-}C(=O){-}N({-}C_6H_4{-}SO_3 \cdot HN(CH_3)_3)$$

where R is polybutenyl (molecular weight of 640).

| Weight Percent Ionic Surfactant | Weight Percent Sulfur | Weight Percent Ash |
|---|---|---|
| — | 2.19 | 10.8 |
| 0.1% TRS-1080 | 2.17 | 10.3 |
| 0.5% TRS-1080 | 2.08 | 10.3 |
| 0.02% Sulfanilate salt | 1.96 | 9.3 |
| 0.25% Sulfanilate salt | 1.63 | 8.5 |
| 0.5% Sulfanilate salt | 1.22 | 7.7 |

## Desulfurization by Microwave Energy

The process described by *P.D. Zavitsanos and K.W. Bleiler; U.S. Patent 4,076,607; February 28, 1978* for coal desulfurization induces thermochemical, in situ reactions between sulfur and other elements in the coal.

The in situ reactions (a reaction within the carbonaceous aggregates among some of the constituent elements) illustrated below allow sulfur to be liberated upon the forming of hydrogen sulfide, sulfur dioxide, or sulfur carbonyl. "Bound" in in these reactions means sulfur bound to iron (as in a pyrite) or sulfur organically bound to carbon (as in a dibenzothiophene). $\Delta H$ is defined as a small amount of energy in the form of activation energy required to induce sulfur gasification. Thus, as shown below, the in situ reactions allow for sulfur to be liberated without a transference of reactant elements to the carbonaceous aggregates.

$$S \text{ (bound)} + H_2O \text{ (adsorbed)} + \Delta H \longrightarrow H_2S \text{ (gas)}$$

$$S \text{ (bound)} + H_2 \text{ (bound)} + \Delta H \longrightarrow H_2S \text{ (gas)}$$

$$S \text{ (bound)} + O_2 \text{ (bound)} + \Delta H \longrightarrow SO_2 \text{ (gas)}$$

$$S \text{ (bound)} + CO_2 \text{ (bound or gas)} + \Delta H \longrightarrow COS \text{ (gas)}$$

Normally, high thermal energy in a range of 600° to 900°C is required to break bonded constituents of the coal (i.e., Fe-S and C-S). However, these processes utilize electromagnetic energy and an accompanying reaction chamber, such as an oven, to deposit microwave energy to sulfur-bearing areas or compounds to effect volatilization of sulfur in a form of a stable gaseous compound. More specifically, microwave thermal energy is deposited in a manner to selectively heat compounds in coal that contain large amounts of highly reactive sulfur, for instance, pyrites, pyritotites and thiophenes. Heating the preceding compounds induces thermochemical in situ reactions between these compounds and other neighboring reactive compounds, namely, $H_2O$, $CO_2$ or bound hydrogen. Further, while the bonds of sulfur-iron and sulfur-carbon are broken, in response to the effects of the microwave initiated reactions, sulfur is released and united to gaseous, reactive elements through molecular bonding.

The frequency of the microwave energy necessary to cause the sulfur to react with gaseous elements present in the coal is of an order of about 2.4 to 10 GHz (gigahertz), while the microwave energy is in a range between 500 and 1,000 watts. Coal at the above frequencies and watt energy is irradiated with microwave energy for a duration of about 40 to 60 seconds at approximately 1 atmosphere or less.

Moreover, the heating energy generated by the microwave energy per gram that is required to induce the in situ reaction is about 3 calories per gram; this is to be compared with 200 calories per gram required to heat coal to around 800°C required for thermal removal of the sulfur. Since the heating value of coal is 10,000 to 14,000 Btu per pound or 5,000 calories per gram, this translated into electrical energy is about 1,500 calories per gram. Hence, by utilizing microwave heat energy, only 0.3 to 1% of the heating value of coal is necessary in order to induce an in situ reaction which favors the volatilization of sulfur in the form of a stable gaseous compound. Additionally, during the reaction the temperature of the coal rises to a modest level of between 50° to 150°C with no significant evolution of hydrogen or carbon-containing matter, which evolution would result in loss of heating value.

Depending on the time duration, particle size, type of sulfur, and origin of coal, around 50% of all the sulfur is liberated. Table 5.1, which follows, shows the effect of desulfurization with microwave energy on bituminous coal containing both organic and pyritic sulfur, under the conditions listed below:

| | |
|---|---|
| Microwave frequency, GHz | 2.45 |
| Sample weight, g | 10 |
| Sample geometry | chunks |
| Radiation time, sec | 20–60 |
| Water content, % | 1 |
| Bed Thickness, cm | 1.5–2 |
| Particle size, cm | 0.1–4 |
| Total sulfur content, wt % | 4.1 |
| Sulfur removed, % | 26.8–53.5 |

## TABLE 5.1:  COAL DESULFURIZATION DATA

| Sulfur Content in Wt.% | | | Sulfur Removed in Wt.% |
|---|---|---|---|
| Before Treatment | After Treatment | Nature of Treatment | |
| 3.94 | | | |
| 3.88 | | | |
| 4.10 | | | |
| 4.45 | | | |
| 4.20 | | | |
| (4.11  Ave.) | | | |
| | 2.85 | | |
| | 2.49 | 20 sec. in | |
| | 2.48 | 1 Atm Air, | |
| | (2.61  Ave.) | Glass Container | 36.5 |
| | 2.53 | 60 sec. in | |
| | 2.18 | 1 Atm Air, | |
| | 2.56 | Glass Container | 41.1 |
| | (2.42  Ave.) | | |
| | 2.50 | 20 sec. in | |
| | 2.67 | 1 Atm Air | |
| | (2.59  Ave.) | No Container | 36.9 |
| | 2.84 | Same as Above | |
| | 2.93 | | |
| | 3.03 | | 28.7 |
| | (2.93  Ave.) | | |
| | 2.77 | 40 sec. in | |
| | 2.43 | 1 Atm Air | |
| | (2.60  Ave.) | No Container | 36.7 |
| | 1.97 | 40 sec. in | |
| | 1.84 | Reduced Air | |
| | (1.91  Ave.) | Pressure (1–5 mm Hg) | 53.5 |
| | 2.38 | 60 sec. in | |
| | 2.51 | Reduced Air | |
| | (2.45  Ave.) | Pressure (1–5 mm Hg.) | 40.4 |

Desulfurization of coal, in accordance with this process, is preferably practiced in a batch process.  In step 1, coal is crushed to size, which size is usually between 0.1 to 5 cm.  In step 2, the crushed coal is transported to a microwave cavity or reactor by a suitable conveyor.  Coal is loaded on the conveyor in a bed thickness, or heights, of around 1 to several cemtimeters.  At step 3, the microwave reactor receives the crushed coal and subjects the coal to irradiation with microwave energy.  The input frequency of the microwave energy is selectively tuned to a particular coupling frequency associated with the sulfur compound in the coal material, and thereby causes the sulfur to react with gaseous elements present in the coal.

Parenthetically, suitable input frequencies are of an order of about 2.4 to 10 GHz. The microwave energy is in a range between 500 to 1,000 watts and the coal is irradiated with microwave energy for a duration of about 40 to 60 seconds at approximately 1 atmosphere of air or less.  At step 4, the resulting or reaction sulfur-containing gaseous compounds, namely $H_2S$, $SO_2$ and COS, are by volume separated, trapped and removed by pumping or forcing the separated volumes into individual compartments. In the compartments, the separated volumes are converted to elemental sulfur or set up for disposal. $H_2S$, for example, may be converted to elemental sulfur by providing a reaction with sulfur dioxide.

Finally, at step 5, the coal subsequent to treatment with microwave energy is removed for further processing. For instance, the coal may be pulverized to a finely divided form with conventional power plant equipment. Or, as another example, the coal may be further burned and/or crushed as required by normal plant procedures to effect a special application for the coal.

# COKE DESULFURIZING PROCESSES

## METALLURGICAL COKE

The most widely used method for smelting iron ores is the reduction of these ores by means of coke in the blast furnace. Coke obtained by the high-temperature carbonization of coal in by-product coke ovens is the type of coke most extensively used. Only certain coals, referred to as coking coals, are suitable for producing coke having the desired physical and chemical properties for metallurgical use. It is essential that the sulfur content of the coke to be used for metallurgical purposes be low. Thus, a specification of the American Society for Testing Materials (ASTM D166-24) requires for the production of metallurgical coke that the composition of the coking coal be such that the dry coke produced therefrom will not contain more than 1.0% of sulfur in the case of foundry coke and 1.3% of sulfur in the case of blast furnace coke.

Under conventional high-temperature coking conditions, where the coking charge is essentially a caking coal of bituminous rank, the sulfur percentage of the coke obtained, based on the weight of the coke, is usually from 80 to 100% that of the sulfur percentage based on the weight of the coal from which it was made. On this basis, coking coals of metallurgical grade may ordinarily not contain more than 1.60% of sulfur. Inasmuch as reserves of such low-sulfur coal are limited, coals containing 1.5% of sulfur or more, must be blended after washing with low-sulfur coals in order to increase the total supply available for producing metallurgical grade coke.

### Manganese Oxide as Acceptor for Hydrogen Sulfide

A desulfurization treatment for carbonaceous solid fuels is described by *E. Gorin, G.P. Curran, and J.D. Batchelor; U.S. Patent 2,824,047; February 18, 1958; assigned to Consolidation Coal Company.* In this process carbonaceous solid fuels are treated in a hydrogen atmosphere in the presence of a solid acceptor for hydrogen sulfide. The sulfur from the carbonaceous solid fuel combines with hydrogen gas to form hydrogen sulfide. The presence of small quantities of hydrogen sulfide in the vapor phase serves to inhibit further reaction between hydrogen and sulfur.  By providing a solid acceptor for the hydrogen sulfide, the inhibiting gas can be removed from the vapor almost immediately upon formation. Thus, the newly formed hydrogen sulfide gas is prevented from exerting an inhibiting influence.  Highly efficient desulfurization of the carbonaceous solid fuels results.

The solid acceptor for hydrogen sulfide thereafter is separated from the desulfurized carbonaceous solid fuels.  The solid acceptor, which contains sulfide, thereafter is regenerated by treatment with air at elevated temperatures.  Oxygen from the air combines with and removes the accepted sulfur of the solid acceptor and restores the hydrogen sulfide accepting property of the solid acceptor.  The regenerated solid acceptor thereafter can be reused for further desulfurization treatments.  Specifically, where solid acceptors containing manganese oxide are used, the reaction in the solid fuels desulfurization zone is as follows:

$$MnO + H_2S \longrightarrow MnS + H_2O$$

The sulfided acceptor thus contains manganese sulfide.  Regeneration of the sulfided acceptor in the presence of air is as follows:

$$MnS + \tfrac{3}{2}O_2 \longrightarrow MnO + SO_2$$

By restoring the active ingredient of the solid acceptor to the manganese oxide form, the solid acceptor is suitable for reuse in further solid fuel desulfurization treatment.

Five other patents by *J.D. Batchelor, G.P. Curran and E. Gorin; U.S. Patents 2,927,063; March 1, 1960; 2,950,229, 2,950,230, and 2,950,231; all August 23, 1960; and 3,101,303; August 20, 1963; all assigned to Consolidation Coal Company* describe the way in which acceptor desulfurization can be carried out.  The generalized flow sheet of Figure 6.1 illustrates the process.

## FIGURE 6.1:  FLOW SHEET

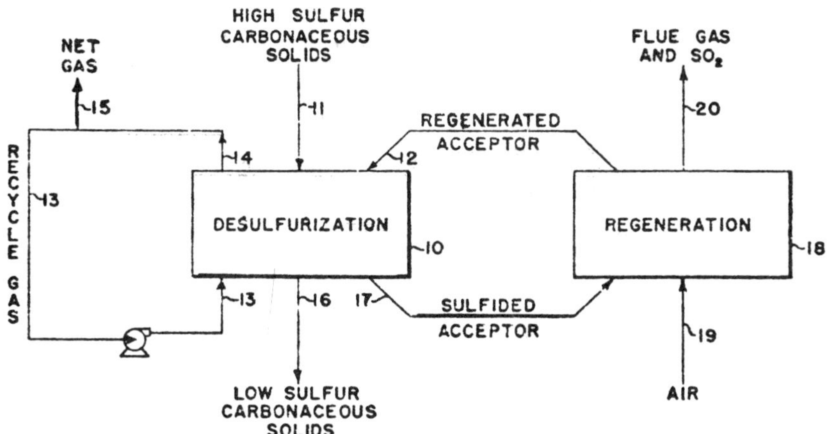


Source:  U.S. Patent 2,950,230

A desulfurization zone **10** receives noncaking carbonaceous solids containing sulfur through a conduit **11** and regenerated acceptor solids through a conduit **12**. In this instance, the active ingredient of the acceptor solids is manganese oxide. A hydrogen-rich treating gas consisting essentially of hydrogen is introduced into the desulfurization zone **10** through a conduit **13**. Additional gases, consisting of hydrogen gas, are autogenously produced through devolatilization of the carbonaceous solids at the elevated temperature of the desulfurization zone **10**. Under preferred operating conditions the autogenously produced devolatilization gases will be in sufficient quantity to provide the full hydrogen requirements for desulfurization so that extrinsic hydrogen production is not required.

The desulfurization zone is maintained at a temperature from 1100° to 1600°F. Below about 1100°F, the desulfurization rate is low. Operation above 1600°F requires excessive heat and also promotes rapid deactivation of the acceptor. The pressure level preferably is high enough to provide a hydrogen gas partial pressure of at least one atmosphere. A total pressure of from one to six atmospheres is preferred.

A typical char (containing sulfur) produced by fluidized carbonization of Pittsburgh Seam coal at 950°F, yields devolatilization gases containing 58.6% hydrogen and 24.8% methane at 1.3 atmospheres and 1350°F. The same char yields devolatilization gases containing 48.7% hydrogen and 32.9% methane at 3 atmospheres and 1350°F. During passage through the desulfurization zone, the treating gases remove sulfur from the carbonaceous solid fuels forming hydrogen sulfide as follows:

$$(-C=S) + H_2 \longrightarrow H_2S + (-C)$$

The $H_2S$, upon formation, is at once absorbed by the solid acceptor and removed from the gase phase.

$$H_2S + MnO \longrightarrow MnS + H_2O$$

Gases are recovered from the desulfurization zone through a conduit **14** and recirculated through conduit **13** for further contact with carbonaceous solids undergoing desulfurization. A net product gas is removed through a conduit **15**. The required residence of carbonaceous solids in the desulfurization zone depends upon the liability of the contaminating sulfur and also upon the level of desulfurization desired. It must be borne in mind that the ultimate sulfur level of the product is determined by the level of $H_2S$ concentration which the manganese oxide will maintain.

Where the hydrogen partial pressure of the treating gases is about one atmosphere or greater, satisfactory desulfurization can be achieved by subjecting the carbonaceous solids to the desulfurization conditions for a period of about 3 hours or less. Increased absolute pressure, as already pointed out, promotes more rapid desulfurization.

Desulfurized carbonaceous solids are removed from the desulfurization zone as product through a conduit **16**. Sulfided acceptor is removed through a conduit **17** and passed to an acceptor regeneration zone **18**. Air is introduced into the regeneration zone **18** through a conduit **19** to raise the temperature of the acceptor through combustion of sulfur along with a portion of the carbonaceous solids commingled therewith and to remove sulfur therefrom through oxidation to sulfur dioxide as follows:

$$MnS + \tfrac{3}{2}O_2 \longrightarrow MnO + SO_2 + heat$$

The temperature within the regeneration zone **18** is maintained at about 1300° to 1600°F. Hot flue gases containing sulfur dioxide are removed from the regeneration zone **18** through a conduit **20**. Excessive oxidation in the regeneration zone **18** should be avoided in order to restrict the quantity of higher oxides of manganese produced. Ideally, some of the acceptor solids recovered from the regeneration zone should be in the form of MnS. By maintaining from about 2 to 15% of the manganese as MnS after regeneration, the oxides of manganese can be maintained principally in the form of MnO rather than as higher oxides such as $Mn_3O_4$ or $Mn_2O_3$.

The presence of higher oxides of manganese in the desulfurization zone undesirably consumes hydrogen gas without accompanying sulfur removal as will be hereinafter described. In general, the amount of oxygen used in the regeneration zone should be within about 20% of the stoichiometric quantity, which would be required for oxidizing all of the MnS to MnO according to the equation above. Regenerated acceptor is returned to the desulfurization zone **10** through the conduit **12** without deliberate cooling to serve therein as a means for removing $H_2S$ therefrom and to supply the heat requirements thereof.

A variety of solid acceptors have been proposed in the above mentioned patents. Pure manganese oxide can be used. U.S. Patents 2,927,063; 2,950,229 and 2,950,230 use manganese oxide impregnated on silica, alumina or silica-alumina carriers as acceptors. Since these acceptors tend to become deactivated at the elevated temperatures used, these three patents describe processes to reactivate the solid acceptors.

In U.S. Patent 2,950,231, the solid hydrogen sulfide acceptor used is a naturally occurring manganese oxide ore containing less than 20% by weight of impurities such as silica, alumina, calcium and iron. Such manganese ores are readily available commercially. Manganese ores of this high degree of purity are mined in portions of Africa. Manganese ores from Morocco, for example, consist essentially of $MnO_2$ in the form of pyrolusite. Analysis of these ores shows the manganese oxide content greater than

The principal impurities are silica, alumina, calcium and iron. Unfortunately, the manganese ores which consist essentially of pyrolusite tend to decrepitate in handling and must receive some preliminary treatment before they will resist abrasion in the described desulfurization process. The natural ore is briquetted at a pressure of 5,000 to 10,000 psig. The briquets are calcined at a temperature of about 2000°F or higher for a period of one or more hours. The calcined briquets thereafter are crushed and screened to produce particles of the desired size for use in the described desulfurization process.

It has been found, however, that the manganese ores consisting essentially of ramsdellite do not require preliminary treatment. The ramsdellite ore can be crushed in its naturally occurring state to produce particles of the desired size which can be used directly as a solid acceptor in the described desulfurization process.

The object of the process described in U.S. Patent 3,101,303 is to minimize the production of higher oxides of manganese in the acceptor regeneration treatment. In this process, a portion of the MnS in the acceptor is allowed to remain in the acceptor throughout the regeneration treatment. By allowing about 2 to 15% of the manganese to remain in the form of MnS, the yield of MnO is unexpectedly selective to the substantial exclusion of the undesirable higher oxides of manganese. The residual MnS does not affect the subsequent $H_2S$ absorption properties of the acceptor since the MnS passes through the desulfurization zone unchanged.

In the preferred embodiment, the acceptor is used not only to remove $H_2S$ from the vapor phase in the desulfurization zone but also to provide the heat requirements for raising the temperature of the carbonaceous solids (undergoing desulfurization) to the desired desulfurization temperature. Thus, there exists an overwhelming stiochiometric excess of MnO in the desulfurization zone, dictated by the heat balance requirements.

**Hydrogenation of Coal Briquets**

The object of the process developed by *J.D. Batchelor, G.P. Curran, R.J. Friedrich and E. Gorin; U.S. Patent 3,117,918; January 14, 1964; assigned to Consolidation Coal Company* is the use of high sulfur caking coal as a raw material for preparing low-sulfur metallurgical fuel of substantially uniform size and shape.

In accordance with this process, heat-processed high-sulfur content coal-char agglomerates are desulfurized to yield metallurgical grade form coke prior to or concurrently with a calcining treatment. In one aspect, a combined desulfurization and calcining treatment of the coal-char agglomerates is conducted in the presence of at least one atmosphere of hydrogen and in an environment having a volume ratio of $H_2S$ to $H_2$ lower than 0.02.

In a second aspect of the process, the coal-char agglomerates are fed at atmospheric pressure in a descending column in countercurrent relation to autogenously produced devolatilization gas consisting principally of hydrogen. In both aspects of the process, the resulting desulfurized and calcined agglomerates are recovered as a metallurgical grade form coke, preferably after being cooled below their atmospheric kindling temperature. The agglomerates are preferably formed either as hot-tumbled coal-char agglomerates, preferably kiln-carbonized at a temperature above 800°F, or as briquetted coal-char agglomerates, preferably containing a pitch binder.

It is preferred to prepare the coal-char agglomerates by briquetting, that is by an operation in which a press is used to effect cohesion and compaction of the fine particles of coal and char. The addition of special binding materials such as pitch to the coal and char particles has been found to result in briquets of substantially improved strength when processed to form coke. The pitch serves particularly as a temporary binder to provide adequate strength to the uncarbonized briquets to permit their ready handling before carbonization.

The coal-char agglomerates contain a caking bituminous coal and a low density char, below about 45 pounds per cubic foot, obtained by a low temperature carbonization, i.e., at a temperature below 1400°F, of a caking bituminous coal.

Either or both the caking bituminous coal and low temperature carbonization
char may utilize low volatile, medium volatile, or high volatile coals, alone or
in suitable admixture. In a particularly preferred feature of this process, char
having a poured bulk density less than about 30 pounds per cubic foot (ASTM
test D292-29) is obtained via a fluidized low temperature carbonization of a high
volatile bituminous coal.

The coal-char briquets are shock heated whereby their surface is almost elevated
instantaneously to a temperature of 900° to 1250°F. The surface of the briquets
is maintained at a temperature of 900° to 1250°F until they have attained
throughout a temperature of 900° to 1250°F. Thereafter the coked briquets are
subjected to a combined desulfurizing and calcining treatment whereby they are
heated to a temperature above 1550°F.

Volatile materials that are evolved from the briquet are recovered as tar vapors
and gases. In the shock heating treatment, tar vapors and a hydrocarbon-rich
gas are the principal evolution products. This latter gas is generated in sufficient
quantity to supply autogenously the hydrogen required in the process, thereby
avoiding need of an extrinsic hydrogen supply.

A preferred embodiment of this process has the following steps. A caking bituminous coal and preferably a high volatile coal are provided as starting material.
A portion of the high volatile coal is processed via fluidized low temperature
carbonization, whereby the coal is converted into gas, tar and a porous finely
divided solid distillation residue termed char. Char production from high volatile coals by fluidized low temperature carbonization processes is swelled and
expanded from the original dimensions of the coal particles into fluffy, rounded
particles.

The sponge-like porous properties of the char particles result in a low bulk density of the material and a correspondingly low physical strength. The bulk density of the material is from 20 to 25 pounds per cubic foot. In general, the
char is sufficiently finely divided, as produced, to pass through a 14-mesh Tyler
standard screen.

The finely divided product char is blended with finely divided caking bituminous
coal to prepare substantially uniform sized briquets. In a preferred embodiment,
the coal employed in the briquet preparation may be the same coal which is
employed in preparing the char via the fluidized low temperature carbonization
process. Alternatively, the coal employed in the briquets may be from a different source provided that the coal possesses highly caking properties. If desired,
binder materials such as pitch may be employed in the briquet preparation stage
to introduce shape-retaining properties to the briquet prior to the subsequent
processing.

The briquets are transferred from the briquet preparation zone into a shock-
heating and holding zone for shock heating whereby their surface is virtually instantaneously elevated to a temperature in the range of 900° to 1250°F. The
surface of the briquets is retained at a temperature of 900° to 1250°F until the
briquets attain throughout a temperature of 900 to 1250°F. The residence time

Although some of the heat required in the shock heating and holding zone may be supplied indirectly, it is preferred to provide the bulk of the heat directly, either by means of hot gases or alternatively by recirculation of a finely divided heat carrying medium. The resulting coked briquets are introduced into a calcining and desulfurizing zone which is maintained under elevated temperatures sufficient to provide at least one atmosphere of hydrogen pressure.

The volume ratio of $H_2S/H_2$ is maintained below 0.02 to avoid suppression of the desulfurizing reaction. The volume ratio of $H_2S/H_2$ can be maintained at a low value by providing sufficient hydrogen gas with external $H_2S$ removal facilities as described in U.S. Patent 2,717,868. Alternatively, the volume ratio of $H_2S/H_2$ may be maintained at a low value by providing in situ solid acceptors for hydrogen sulfide, such as manganese oxide, as described in U.S. Patent 2,824,047.

In this latter embodiment, the solid acceptors for hydrogen sulfide may serve also as the heat carrying material. Concomitant devolatilization of the briquets occurs in the calcining and desulfurizing zone to produce additional hydrogen gas which may be employed by appropriate recycle techniques in the process. The desulfurization process proceeds at an optimum rate in the temperature range of 1300° to 1500°F. Prolonged exposure of the briquet at temperatures above 1500°F adversely affects the attainable desulfurization reaction rate.

Where desirable high strength properties in the solid product are sought, a final calcining temperature above 1500°F is required. Thus, optimum results are obtained by conducting the desulfurization reaction principally while the briquets are being raised in temperature within the combined desulfurizing and calcining zone.

The resulting calcined desulfurized briquets are gradually cooled directly by means of a coolant medium in a cooling zone. The cooling treatment preferably is conducted at atmospheric pressures and may be effected in part simultaneously with the transfer of calcined desulfurized briquets from the superatmospheric pressures existing in the calcining and desulfurizing zone by means of a pressure seal and apparatus.

*Example 1:* Briquets which were prepared as described from Arkwright coal, a typical Pittsburgh seam bituminous coal, contained 2.20% sulfur. The briquets had been prepared from the following formulation in cylindrical shapes, 2 inches diameter by 2⅜ inches high.

| Ingredient | Percent by Weight |
|---|---|
| Arkwright coal | 25.0 |
| Char produced by fluidized low temperature carbonization of Arkwright coal | 58.5 |
| Low temperature carbonization pitch | 5.3 |
| High carbon content pitch | 6.2 |
| Recycle coke particles | 5.0 |

The briquets were shock heated and held at 950°F in a fluidized bed of sand for 30 minutes. Thereafter, the briquets were heated at 6° to 8°F per minute to a temperature of 1600°F in an atmosphere of 30 psig hydrogen. The product briquet had a sulfur content of 1.24% after one hour exposure at 1600°F.

*Example 2:* Briquets having one inch diameter were prepared from Arkwright coal and char obtained by fluidized low temperature carbonization of Arkwright coal in the proportions shown in Example 1. The briquets contained 2.30% sulfur. Following shock heating at 1100°F in a fluidized bed of sand, the briquets were retained at 1100°F for 30 minutes and thereafter heated to 1600°F at 8° to 10°F per minute. The briquets were retained at 1600°F for one hour at a pressure of six atmospheres absolute of hydrogen gas. The product briquets had a sulfur content of 0.51%. The product briquets had a cold tumbler index, +½" (a measure of strength), of 83% which is considered good.

## Treatment with Fused Salt

The process developed by *P.X. Masciantonio; U.S. Patent 3,166,483; Jan. 19, 1965; assigned to United States Steel Corporation* provides a method of extracting both inorganic and organic sulfur from coal in which coal particles are agitated in a fused salt bath at an elevated temperature for a brief period, the desulfurized coal particles are separated from the fused salt and washed, and the fused salt reused.

Crushed coking coal and fused salt are fed to a heated treatment vessel in a ratio of at least 4 parts by weight of salt to each part by weight of coal. The coal should be crushed to minus ½" or finer. The fused salt is one which produces a strongly alkaline reaction and melts at a temperature sufficiently low to produce no significant adverse effect on the coal. Examples of suitable fused salts are NaOH, KOH, anhydrous sodium or potassium acetate, or mixtures of the foregoing.

Lime, sodium silicate or common salt melt at temperatures which are too high for them to be used alone, but they can be added as diluents. There may also advantageously be added sodium metal to the fused salt. A suitable temperature range for the fused salt treatment is 200° to 450°C or preferably 300° to 350°C, since higher temperatures tend to destroy coking properties of the coal. The term fused salt is used in accordance with the definition recognized in high-temperature chemistry, that is, a molten state of primarily ionic compounds.

The mixture of coal particles and fused salt is agitated in the treatment vessel for a period up to 45 minutes. The fused salt reacts with sulfur present in the coal as iron pyrite ($FeS_2$) or as organic sulfur to form an alkali metal sulfide ($Na_2S$ or $K_2S$) which dissolves in the fused salt. In this manner, the sulfur content of a typical coking coal can be lowered from 1.6% to about 0.6%. Such coal goes into a semifluid plastic condition in which the fused salt can reach the sulfur-bearing compounds, but retains suitable coking properties on cooling, provided the treatment is not prolonged beyond 45 minutes.

Next, the coal particles are allowed to settle and the fused salt decanted therefrom. The decanter is heated or insulated to maintain the salt in fused condition while the coal particles settle. Coal particles recovered in the decanter are washed with clean water to thereby wash away salt from the surfaces of the coal particles and also cool the material to ambient temperatures. It is important to wash the coal substantially free of alkali metal compounds, since they are objectionable when coke made from the coal is used in a blast furnace.

As the fused salt repeatedly reacts with sulfur, its sulfur content of course builds up. Ultimately sulfur in the fused salt reaches a concentration at which the salt is incapable of extracting any more sulfur from carbonaceous material because of equilibrium considerations. However, this equilibrium is reached only after the fused salt has a sulfur content of 5% by weight. For a typical coal, the fused salt can be reused up to about twenty times before its sulfur content builds up to this value. The following specific examples demonstrate how the process may be practiced.

*Example 1:*  100 grams each of KOH and NaOH are introduced into a one-liter iron pot and the mixture is heated to 300°C, at which temperature it was molten. The rapidly stirred fused salt has slowly added to it 50 grams of minus 40 mesh Robena coal which had a sulfur content of 1.64%. The resulting melt is stirred for 45 minutes at 350°C, and thereafter the fused salt is decanted from the coal particles which are washed and dried. The sulfur content of the coal after this treatment was about 0.60%, and the coal retained good coking properties. The yield of treated coal was 45 grams (90%). The fused salt was ready for reuse.

*Example 2:*  The procedure described in Example 1 was repeated, except that the salt was heated to 400°C and 3 grams of metallic sodium was added to the melt. A vigorous reaction immediately after adding the sodium was observed, whereupon the coal particles were separated. After washing and drying, the coal had a sulfur content of 0.60%, with a yield of 93%, and it retained its coking properties.

**Simultaneous Coking and Sulfur Removal**

*H. Loevenstein; U.S. Patent 3,272,721; September 13, 1966; assigned to Harvey Aluminum Incorporated* describes a coking process in which the coal from which the coke is produced is desulfurized as an integral part of the coking operation.

Briefly, the coking process is one in which the temperatures used are such that the heating of the coal is retarded during the early stages of the process. More particularly, the process comprises gradually heating the coal to be desulfurized and coked from a temperature of 500°C to a temperature of 800°C for two to five hours. It has been found that when this retarded heating procedure is used, a highly efficient desulfurization of the coal results.

It may be theorized that this efficient desulfurization of the coal occurs because the retarded heating of the coking process does not remove the volatiles contained in the coal too rapidly. Surprisingly, it would appear that retention of a substantial amount of volatiles in the coal during the desulfurization step substantially improves the efficiency of this step. It would appear that the volatiles in the coal react with the sulfur, forming hydrogen sulfide which is driven off along with a portion of the volatiles, thus eliminating or greatly reducing the sulfur content.

The coal ordinarily used for coking normally contains at least 10% volatiles. The coal used in the process should contain at least 3% volatiles. Thus conventional coking coal is well suited for the process. The rate of reaction between the volatiles and the sulfur increases as the temperature is increased, but the speed of removal of the volatiles is also increased with increases in temperature. Therefore, in conventional coking processes wherein the temperature of the coal is rapidly raised to temperatures on the order of 800°C and higher, the sulfur content of the coal is not reduced by an appreciable amount.

Furthermore, contrary to the widely accepted opinion of those skilled in the art, it is not necessary or desirable to rapidly heat the coal to the optimum desulfurization temperature and then attempt to keep it at that fixed temperature for the desired period of time. Rather, it is a significant contribution of the described process that highly effective desulfurization can be obtained by slowly heating the coal in order to maintain a volatiles content which is sufficiently high to effect removal of the desired or optimum amount of sulfur. This phenomenon permits desulfurization in existing retorts, which is quite difficult when operating for many hours at constant temperature. The desulfurization in the retorts is an advantage of this process.

Additional gases may be used to improve desulfurization. Hydrogen and hydrogen-containing gases such as ammonia, coke oven gas, etc., which react with the sulfur in the coal thereby forming hydrogen sulfide may be used. In addition, neutral gases such as nitrogen which retard the degasification by lowering the partial pressure of the volatiles, thereby causing a more efficient reaction between the volatiles and the sulfur, may also be used. However, once the volatiles have been reduced below a minimum level, even the addition of hydrogen has little effect on sulfur removal. The volatiles contained in the coke are believed to be comprised mainly of hydrocarbons with, perhaps, some other organic compositions.

It is another contribution of this process that desulfurization of the coal may be performed without grinding it. In the past, it has been considered highly desirable, if not essential, that the coal be ground in order to obtain effective desulfurization. The following specific examples are illustrative of the process described above but it is to be understood that the process is not to be limited to the details thereof. All proportions in these examples are set forth in parts by weight unless otherwise stated.

*Example 1:* Three 10 gram samples of a coal containing 0.63% sulfur and 17.83% volatiles were heated continuously to bring them to the temperatures indicated in the table. This heating was performed in the presence of hydrogen, the flow rate of the hydrogen being about 350 cc per minute.

| Heating Time | Temperature Range, °C | Sulfur Removed, % |
|---|---|---|
| 22 min | 500–800 | 15.1 |
| 1 hr, 13 min | 500–800 | 29.1 |
| 4 hr, 15 min | 500–750 | 41.2 |

It is apparent from the table that retarded heating, i.e., heating with less intensity, causes a substantial improvement in the efficiency of desulfurization.

*Example 2:* It has been found that when additional hydrogen is supplied during desulfurization, the hydrogen flow need not be initiated until the temperature is in the range of from 500° to 700°C. This is of substantial importance since it eliminates the danger of explosion which is always present when hydrogen is introduced into a furnace atmosphere which still contains oxygen. At temperatures above 500°C, the furnace atmosphere consists essentially only of hydrocarbon vapors and hydrogen sulfide. If desired, heating may be initiated in an atmosphere of nitrogen and the atmosphere then converted to hydrogen when a temperature of about 500° to 700°C is reached. In general, it is preferred to convert the atmosphere to hydrogen since this will result in a more efficient desulfurization.

In the present example, coal having a sulfur content of 1.71% and a volatiles content of 33.10% was slowly heated to 800°C. When no additional gas was introduced until the temperature reached 700°C, and then hydrogen was introduced at a flow rate of 300 cc per minute, the sulfur in the coal was reduced by 64.1%. When nitrogen at a flow rate of 300 cc per minute was used until the temperature reached 700°C and the atmosphere was then converted to hydrogen at a flow rate of 300 cc per minute, the total amount of sulfur removed was 76.3%. When only nitrogen at a flow rate of 300 cc per minute was used throughout the process, the total amount of sulfur removed was 59.4%.

## Calcination of Agglomerates

Many processes are known, and some are in commercial use, for making agglomerates from carbonaceous solids and binders of diverse kinds. Briquetting presses, rotary retorts, and other equipment have been employed in such processes to make agglomerates which, upon appropriate calcination, make coke suitable for use in metallurgical operations. The term "form coke" is sometimes used to describe agglomerates in a calcined state and is so used herein.

The strength of form coke must be adequate to sustain the burden of the ore reduction furnace. In the case of the blast furnace, the strength of the form coke should be very high to minimize formation of fine particles which decrease furnace burden permeability. Calcination of the agglomerates plays a critical role in producing form coke of the requisite strength. It also serves to regulate the volatile matter content of the form coke. Still further, calcination may be a factor in the desulfurization of the agglomerates to produce a low sulfur form coke.

The particular process selected for achieving one or more of these objectives is also a function of the composition of the green (i.e. uncalcined) agglomerates. Some agglomerates require shock heating to prevent grape clustering. Shock heating consists of sudden exposure of the agglomerates to a high temperature. Grape clustering is the adhesion of individual agglomerates to one another to form grape-like clusters. Other agglomerates require carefully controlled low heating rates, at least in certain temperature ranges to achieve the desired strength of the form coke.

The primary object of the process developed by *E. Gorin; U.S. Patent 3,671,401; June 20, 1972; assigned to Consolidation Coal Company* is to provide an improved, continuous, commercially feasible process for calcining carbonaceous agglomerates. A secondary object is to provide a process wherein desulfurization is also effected. According to the broadest aspects of this process, the green carbonaceous agglomerates are conducted through two different heat transfer zones, in the first of which gas is the principal source of heat and in the second of which solids are the principal source of heat. The first heat transfer zone is essentially a preheating zone. In this preheating zone, the heat is largely supplied by the hot effluent gas from the second heat transfer zone. The second heat transfer zone is the calcination zone proper. In this second zone, the preheated agglomerates from the first zone are calcined at a temperature above 1400°F. The heat is supplied in the second zone by a stream of hot heat carrier solids.

More specifically, the transfer of heat to the agglomerates in the second zone, that is, the calcination zone proper, is accomplished as follows: A downwardly moving bed of the preheated agglomerates from the preheating zone is established and maintained. Then a stream of the hot heat carrier solids is showered down over the agglomerates and through the interstices of the downwardly moving bed at a velocity which is greater than that of the bed itself. At the same time, a stream of nonoxidizing gas is circulated upwardly through the bed in countercurrent flow relationship to the bed and to the hot heat carrier solids.

The latter supply the heat required to calcine the agglomerates either by direct transfer to the agglomerates or by transfer to the gas and then to the agglomerates. The stream of upflowing gas aids in maintaining uniformity of heat transfer throughout the bed of agglomerates by reason of its dispersal of the downwardly showering heat carrier. In the process, its own temperature is raised to the point where it can serve as the preheating gas used in the first heat transfer zone.

The agglomerates are maintained in the calcination zone long enough to produce the desired calcination. Then the heat carrier solids are separated and recycled after appropriate regeneration which simply consists of reheating if the heat carrier is an inert solid such as sand. However, the heat carrier may be an $H_2S$ acceptor such as manganese oxide. By use of such an acceptor as a heat carrier, desulfurization may be concurrently effected in the case of agglomerates derived from a sulfur-containing coal.

In the calcination of such hydrocarbonaceous solids, some hydrogen is evolved which reacts with the sulfur to form $H_2S$. The acceptor reacts with the $H_2S$ to form the sulfide. The latter may be regenerated in any suitable manner, as well as reheated, before recycle to the calcination zone. Accordingly, as used herein, the term regeneration means reheating and restoring the heat carrier to its effective state for reuse in the calcination zone.

In accordance with the preferred embodiment, the process is applied to hydrocarbonaceous agglomerates which require a controlled heating rate from 1100° to 1450°F to produce form coke of the requisite strength. The hot effluent gas from the calcination zone is passed upwardly and countercurrently through a downwardly moving bed of the agglomerates in the preheating zone. The relative temperatures of solids and gas in the preheating zone, as well as their respective velocities, are regulated to maintain a heating rate which does not exceed 10°F per minute over the range of 1100° to 1450°F.

**Treatment with Iron and/or Iron Oxide plus Metal Chloride**

The process developed by *R.H. Long and M.C. Sze; U.S. Patent 3,873,427; March 25, 1975; assigned to The Lummus Company* relates to the desulfurization of coke, and more particularly to an improved process for producing a coke having a sulfur content of less than 0.85%. When the coke is derived from coal, desulfurization is effected with magnesium, calcium, ferrous or ferric chloride plus either iron and/or ferrous or ferric oxide. The coke, in admixture with the desulfurizing additive, is heated to a temperature of at least 2,100°F for a time sufficient to produce a desulfurized coke having a sulfur content of no greater than 0.85%, and preferably no greater than 0.5%.

The desulfurizing additive may be admixed with the coal prior to coking, in which case the coking operation is completed by desulfurization at temperatures of at least 2100°F. Alternatively, the desulfurizing additive may be added to the coke, and the coke subsequently desulfurized at temperatures of at least 2100°F. It is therefore to be understood that desulfurization may be effected as part of the coking operation or subsequent to the coking operation, with the desulfurizing additive being combined with the coking feedstock prior to coking or combined with the coke from a coking operation.

The ferruginous material (either iron and/or ferrous oxide and/or ferric oxide) may be employed in an impure state. Thus, for example, it may be added as taconite fines, iron ore, iron etc. In general, it is employed in a finely divided state (e.g., a particle size of −100 mesh with 50% or more being −325 mesh) in an amount to provide a ferruginous material to coke weight ratio of from 1:4 to 1:30, and preferably from 1:10 to 1:16. In the case where the coke is produced from a coal containing pyrites in an amount of 0.5 weight percent or greater, it is possible to effect the required desulfurization without the use of a ferruginous material, i.e., the metal chloride salt is used as the only desulfurization additive.

Additional beneficial results are achieved, however, by using the ferruginous additive in combination with the metal chloride salt, e.g., use of lower amounts of metal chloride salt, lower temperature, shorter periods for achieving the desired desulfurization and/or increased desulfurization. In some cases, even with the presence of pyrites, the ferruginous material may be required to achieve the desired degree of desulfurization.

The chloride salt (ferrous chloride and/or ferric chloride and/or calcium chloride and/or magnesium chloride) is generally employed in an amount to provide a chloride salt to ferruginous material weight ratio of from 4:1 to 1:100, and preferably a chloride salt to ferruginous material weight ratio of from 0.5:1 to 4:1 for ferrous and ferric chloride, and from 1:100 to 3:100 for the other chlorides. In the case where the chloride salt is used as the only desulfurizing additive (coke produced from coal containing pyrites), the chloride salt is employed in amounts to provide a chloride salt to coke weight ratio ranging from 0.0003:1 to 1:1.

The desulfurization is effected in a reducing atmosphere at a coke temperature of at least 2100°F, with the temperature of the coke generally not exceeding 2600°F. The desulfurization is preferably effected at a temperature from 2100° to 2300°F. (The furnace or oven in which the desulfurization is effected is generally at a temperature which is 100°F higher than the temperature of the coke). The coke is maintained at the desulfurization temperature for a time sufficient to reduce the sulfur content of the coke to no greater than 0.85%, and preferably no greater than 0.5%. The precise time required for such desulfurization will vary with the amount of sulfur originally present in the coke and the distribution, as to type, of the sulfur, i.e., pyritic, sulfate, sulfide, organic. In general, the time is at least one-half hour, with the time period generally not exceeding 16 hours; typically the process takes from 1 to 8 hours.

Figure 6.2 illustrates a processing sequence for producing desulfurized coke from coal. Coal is ground to 100% –10 mesh, and if required, is intimately mixed with ground ferruginous or iron ore concentrate. The ground coal is then blended with an aqueous solution of metal chloride by either spraying or pugging and the mixture is dried. Alternatively, the metal chloride may be added by dry blending.

## FIGURE 6.2: DESULFURIZED COKE FROM COAL

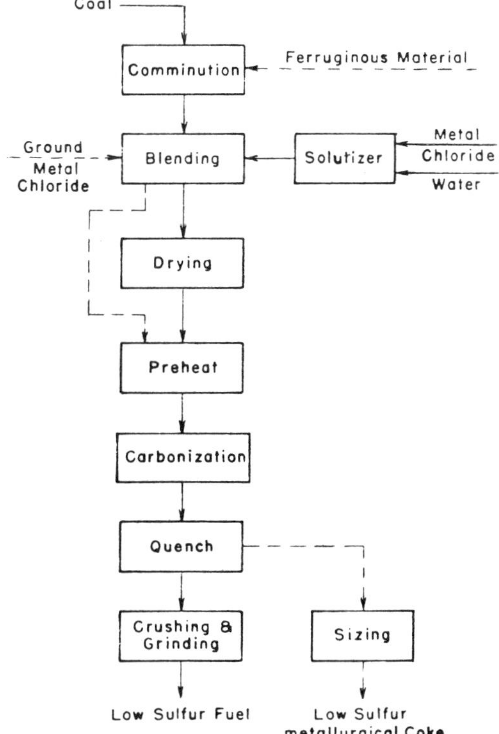


Source: U.S. Patent 3,873,427

The mixture of coal and metal chloride, which may further include added ferruginous material, is preheated and partially oxidized followed by carbonization, including desulfurization at a temperature of at least 2100°F. The desulfurized coke is quenched and may be crushed and ground for use as a low sulfur fuel or sized for use as a low sulfur metallurgical coke.

The process will be further illustrated by the following example.  Unless otherwise indicated all parts and percentages are by weight.

*Example:*  An Illinois No. 6 coal (4.17% sulfur) is coked and desulfurized in accordance with the process as tabulated in the table.

| Run No. | Time at 2200°F (hr) | Additive | Weight Percent* | Sulfur in Product (wt %) |
|---|---|---|---|---|
| 1 | 16 | none | — | 1.08 |
| 2 | 16 | CaCl$_2$ | 2 | 0.27 |
| 3 | 16 | MgCl$_2$·6H$_2$O | 2 | 0.65 |
| 4 | 16 | FeCl$_3$ | 10 | 0.57 |
| 5 | 4 | FeCl$_2$·4H$_2$O | 6 | 0.7 |
| 6 | 16 | Taconite | 5 | 0.22 |
|   |   | FeCl$_3$ | 5 |   |
| 7 | 16 | Taconite | 5 | 0.47 |
|   |   | FeCl$_2$·4H$_2$O | 10 |   |
| 8 | 4 | CaCl$_2$ | 2 | 0.72 |
| 9 | 4 | Taconite | 3 | 0.55 |
|   |   | CaCl$_2$ | 2 |   |

*Based on coal.

## Treatment with Phosgene or CO plus Cl$_2$

In accordance with the process provided by *R.H. Long and M.C. Sze; U.S. Patent 3,878,051; April 15, 1975* the coke is contacted with a desulfurizing gas, with or without the presence of other additives, at a temperature from 1200°F to 1800°F for a time sufficient to produce a coke having a sulfur content of no greater than 0.85%, and preferably no greater than 0.5%.

The desulfurizing gas is either a mixture of carbon monoxide and chlorine or phosgene, and generally also includes a diluent gas, such as nitrogen, in order to minimize the concentration of phosgene (phosgene is also generated when using a mixture of carbon monoxide and chlorine).  A diluent gas need not be present although the use of a diluent is preferred.  Although the desulfurization is effected under reducing conditions, oxygen can also be present in the gas.

In employing a mixture of carbon monoxide and chlorine, the relative proportions of the two materials can vary over a wide range, in that it is believed that the desulfurization is effected by in situ generation of phosgene.  As should be apparent, it is most economic to employ the carbon monoxide and chlorine in amounts approximating equimolar amounts.  Typically, the ratio of one of the two components to the other ranges from 0.5:1 to 1.5:1.

In the case where the coke is derived from coal containing pyrites, in an amount of at least 0.5% by weight of the coal, improved results are obtained by the use of a desulfurizing amount of a ferruginous material.  If the coal used in preparing the coke does not contain pyrites, then a ferruginous material is also used in effecting the desulfurization.

The ferruginous material is either iron and/or an oxide of iron (ferrous oxide and/or ferric oxide) and may be added as taconite fines, iron ore, iron, etc.  In general, the ferruginous material, if employed, is in a finely divided state, e.g., a particle size of −100 mesh with 50% or more being −325 mesh and is present in an amount to provide a ferruginous material to coke weight ratio of from 5:95 to 25:75, and preferably from 10:90 to 15:85.

In accordance with a particularly preferred embodiment, in addition to the ferruginous material, the coke is admixed with a desulfurizing amount of promoting additive which is either sulfuric acid, ferric sulfate or mixtures thereof in that the presence of such an additive has been found to promote desulfurization of the coke with the desulfurizing gas.

The sulfuric acid and/or ferric sulfate additive is generally employed in an amount from 1 to 8%, and preferably in an amount from 3 to 8%, all by weight, based on coke. In some cases, i.e., cokes of higher sulfur content produced from petroleum feedstocks, in order to achieve the desired amount of desulfurization, it may be necessary to use sulfuric acid and/or ferric sulfate as a promoting additive.

The desulfurization is effected in a reducing atmosphere at a coke temperature from 1200° to 1800°F and preferably at a temperature from 1400° to 1800°F with a temperature of 1500°F being generally preferred. (The furnace or oven in which the desulfurization is effected is generally at a temperature which is 100°F higher than the temperature of the coke). The desulfurization with phosgene or a mixture of carbon monoxide and chlorine can not be effectively employed at coke temperatures in the order of 1900°F and higher and, accordingly, in general, the temperature of the coke does not exceed 1800°F.

The coke is maintained at the desulfurization temperature for a time sufficient to reduce the sulfur content of the coke to no greater than 0.85%, and preferably no greater than 0.5%. The precise time required for such desulfurization will vary with the amount of sulfur originally present in the coke and the distribution, as to type, of the sulfur, i.e., pyritic, sulfate, sulfide, or organic. In general, the time is at least one-half hour, with the time period generally not exceeding 16 hours. Typically, the required desulfurization time is in the order of 1 to 8 hours.

*Example:* Coal (100%, –10 mesh) is contacted with the gaseous reactant in a fixed bed in a tubular reactor as reported in the table below. The coal is Illinois No. 6 having 4.17% sulfur and the space velocities are in the order of 1 to 3. Run 4 is not effected in accordance with the described process.

| Run No. | Coal Mix, wt %. | | Reactant | | .. Run Conditions .. | | |
|---|---|---|---|---|---|---|---|
| | Coal | Additive | Gas | Mol % | Temperature (°F) | Time (hr) | Sulfur in Product (%) |
| 1 | 100 | — | CO | 12.5 | 1500 | 4 | 0.4 |
| | | | $Cl_2$ | 12.5 | | | |
| | | | $N_2$ | 74 | | | |
| | | | $O_2$ | 1 | | | |
| 2 | 100 | — | CO | 12.5 | 1500 | 4 | 0.5 |
| | | | $Cl_2$ | 12.5 | | | |
| | | | $N_2$ | 75 | | | |
| 3 | 90 | 10-$H_2SO_4$ | CO | 12.5 | 1500 | 4 | 0.26 |
| | | | $Cl_2$ | 12.5 | | | |
| | | | $N_2$ | 74 | | | |
| | | | $O_2$ | 1 | | | |
| 4 | 100 | — | — | — | 1500 | 4 | 1.25 |

### Treatment with Sulfur Vapor

In the process developed by *D. MacGregor; U.S. Patent 4,011,303; March 8, 1977; assigned to William H. Sayler* sulfur-bearing coke is desulfurized by contacting it with a dilute mixture of elemental sulfur vapor and a gas carrier which is substantially inert with respect to the coke being treated, the coke being first heated to a temperature at which it will react with the dilute elemental sulfur vapor to form carbon disulfide.

The reaction proceeds utilizing sulfur components of the coke along with the introduced sulfur. Inorganic compounds of sulfur are not removed by this procedure, but, since these do not give rise to air-polluting gases when the coke is burned as a fuel, it is of no consequence. Air polluting forms of sulfur are removed, and this results in a relatively clean-burning fuel product.

The reaction takes place at approximately 1150°F and on up to the dissociation temperature of carbon disulfide. Heating of the coke and reacting it with dilute elemental sulfur vapor may be accomplished in a variety of different types of apparatus. Reaction times will vary, depending upon the type of apparatus employed and the concentration of the sulfur vapor. Sulfur concentrations of between 1 and 10% by volume in an inert gas are preferred. Satisfactory results are usually obtained in one-half to one hour with sulfur vapor concentrations of 2 to 3%. The carbon disulfide formed is removed as a vapor and is recovered as a useful by-product by means well known in the art.

Figure 6.3 shows schematically a satisfactory form of furnace and reaction chamber for carrying out the process. Sulfur-bearing coke is fed into reaction chamber **10** of a reaction vessel **11** through an entryway **12** suitably arranged to prevent the escape of gases. During initial start-up of the apparatus, the coke entering the apparatus will fall through the reaction chamber into a lower cooling chamber **13** and will lodge against and build up upon closed exit port **14**. Coke will be fed into the apparatus until both the cooling chamber and the reaction chamber are filled.

Heating means, such as a tubular furnace **15**, surrounds the reaction chamber, the operation being such as will raise the temperature of the coke to a point within the range of preferably 1800° to 2200°F and keep it at that temperature for the time required to effect the desired reaction.

Provision is made for introducing dilute elemental sulfur vapor into the reaction chamber so it will permeate the column of coke therein. As illustrated, an inert carrier gas is introduced into the apparatus at or near the bottom of the cooling chamber through a port **17**. Nitrogen has been found to be particularly satisfactory as the carrier gas.

Such carrier gas flows upwardly through the coke in the cooling chamber, simultaneously cooling the coke and heating the gas. The so-heated gas mixes with elemental sulfur vapor that is injected into the apparatus through a manifold **16** at the bottom of or just below the reaction chamber, and both carrier gas and sulfur vapor move upwardly through the heated coke. Instead of sulfur vapor, sulfur solids, preferably in powder form, may be injected through manifold **16** into the lower part of the reaction chamber.

## FIGURE 6.3:  TREATMENT OF COKE WITH SULFUR VAPOR

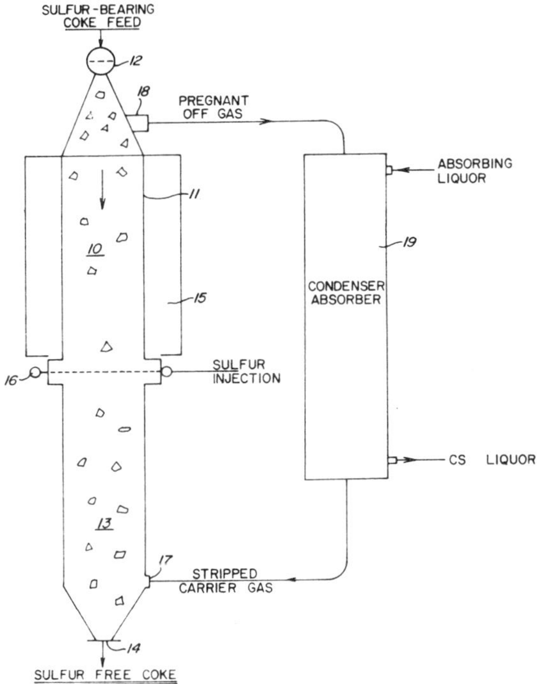


Source:  U.S. Patent 4,011,303

The sulfur solids are vaporized therein by the heat of the apparatus, the coke
material and the carrier gas, and the vapor is carried upwardly through the heated
coke in the reaction chamber by the carrier gas.  In the reaction chamber, the
elemental sulfur reacts with the heated coke to produce carbon disulfide.  The
general formula for the reaction is:  $C + 2S \rightarrow CS_2$.  Carbon for the reaction
comes from the coke, while the sulfur comes from both the elemental sulfur
introduced into the chamber and sulfur contained by combustible portions of
the coke.

Although the sulfur vapor appears to preferentially react with the sulfur-carbon
complexes to form carbon disulfide, if there is an excess of sulfur vapor present,
the sulfur will react with the unbonded carbon in the coke to produce additional carbon disulfide.  Since it is not the purpose of the process to produce
carbon disulfide, any unnecessary consumption of carbon is wasteful and highly
undesirable.  To insure that excessive sulfur vapor is not present, the sulfur vapor
is diluted to a substantial extent by the carrier gas.  It has been found that the
best reaction control is obtained when the sulfur vapor is kept between 1 and
10% by volume of the sulfur vapor-carrier gas mixture, with the preferred concentration of sulfur vapor being between 2 and 3%.  When larger concentrations

of sulfur vapor are used, some of the sulfur reacts with unbonded carbon, as explained above, to form carbon disulfide. Also, with larger concentrations of sulfur, the carbon disulfide reaction proceeds rapidly so that it is difficult to stop the reaction at the point at which the constituent sulfur in the coke is substantially all consumed. If the reaction continues beyond the point at which the sulfur in the coke is consumed, carbon disulfide will continue to be produced on the basis of the supplied sulfur and the unbonded carbon in the coke. This, obviously, creates the same problem as an excess of sulfur vapor and unnecessarily consumes the carbon in the coke.

It is also important for the reaction to be carried out in an inert (except for the sulfur vapor) atmosphere. If other active elements are present, the carbon will react and further reduce the amount of coke output. If oxygen is present, for example, products such as COS, CO, $SO_2$, and $CO_2$ will be formed. The carbon disulfide gas produced by the reaction taking place in the reaction chamber rises in the chamber with the carrier gas and is drawn off at or near the top, for example through an outlet port **18**. Liquid carbon disulfide is recovered by passing the $CS_2$ gas and the carrier gas through a condenser-absorber **19**.

Experiments demonstrating the process were conducted in an apparatus comprising a horizontal quartz tube and an electric resistance furnace surrounding the center portion of the tube. Coke samples were placed in the portion of the tube surrounded by the furnace, porous silica plugs being placed on either side of the sample. Powdered sulfur was placed in a portion of the tube outside the furnace. Nitrogen gas was passed through the tube. The coke was heated to a reaction temperature of 2,000°F. Once the coke was at reaction temperature, the portion of the tube containing the sulfur was heated with a bunsen burner to vaporize the sulfur. The sulfur vapor was carried by the nitrogen carrier gas through the heated coke.

The process was carried out on a number of coke samples for different periods of time. Each timed run was made with a fresh sample of coke. Some of the results are shown in the following table.

### Coal Coke from Geneva Steel Export Batch with Combustible Sulfur Content of 2.37%, 10 Gram Samples

| REACTION TIME | % COMBUSTIBLE SULFUR REMAINING IN REMAINING COKE | % COMBUSTIBLE SULFUR REMOVED FROM COKE |
|---|---|---|
| 5 Minutes | 2.35 | 0.84 |
| 10 Minutes | 2.24 | 5.48 |
| 15 Minutes | 1.38 | 41.77 |
| 20 Minutes | 0.62 | 73.84 |
| 25 Minutes | 0.38 | 83.97 |
| 30 Minutes | 0.24 | 89.88 |
| 60 Minutes | 0.03 | 98.74 |

As can be seen from the table, almost 90% of the combustible sulfur content of the coke is removed within 30 minutes with over 98% being removed within one hour.

## ANODE COKE

In the production of aluminum metal by the electrolysis of aluminum oxide, there is employed a carbon anode conventionally made from petroleum or coke-oven pitch coke. The carbon anode is consumed during the electrolysis, causing many of the impurities present in the carbon, such as silicon, vanadium and iron, to pass into the aluminum.

Generally, a purchase specification of 1% total ash is imposed on coke used in the manufacture of electrodes for the aluminum industry. An auxiliary specification of less than 2% sulfur in the calcined coke is also imposed. This severely limits the raw materials that can be used in the manufacture of electrodes, and suitable materials heretofore have been confined to coke produced from certain grades of petroleum residuum, coke produced by carbonization of coal tar or coke-oven pitch, and coke produced by the carbonization of gilsonite.

There are reported instances where cokes of marginal purity have been produced from certain seams of coal by froth flotation and chemical cleaning, but there are no major seams of coal in the United States that can be cleaned adequately by froth flotation to produce coke that will meet the specification as to ash content of anodes for use in the aluminum industry. Chemical cleaning of coals, for example, with mineral acids, is an expensive procedure which limits its application to emergencies; and few seams of coal are available which are amenable to this process.

Because of its relatively high purity and availability, petroleum coke has generally been the material of choice for the preparation of anodes of the type heretofore discussed, but not all petroleum crudes yield a residuum of satisfactory quality for these purposes. In fact, the availability of crudes which yield a satisfactorily low sulfur and vanadium content for the production of carbon of desired purity is so limited that sources of pure carbon for future expansion of domestic aluminum production are uncertain.

### Solution of Coal in Creosote-Type Coal Tar Fraction

The process described by *V.L. Bullough and W.C. Schroeder; U.S. Patent 3,240,566; March 15, 1966; assigned to Reynolds Metals Company* relates to the production of high purity carbon of the type employed in carbon and graphite electrodes, and it deals more particularly with a process for the removal of impurities from bituminous coal to produce a substantially ash-free carbon which can be used in the manufacture of carbon electrodes suitable for the electro-metallurgical industries, especially in the aluminum industry.

In accordance with the described process, it has been found that the soluble constituents of coal, particularly bituminous coal, can be dissolved in an aromatic hydrocarbon oil of relatively low viscosity and boiling range, under superatmospheric pressure, and at a temperature maintained at above the temperature of maximum solubility of the coal constituents.

The aromatic hydrocarbon oil which is advantageously employed for the solution treatment of coals, in accordance with the process, is preferably a lower boiling fraction obtained by the fractional distillation of coal tar or coke oven tar. Preferably the fraction is a creosote type fraction from which light oils have been substantially removed, i.e. benzene, toluene, and mixed xylenes, so that the creosote type oil has an initial distillation temperature of about 180°C at atmospheric pressure.  The creosote type extraction oil which has been found suitable for these purposes may, however, depending upon the selection of operating conditions, include high boiling anthracene oil fractions, and in fact, any make-up to the process necessitated by loss or removal from the system of the solvent-valuable lower boiling creosote type oil fractions can be supplied from relatively inexpensive heavy residue creosote oil.

Thus, in accordance with a first aspect of the process, bituminous coal can be extracted under critical and carefully controlled conditions, up to a solubility of more than 80%, in an aromatic liquid hydrocarbon creosote oil-type solvent having a relatively low viscosity and a low initial boiling point of about 180°C. Moreover, this aromatic liquid hydrocarbon extractant can be generated in the sustained operation of the process, so that bituminous coal becomes the only raw material needed in the process.

In the practice of the extraction aspect of the process, there is selected a suitable bituminous coal, or other carbonaceous material of similar composition. Preferably freshly mined coal is used, and the raw coal is crushed and pulverized to optimum particle size.  Examples of suitable bituminous coals include Alabama and Kentucky high volatile types.  Coals of this type may analyze, for example, on the as-received basis, from 35 to 45% volatile matter, 72 to 84% carbon, and 2 to 10% ash, depending upon quality.

The coal is advantageously crushed and pulverized, then preheated, if desired, to reduce its moisture content.  The crushing may be carried out in conventional equipment, for example, an air-swept ball mill with air classifier, and the moisture reduced by a flow of heated flue gas through the ball mill to a level of 1%.

The crushed and dried coal is then fed to a suitable mixing tank and mixed with a suitable amount of a creosote type extraction oil of the character described, or with an extraction oil fraction generated from within the process system.  The coal extraction oil mixture is then transferred to a pressure digester.  Alternatively, the coal and extraction oil can be fed directly into the pressure vessel, which serves both as mixer and digester.  The pressure vessel comprises an autoclave equipped with heating elements, a cooling coil, a stirrer, and pressure and temperature indicating devices.

The finely divided coal is heated in the digester with the solvent at a temperature slightly in excess of the temperature required to achieve a maximum solubility of the coal constituents in the extraction oil, preferably at a temperature from about 5° to 50°C in excess of the maximum solubility temperature.  Where the coal and oil are premixed, the mixing may be carried out at a temperature of about 50°C if the mixing tank is open, up to about 150°C when the tank is closed.

The mixture of coal and extracting oil in the digester is brought to a temperature suitable for maximum solubility, as previously indicated.  This temperature will generally lie in the range from 380°C to about 450°C, depending upon the type of coal.  The pressure in the extraction vessel at these temperatures will rise to 750 to 2,000 psig.  The mixture is stirred at a velocity sufficient to maintain the solids in suspension. The mixture is held at the extraction temperature for a period of time from 2 to 60 minutes but a retention time of 10 to 30 minutes is preferred. The temperature of the mixture is then reduced by cooling to between about 160° and 220°C. Any gaseous or liquid hydrocarbon material present having a boiling temperature below 160° to 220°C is distilled off, and collected for use as fuel or as a valuable light oil by-product of the process. The charge is then transferred to a centrifuging system, described below, for separation of insolubles. The weight ratio of extracting oil to coal may range from 1:1 to 6:1, but is preferably maintained between 2.5:1 to 4:1.

The temperature to which the coal and extracting oil mixture is heated has been found to have a marked influence in the purity of the carbon or coke ultimately obtained. This temperature must be above that of maximum solubility of the coal constituents to produce a final carbon or coke which is substantially ash-free. It is thought that finely disseminated particles of ash in the hot extraction mixture at these elevated temperatures may act as nucleating sites for carbonization, and thus some material which would otherwise contribute to ash content in the final carbon product is trapped within a partially carbonized matrix and removed by subsequent processing.

## Solvent Refining Process

Mineral matter, fusain, volatile matter and moisture primarily constitute the remaining components present in coal. Mineral matter deposited in the sedimentary deposits by infiltration of ground waters during coalification and remaining after coal has been burned is called ash. Fusain, which is substantially consumed during the combustion of coal, may be considered a mineral charcoal. The rank of coal (i.e. the degree of coalification) is determined by its carbon content which increases with the natural series of lignite, subbituminous coal and bituminous coal. In this series, the fixed carbon content generally increases whereas the moisture and oxygen content decreases.

The presence of fusain, mineral matter (analyzed as ash) and combined sulfur is the primary reason for the lack of utilization of coal for preparing coke suitable for making anodes to be used in the production of aluminum. Sulfur is usually present as organic sulfur, pyritic sulfur and/or inorganic sulfate. It is known that the ratio of inorganic sulfate and pyritic sulfur to organic sulfur increases with coals of increasing total sulfur content.

The object of *W.J. Bloomer and S.W. Martin; U.S. Patent 3,379,638; April 23, 1968; assigned to The Lummus Company and Great Lakes Carbon Corporation; W.J. Bloomer; U.S. Patent 3,375,188; March 26, 1968; assigned to The Lummus Company; and W.J. Bloomer and S.W. Martin; U.S. Patent 3,109,803; Nov. 5, 1963; assigned to The Lummus Company* is the provision of substantially ash-free coke from a coal selected from the group consisting of bituminous coal, subbituminous coal and lignite.

In accordance with this process, it is proposed to digest and/or dissolve the extractable carbonaceous matter (which excludes fusain) present in the pulverized raw coal under conditions of elevated temperatures and pressures utilizing a high boiling liquid solvent of high aromaticity thereby forming a solution of such extractable carbonaceous matter (hereinafter referred to as a coal solution). Fusain and the mineral matter or ash are substantially unaffected by the solvent and are suspended in the coal solution. The coal solution while in a free-flowing state is filtered to separate the suspended solids (including undissolved extractable carbonaceous matter) and is thereafter heated to a temperature above incipient coking and passed to a coking unit. Product coke is withdrawn from the coking unit and passed to subsequent processing units including coke handling and calcining facilities.

The vaporous products are withdrawn from the coking unit and are introduced into a fractionating unit wherein they are separated into various fractions, such as, for example: a light gas and gasoline fraction; a medium and heavy middle oil distillate fraction; and a heavy bottoms distillate fraction. A portion or all of the high boiling liquid solvent of high aromaticity utilized for dissolving or digesting the extractable carbonaceous matter present in the raw coal is derived from a distillate fraction recovered from the fractionating unit.

Referring to Figure 6.4, ground or pulverized coal selected from the group consisting of bituminous coal, subbituminous coal and lignite, and/or mixtures thereof are collected in a hopper **1** from which it is continuously distributed at a desired rate by a conveying mechanism **2** into a solutizer tank **3** maintained at a pressure of from 1 to 7–8 atmospheres.

**FIGURE 6.4: SOLVENT REFINING PROCESS FOR PRODUCING COKE**

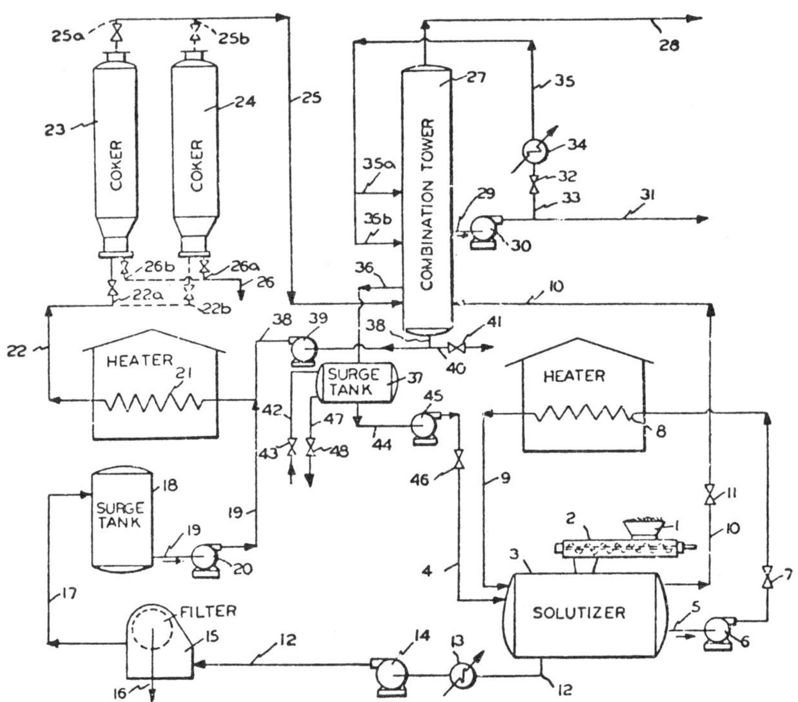


Source: U.S. Patent 3,379,638

As illustrated, conveying mechanism **2** is a screw type feeder which introduces the coal into solutizer **3** without loss of pressure therein.  Any conventional means of mechanical transfer may suffice, providing the means allows for positive transfer of the coal into the solutizer without a substantial loss of pressure therein.  The high boiling liquid solvent of high aromaticity is introduced into the solutizer through line **4** at a rate so as to provide a ratio of solvent to coal of from ½:1 to 6:1.  Normally, a solvent to coal ratio of from 2:1 to 3:1 is preferred, since effective extraction rates are obtained within this ratio range while minimizing filtration costs.

The solutizer is maintained at a temperature of from 600 to 850°F, preferably of from 750° to 825°F and above the final decomposition temperature of the initial coal whereby a substantial portion of the extractable carbonaceous matter in the raw coal is thermally depolymerized.  The products of depolymerization are soluble in the highly aromatic solvent and thereby form, with the solvent, the coal solution.  Undissolved and insoluble solids including undissolved extractable carbonaceous matter, and insoluble mineral matter or ash and mineral charcoal or fusain are suspended in the coal solution.  An agitator (not shown) may be provided to agitate the coal-solvent mixture during solvation.

Solvation temperatures are maintained in solutizer **3** by withdrawing and circulating a portion of the coal solution and coal-solvent mixture through an external heating system.  The withdrawn portion is passed through line **5** by pump **6** under the control of valve **7** to heater **8** and thereafter reintroduced through line **9** into the solutizer.  In this manner, solvation temperatures are maintained within the solutizer without the necessity of a high temperature and pressure heating system, which would be the case, if the solvent was preheated to a temperature sufficient to maintain solvation temperatures within the solutizer.

A through-put time of the raw coal of from about 5 to 120 minutes is normally sufficient to dissolve or digest effectively and efficiently the extractable carbonaceous matter.  Since the solvent may have an initial boiling temperature (converted to one atmosphere) as low as 650°F, whereas solvation temperatures may be as high as 850°F, the solutizer is provided with vent line **10** under the control of valve **11** to permit the withdrawal of the lower boiling components of the solvent and any volatile matter vaporized from the raw coal.

In this respect, it has been observed that the quantity of volatile matter is practically negligible.  A substantially uniform coal solution, wherein undissolved and insoluble solids are suspended, which include mineral charcoal or fusain and mineral matter or ash, is withdrawn through the bottom draw-off **12** and is drawn through cooler **13** by pump **14** and passed to a continuous rotary filter **15**.

The coal solution is cooled to a temperature of from 400 to 700°F during its passage through the cooler.  Preferably, the rotary filter is precoated with conventional filter aids and is normally operated at a pressure of 40 to 60 psig to effect efficient removal of substantially all of the suspended solids including undissolved carbonaceous matter.  The filter cake is washed and dried to recover absorbed solvent and is withdrawn from the filter through line **16**.  The substantially de-ashed coal solution is passed through line **17** to surge drum **18** from which it is passed through line **19** and pump **20** to heater **21** (a suitable coil heater).  The coal solution is heated to a temperature of from 850° to 1050°F in heater **21** and is passed therefrom through line **22** to a coking unit.  The coking unit, as illustrated, is a delayed coker and is comprised of coke drums **23** and **24**.  The heated coal solution in line **22** is introduced through line **22a** into coker **23** wherein the charge is decomposed into coke and a vaporous effluent.  The coker overhead in line **25a** is passed through line **25** to a fractionating unit.  While coker **23** is being filled with coke, coker **24** is being decoked with product coke withdrawn through lines **26a** and **26** for subsequent processing.

In normal operation, cooling and decoking of coker **24** is completed prior to the filling of the coker **23**.  With decoking completed on coker **24** and having filled coker **23** to a predetermined level, the coker charge is diverted to coker **24** through line **22b**, with the vaporous effluent in line **25b** being passed to the fractionating unit through line **25**.  After cooling, coker **23** is decoked, with product coke being passed through lines **26b** and **26** for subsequent processing.

The coker overhead in line **25** is introduced into a fractionating or combination tower **27** which is provided with suitable fractionating decks (not shown).  Introduction of the effluent into the tower may result in some foaming.  This may be effectively inhibited by the addition of a small amount of an antifoam agent, at the point of introduction or at some elevated point in the tower.  The hereinbefore mentioned distillate and volatile matter in line **10**, which are evolved during solvation, are introduced into the lower portion of the tower.

The combination tower overhead products in line **28** comprising condensible and noncondensible components are passed to conventional processing units to separate the condensible components from the noncondensible components.  A medium and heavy middle oil is withdrawn from an intermediate point on the tower through line **29** by pump **30** and is passed through line **31** to refining units (not shown).  A portion of the middle oil in line **29**, under the control of valve **32**, is passed through line **33**, waste heat boiler **34**, and line **35** and is thereafter split into two portions (lines **35a** and **35b**) for introduction as reflux into combination tower **27**.

Tower bottoms in line **36** are passed to surge tank **37**.  By properly controlling the temperature level within the tower, the distillate fraction in line **36** has an initial boiling temperature (converted to one atmosphere) of from 650° to 850°F and represents all or a portion of the solvent to be used for dissolving or digesting the extractable carbonaceous matter in the pulverized raw coal feed.  As hereinbefore mentioned, when processing bituminous coal, solvent requirements may usually be satisfied by operating the fractionating unit so as to obtain a distillate having an initial boiling temperature (converted to one atmosphere) of 750°F.

If additional solvent is required (over the 750°F+ distillate) the fractionating unit may be operated to provide a distillate having an initial boiling temperature (converted to one atmosphere) as low as 650°F, or if desired, the fractionating unit may be operated to provide the latter distillate, withdrawing excess solvent over solutizer requirements for utilization in other processes.

Start-up or make-up solvent is introduced into surge tank **37** through line **38** under the control of valve **39**.  The quantity of solvent necessary to provide a solvent to coal ratio of from ½:1 to 6:1 is withdrawn from the tank through line **44**, and passed by pump **45** under the control of valve **46** into line **4**.  After start-up, should the quantity of captive solvent exceed solvation requirements, excess solvent may be withdrawn from the tank **37** through line **47** under the control of valve **48**.  It is generally contemplated that solvent requirements for the solvation of bituminous coal may be fulfilled by controlling the operating pressure and temperature of the combination tower so as to permit the direct introduction of the distillate fraction in line **36** to solutizer **3**, in which case, surge tank **37** is superfluous.

*Example:*  A bituminous coal having the following proximate analysis was used:

|  | Percent by Weight |
|---|---|
| Moisture | 1.4 |
| Volatile matter | 36.6 |
| Fixed carbon | 56.6 |
| Ash | 5.4 |

It was crushed and ground to a particle size distribution whereby 37.3% was retained on a 100 mesh screen.  For comparison purposes, combined sulfur was analyzed as being 1.46 wt %.  The crushed coal and a tar distillate having an initial boiling temperature of about 750°F (converted to one atmosphere) were introduced into the solutizer zone to provide a 3 to 1 ratio of solvent coal.  The mixture was agitated while maintaining a temperature of 800°F and a pressure of 5 atmospheres.  The resulting coal solution was cooled to a temperature of 450°F and passed through a filter precoated with a standard filter aid.  The filtered coal solution represented an 86.3% recovery of extractable carbonaceous matter based on the crushed coal and had the following properties:

| | |
|---|---|
| SG (100°/100°F) | 1,2407 |
| Softening point, °F | |
| (B.&R.) | 152 |
| Sulfur (wt %) | 0.46 |
| Carbon residue (wt %) | |
| Ramsbottom | 30.9 |
| Conradson | 31.3 |
| $CS_2$ solubility (wt %) | |
| Bitumen | 76.66 |
| Ash | 0.02 |
| Difference | 23.22 |

The filtered coal solution was heated to 910°F and coked to provide a coke which was recovered from the coker having the following properties:

|  | Percent by Weight |
|---|---|
| Volatile matter | 10.9 |
| Sulfur | 0.27 |
| Ash | 0.13 |
| Iron | 0.036 |
| Silicon | 0.009 |
| $R_2O_3$ | 0.112 |
| Nickel | 0.0073 |
| Titanium | 0.013 |
| Vanadium | 0.00024 |

## Combination Coal Deashing and Coking Process

According to *W.C. Bull and V.L. Bullough; U.S. Patent 3,997,422; Dec. 14, 1976; assigned to Gulf Oil Corporation and to Reynolds Metals Company* coal is first solvent deashed and desulfurized in the presence of added hydrogen to produce a low hydrogen-content deashed coal which is solid at room temperature. A low-ash low-sulfur solid coal is the primary solvation product, but a smaller quantity of normally liquid product is also produced, together with some gases.

Some or all of the normally solid deashed coal is passed to a delayed coker which is operated at a higher temperature than is employed in the solvent deashing process and without added hydrogen. In the coker, the solid coal is subjected to thermal cracking to produce normally liquid hydrocarbon, a hydrogen-containing gas and low-ash low-sulfur coke. The coke can be passed to a calciner for conversion to a calcined coke which meets commercial metal and sulfur specifications for use as an electrode in the manufacture, for example, of aluminum.

The coke recovered from the thermal coker has a lower hydrogen content than the deashed solid coal supplied to the coker, accounting for the hydrogen made in the coker. The hydrogen produced in the coker constitutes at least a partial recovery of the gaseous hydrogen combined with the deashed coal in the solvent deashing step and may also include hydrogen present in the raw coal. The coker hydrogen is purified and recycled to the solvent deashing process and can typically constitute up to 30%, or more, of the hydrogen requirement of the solvent deashing step.

Substantially the entire coker hydrocarbonaceous liquid product boiling in the deashing solvent range is also recycled to the deashing process, wherein it constitutes a portion of the total process solvent. If this portion of the total quantity of liquid solvent were produced in the solvent deashing step, its production would require a net consumption, rather than a net production of hydrogen.

This process takes advantage of the production of liquid boiling in the deashing solvent range in a thermal coking step, and of the fact that this liquid has good hydrogen donor qualities as required by a solvent in the deashing step, by recycling this solvent boiling range liquid to the solvent deashing step and by modifying operation of the solvent deashing process to integrate the coker and the solvent deashing operations into highly interdependent combination processes which together produce substantially no more than the total requirements of solvent boiling range liquid.

Figure 6.5a shows the sulfur content in the deashed coal as a function of total preheater and dissolver residence time at various maximum preheater temperatures. Figure 6.5a shows that residence time exerts a greater effect on sulfur level in the vacuum bottoms at high temperatures than at low temperatures. Figure 6.5a shows that if significant sulfur is to be removed without utilizing relatively high temperatures, a prolonged residence time must accompany low temperature operation.

Figure 6.5b shows the process schematically. As shown in the figure, pulverized coal is charged to the process through line **10**, is contacted with recycle hydrogen from line **40**, and forms a slurry with recycle solvent which is charged through line **14**.

## FIGURE 6.5:  COAL DEASHING AND COKING PROCESS

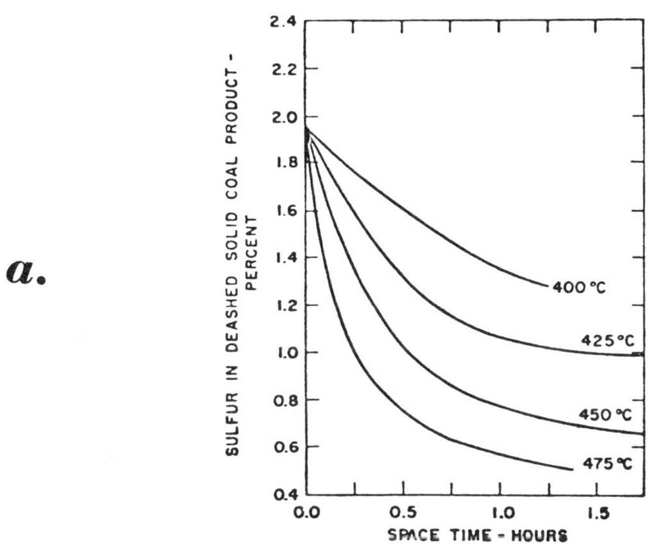


*a.*

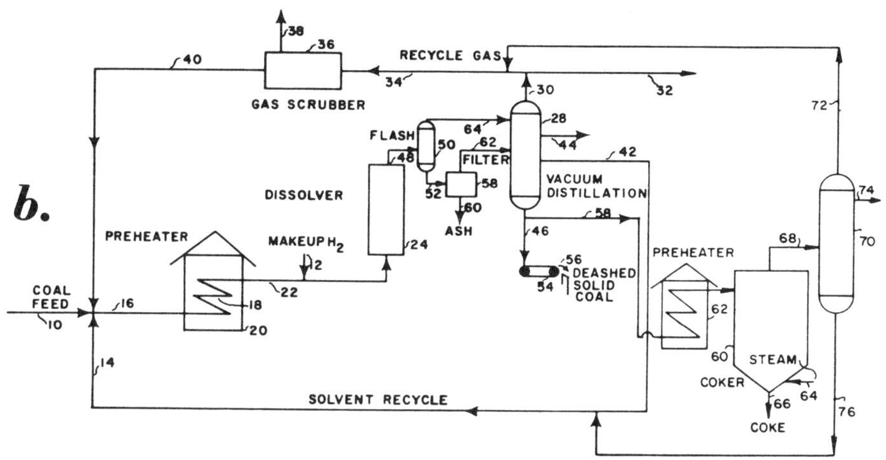


*b.*

(a)   Sulfur content as a function of preheater and dissolver
      residence time
(b)   A schematic of the coking process

Source:  U.S. Patent 3,997,422

The slurry passes through line **16** to solvent deashing preheater tube **18** having a high length to diameter ratio which is greater than 100, generally, and, preferably, greater than 1,000 to permit plug flow. The preheater tube is disposed in a furnace **20** so that in the preheater the temperature of a plug of feed slurry increases from a low inlet value to a maximum temperature at the preheater outlet.

The high temperature effluent slurry from the preheater is then passed through line **22** where it is cooled before reaching dissolver **24** by the addition of cold makeup hydrogen through line **12**. Other methods for cooling can include water injection, a heat exchanger or any other suitable means. The residence time in the dissolver is substantially longer than the residence time in the preheater by virtue of the fact that the length to diameter ratio is considerably lower in the dissolver than in the preheater, causing backmixing and loss of plug flow. The slurry in the dissolver is at substantially a uniform temperature whereas the slurry in the preheater increases in temperature from the inlet to the exit end thereof.

The slurry leaving the dissolver passes through line **48** to flash chamber **50** from which a lighter overhead stream passes through line **64** to vacuum distillation column **28** while ash-containing heavy fuel is removed as flash residue through line **52** and passed to filter **58**. Ash is removed from the flash residue through line **60** while the deashed residue is passed to the vacuum distillation column through line **62**. Gases, including hydrogen for recycle, are removed overhead from the distillation column through line **30** and are either withdrawn from the process through line **32** or passed through line **34** to scrubber **36** to remove impurities through line **38** and prepare a purified hydrogen stream for recycle to the next pass through line **40**.

All the distillate liquid boiling above 200° or 260°C produced in the solvent deashing process is removed from a mid-region of the distillation column through line **42**. A liquid fraction, including naphtha, boiling lower than the solvent boiling range is removed from the distillation column through line **44**. Since the solvent deashing process produces insufficient solvent liquid for the next pass, not only is the entire volume of solvent range liquid in line **42** recycled through line **14**, but this volume is blended with a coker solvent stream, entering through line **76**.

Vacuum bottoms is removed from the distillation column through line **46** and can be, in part, passed to moving conveyor belt **54**. On the conveyor belt the bottoms product is cooled to room temperature, at which temperature it solidifies. Deashed solid coal containing as low an ash content as is practical is removed from the conveyor belt by a suitable belt scrapper means, as indicated at **56**.

A portion of the vacuum bottoms flowing in line **46**, which can be a major or a minor portion, or about one-half, is diverted through line **58** through preheater **62** to delayed thermal coker drum **60**. If desired, all of the vacuum bottoms in line **46** can be passed to coker **60**. In the preheater, the vacuum bottoms is heated to a temperature between about 485° and 510°C before reaching the coker. This temperature is higher than the maximum temperature used in the dissolver stage. Steam is added to the coker through line **64** and coke is removed through line **66**. Hydrogen is not added to the coker. The coker is operated at any convenient low pressure, such as 1.7 to 2.8 kg/cm$^2$. Liquid and gaseous product from the coker is passed overhead through line **68** at a temperature of 425° to 460°C to distillation column **70**.  Distillation column **70** discharges a hydrogen-containing gaseous stream overhead through line **72** to line **34** and gas scrubber **36** to supply hydrogen produced in coker **60** for recycle to the solvent deashing process.  30% of process hydrogen requirements can pass through line **72**.  A liquid product, including naphtha, boiling below 200° to 260°C is removed from the process through line **74**.

The liquid produced in coker **60** boiling in the solvent range, which is typically distillate boiling at 200° to 260°C+, is passed in its entirety through line **76** to line **42** for blending with the solvent liquid in line **42** and for recycle as solvent in the deashing process.  The solvent blend returning to the solvent deashing process through line **14** includes substantially the total solvent boiling range liquid recovered from both the solvent deashing process and the coking process excluding, of course, mechanical and other losses.

## REDUCTION OF SULFUR CONTENT IN USSR COALS

> The material in this section was taken from a report by V.S. Kaminskii and T.K. Yagodkina (ORNL-tr-4330; for details see bibliography on page 301), and translated by L. Kobylenski.

Sulfur is one of the most harmful mineral deposits in coals.  This necessitates a significant reduction of the sulfur content in coal which is to be used as metallurgical coke.  In the USSR the sulfur content in coals covers a wide range—from 0.2 to 10%.  In the case of coking coals it runs up to 6.0%.  Most coal deposits in the Soviet Union are low in sulfur, however, these resources are concentrated for the most part in Asiatic USSR.  In European regions of the USSR, where major industrial coal mining is concentrated, more than half of the coal extracted contains more than 2.5% sulfur.

The coals of the Kizlovsky, Moscow, Dnieper, Irkutsk and Donets Basins are known to have the largest sulfur content.  Coal from the Kizlovsky Basin (sulfur content up to 6%) is basically used in the production of coke for nonferrous metallurgy and in other branches of industry, where the presence of sulfur is not a significant detriment.  Coal from the Moscow, Dnieper and Irkutsk Basins (average sulfur content 3 to 4%) is used in energy projects. The Donets Basin is considered to be the main supplier of coking coal in the south.

The level of sulfur in run-of-mine coking coal from the Ukraine is found to be 2.60%.  High-level sulfur coal deposits (more than 2.5%) can contain as much as 58%.  The question of reducing the sulfur content in coal and its efficiency must be settled by combination measures developed to guarantee desulfurization at the following stages:  in the mining process (the mining of coal strata); in the process of beneficiation (during coke preparation); in blast furnace production; in-the-blast-furnace desulfurization of pig iron, and also in stages of metallurgical conversion; and in purification of power plant exhaust.

The significance of all the methods above is obvious.  In this discussion only one of the ways of solving the problem is examined, namely, desulfurization of coking coal during its beneficiation. The possibility of removing sulfur in the process depends upon a series of factors. The main ones are: the average amount of sulfur in coal; the relationship between different types of sulfur; the nature of the iron disulfide dispersement; the physical and chemical characteristics of iron disulfides (porosity, hardness, electric conductivity, magnetic susceptibility, degree of contamination by different impurities, crystalline structure, etc.); distribution of sulfur by density and thickness; and technology applied in coal beneficiation plants.

Sulfur in coal appears in the form of organic and inorganic compounds; little is known about its chemical structure. In the inorganic group are these sulfides which are basically disulfides of iron (pyrite, marcasite and possibly melnikovite), sulfates, and also simple sulfur.

When the desulfurization of coals during beneficiation is being considered, attention is usually paid only to the reduction of the total sulfur present. Meanwhile the most harmful sulfur for blast furnace operation is organic sulfur, since it is the organic sulfur compounds of coal that interact with the decomposition products of pyrite during the coking process and therefore remain in the coke. For this reason, special emphasis is placed upon the removal of organic sulfur as well as sulfur during the enrichment process.

Depending on its origin, organic sulfur in coals is divided into types, primary and secondary. Secondary organic sulfur may be of two kinds, intermolecular and bridged; the latter, genetically connected with iron disulfides and concentrated in layers directly adjacent to the surface of disulfide granules is known as secondary organic sulfur A; and the former, chemically related to all organic coal stocks is called secondary organic sulfur B.

Iron disulfides, depending upon conditions and time of origin, may be subdivided into two basic types, biogenetic and abiogenetic. An analysis of the history of sulfur formation in coals shows that a variety of conditions during the period of matter accumulation caused the formation of various iron disulfide modifications (pyrite and marcasite), with somewhat wider dispersement of marcasite, in the authors' opinion, than had previously been considered.

Gravitational methods of beneficiation are not applicable for coals with a significant amount of finely dispersed disulfide impurities. However, a significant portion of sulfidic sulfur and a certain part of secondary sulfur (intermolecular organic sulfur) may be removed from fine or crushed coal with the help of certain beneficiation methods.

The study of sulfur distribution over density fractions showed a uniform distribution of sulfur into fractions for Donets Basin coals containing an insignificant amount of inorganic sulfur, and an unequal distribution for coals high in sulfur. The sulfur content in intermediate density fractions and sulfur free coals is considerably higher than in lightweight fractions. It should be noted that in the majority of cases a sharp rise in sulfur content is observed during transition from a density fraction of 1,300 (1,350) kg/m$^3$ to one of 1,400 kg/m$^3$.

A uniform distribution occurred for sulfur coals of the Donets Basin when these were analyzed as to their coarseness. The sulfur content for coarse-type coals was somewhat higher than for fine-type coals.

Experience has shown that when conventional beneficiation methods were applied to charred coals in Soviet factories, the overall sulfur content in clean coals of all types was reduced to no more than 18 to 20%.

Lately in the USSR and other countries a method for enriching coal fines in concentration tables is gaining popularity, whereby absolute sulfur content reduction is generally greater than in jigging machines. However, examination of a concentrate bed, SKPM-6, at the Gorlovsky Central Concentration Plant revealed that during highly sulfurous sludge (3.24%) enrichment, sulfur reduction in clean coal reached only 16.8%. At the Soviet Central Concentration Plant, sulfur extraction by means of a SKPM-6 reached 31.5%.

By using a greater force field instead of a gravitational one, a finer classification of coal fines may be obtained. Along this line special attention should be paid to an experimental Soviet method of centrifugal beneficiation in heavy liquid and subsequent simplification of its development—centrifugal hydroseparation in an aqueous medium. In centrifugal enrichment of coal in heavy liquid, the degree of desulfurization attained is 50%, in hydroseparation, 40%. Enrichment of fine coal for desulfurization in suspended cyclones and that of thin fines in an aqueous substance in hydrocyclones has widespread appeal.

Significant studies are also being conducted in the area of magnetic and electromagnetic field application in coal desulfurization. The possibility of reducing the sulfur content of a unique Moscow area coal containing 8.4% sulfur by cooling it in a magnetic field was researched at the Moscow Mining Institute. A reduction of over 60% sulfur was acheived.

Studies were conducted on the use of magnetic hydrodynamic separation for the enrichment of mineral resources by various organizations. The drawback of this process is the need for the use of aqueous solutions of salts, acids and alkalis as an electrolyte, which, as in the case of enrichment by heavy liquids, necessitates a washing-off of the enrichment products and a regeneration of the electrolytes. In addition, the effectiveness of the separation process is reduced during transition to finely granulated materials. It can be deduced from this that the outlook for magnetic hydrodynamic separation in reducing the sulfur content in coals is not promising.

Studies of coal beneficiation in a magnetic liquid, a process which does not waste so much electric energy, have shown that it is possible to obtain a low-ash concentrate with a reduced sulfur content from coal sludge and crushed intermediate products. However, application of paramagnetic salts with high magnetic susceptibility is limited, due to the high cost of the salts.

Research into the application of superconductors for magnetic separation of mineral resources is being conducted by the Baikov Metallurgical Institute AN-USSR, MGI and DGI. Laboratory experiments at Uzlovsky TCF confirmed that the sulfur extraction in residue could reach 69.5% in concentrated beneficiation with a magnetic field of superconductive solenoid. In testing beneficiation for possible desulfurization in Moscow and Donets coal by the magnetic hydrodynamic separation method in a super-strength magnetic field, it was shown that the sulfur, which was combined with pyrite, could be reduced by 70 to 88%. It should be noted that the use of superconductive magnets increases the cost of the process.

The mining industry in the United States has experimented with removing pyrites from sulfur coals by a method of two-stage flotation. The first stage removes the high-ash waste. The foamy product is diluted with clean water and
directed to the second stage, where the pulp is treated with an organic colloid
for quenching. After that it is fed into a collector (xanthate) and foaming agent
for removal of pyrites. The experiments were conducted with coal pulverized
to 0.5 mm.

A significant reduction of sulfide sulfur occurs in the first stage. In the second
stage the reduction of sulfur content is achieved at the expense of the concentrate. The high-sulfur wastes (5.84 to 5.90%) are low in ash (13.4 to 13.7%).
This is due to a breakdown in selectivity during reverse flotation. The result
is natural since xanthate is a collector for pyrites as well as for coal. It should
be noted that when marcasite is present in coals its flotation would be even
more difficult than that of pyrites in the two-stage system.

Contradictory opinions are held by scientists regarding desulfurization by flotation, and not one of the systems mentioned above has been applied in industry.
Studies which revealed the presence of sulfur in flotation products issuing from
enrichment plants in the Soviet Union showed that with flotation the sulfur content in concentrates was reduced 0.1% at most.

Tests using centrifugal hydroseparation on highly sulfurous sludges and residues
in run-of-mine coals entering Donets Basin plants with a 3 to 4% sulfur content,
showed a sulfur reduction of 30 to 40%. In the case of coals with a low sulfur
content the effects of desulfurization are reduced.

While studying the reduction of various types of sulfur by centrifugal hydroseparation scientists have confirmed that it is possible to remove secondary organic
sulfur A (intramolecular) in wastes together with granules of iron disulfides and
aureoles of an organic coal compound surrounding them (which is the richest
type of sulfur) and fine-grained disulfide impurities. Secondary organic sulfur
was reduced 19 to 31%.

Comparative tests on centrifugal hydroseparation and other enrichment methods
(separation in concentration tables, centrifugal enrichment in heavy solutions)
confirmed that hydroseparation is an effective method in desulfurizing coals and
that it is possible to obtain properties for sulfur reduction that approximate the
theoretical. While the technological experiments were being conducted, the hy-
drodynamics of intrarotor liquid flows in a centrifugal hydroseparator were being studied. The effects of hydroseparator volume, rotor rotation rate, the
screw conveyer and its operation system on the intensity of liquid displacement
from the rotor, depth of the removable liquid layer and on the amount of average axle flow rate was determined.

A centrifugal hydroseparator called the NOGSH-1100V was built to carry out
the hydroseparation process on an industrial scale. The experimental model
is used for industrial experiments at the Kalininsky TCF Lonets-Enrichment
board.

Two technological techniques of removing sulfur from coking coals during beneficiation warrant attention. The first type allows the coal to be beneficiated by
the usual methods, but before delivery to the flotation process, the siftings

(undersized sludge waters of dehydrating and desludging sieves as well as the centrifugal effluents of dehydrating centrifuges) are subject to preliminary enrichment for removal of iron disulfides and prevention of their entry into the froth.  Beneficiation may be carried out in concentration tables, in hydrocyclones, or by the hydroseparation method.  This technique is suitable for use in operating plants without excessive remodeling being necessary, but only a relatively small part of the coal is subject to special desulfurization.

For more extensive sulfur removal it is necessary to grind lump coal for exposure of pyrite granules and to beneficiate the fines resulting from this method.  For reducing the amount of ground material, a second technological technique offers the removal of lightweight low sulfur fractions (separation density 1,300 to 1,350 kg/m$^3$) and of heavyweight high sulfur fractions (separation density 1,800 to 2,000 kg/m$^3$) from lump coal in mineral suspensions.  The intermediate product is broken down to expose adhesions and is rebeneficiated by any of the above-mentioned methods.

It was shown that in fine grinding of up to 0.04 mm, a maximal exposure of adhesions may be achieved.  However, such an operation creates insurmountable difficulties during beneficiation and dehydration.  Grinding mechanical type coal without first removing from it lightweight fractions containing the least amount of sulfur and then breaking down the intermediate products to less than 1 to 3 millimeters is not recommended.  A study was made of the effect that the grinding of an intermediate fraction (1,350 to 2,000 kg/m$^3$) had on a supplementary yield of low sulfur products from the Kalinin mine (fractions $>$1,300 kg/m$^3$), which had undergone previous removal of a lightweight fraction ($<$1,300 kg/m$^3$) from lump coal.  A significant reduction of sulfide sulfur (an average of 70%) was noted.

As was noted above, desulfurization of natural residues and fragmented coal may be effected by various methods (in hydrocyclones, suspension cyclones, on concentration beds and in hydroseparators).  Consequently variations on these enrichment methods were tested in conjunction with the Central GIPRO mine in four plants where enrichment of residues takes place and in the design of a new enrichment plant which will be based on extensive desulfurization techniques of fragmented coal.

Since the economics of desulfurizing coking coals is determined by comparing costs of the removal of sulfur, costs were determined for the following:  0.1% sulfur reduction in a carbon mixture entering the coking process, a 0.1% sulfur reduction in metallurgic coke, and for 550 kg of coke to melt 1 ton of pig iron. It was discovered that the greatest reduction in expenditures in melting pig iron in order to further reduce desulfurization costs was attained by hydroseparation.

# OTHER DESULFURIZING PROCESSES

## DESULFURIZING CHAR

### Slurrying with Phosphoric Acid

The principal purpose of the process provided by *J.G. Santangelo and T.P. Dorchak; U.S. Patent 3,812,017; May 21, 1974; assigned to Kennecott Copper Corporation* is to remove sulfur during the production of char from coal.

The term "char" as used herein, in contradistinction to "coke," is characterized as the solid carbonaceous residue of coal which has been distilled between 400° and 800°C and containing residual volatile matter of at least 5% by weight. On the other hand, coke normally contains less than 2% by weight volatile matter and is the product of destructive distillation of coal at temperatures above about 1000°C.

The process comprises the steps of mixing crushed coal with an acid selected from phosphoric acids, phosphorous acids and mixtures thereof, removing the excess acid, if any, from the mixture and heating the mixture to a temperature of between 400° and 1100°C for at least 15 minutes.

Using this process, experiments have shown that sulfur could be reduced to as low as 1.7% from a 3.0% sulfur Western Kentucky No. 11 coal. From a 4.3% sulfur-containing Illinois No. 6 coal, a char of 2.76% sulfur content can be produced. From a 6.7% sulfur–containing Missouri coal, a char of 3.3% sulfur content can be produced.

In a preferred embodiment a 5 or 10% aqueous solution of phosphoric acid is slurried with crushed coal. The slurry is then filtered to remove excess acid. After filtration the coal will contain from 10 to 30% liquid based on the weight of the coal. 20 weight percent liquid on the crushed coal is preferred. Experiments have shown that as little as 0.25 weight percent phosphoric acid based on the coal weight can be used in effecting the removal of sulfur from the coal during the carbonization thereof. Amounts of phosphoric acid up to 25% by weight on the coal or more may be used. However, the preferred range for economic consideration is between 1 and 5% based on the weight of the coal. The coal is crushed in any convenient manner in crushers or grinding mills. Preferably the coal will be reduced to a size between 100 and 10 mesh U.S. Sieve series. Larger particles of coal may be desulfurized using this process but the time at temperature is thereby increased.

The coal, after it has been slurried with phosphoric acid and the excess acid, if any, removed therefrom, is fed to a suitable vessel for heating. The coal-phosphoric acid mixture may be heated rather slowly to a range of between 400° and 1300°F, with the preferred range being between 700° and 1200°F. It is believed that the most rapid sulfur removal occurs at about 700°F. At this temperature the evolution of volatile matter, including some organic sulfur compounds, begins, together with pyrite decomposition to pyrrhotite. Once the selected temperature has been reached it should be maintained for a period of time of at least 15 to 30 minutes or longer.

During the heating of the coal-phosphoric acid mixture it is preferred to maintain an inert or a mildly oxidizing atmosphere in the heating vessel. The inert atmosphere may be carbon dioxide or an atmosphere of spent gas from a coal combustion. Once the selected temperature has been reached, a reducing atmosphere is introduced into the vessel. Preferably the reducing atmosphere is a continuously changing one in order to remove the sulfur compounds as they are released from the charring coal. The reducing atmosphere may be any combination of hydrogen, carbon monoxide, methane and an inert gas such as nitrogen. A convenient way to produce a reducing atmosphere is to feed steam and carbon monoxide simultaneously into and through the heating vessel.

During the heat-up, charring and cool-down periods, it is preferred to use a flowing gas in the reactor vessel. More efficient sulfur removal is obtained since the volatile sulfur compounds such as sulfur dioxide, hydrogen sulfide, etc. are carried off in the gas stream. Preferably the gas in the reaction vessel will be changed every few seconds such that the gas space velocity will be between 0.1 and 1.0.

*Example 1:* A 12 g sample of Illinois No. 6 coal containing 4.33% by weight sulfur was crushed and slurried with 15 ml of 10% phosphoric acid. Excess acid was filtered off and the sample washed with water and dried. The sample was not subjected to the thermal treatment but was analyzed for sulfur. The final total sulfur content was found to be 4.16%.

This example shows that only an insignificant quantity of the sulfur is removed by the phosphoric acid alone. Thus it follows that the phosphoric acid pretreatment materially assists in removing the sulfur during the later thermal treatment.

*Example 2:* 15 g of Western Kentucky No. 11 coal was ground to a 100 mesh and slurried with 12 ml of an aqueous 5 or 10% phosphoric acid solution for 15 minutes. The slurry was then filtered to remove excess acid and approximately half of the sample was weighed into a silica boat inside a tubular furnace. The furnace train consisted of compressed gas tanks, a regulator, a flow meter, a tube reactor, tar traps and a cadmium scrubber. The flowing gas was set to 360 ml/min through the 18 x 0.8 inch inside diameter quartz tube reactor. The inert gas was prepurified nitrogen and the reducing gas consisted of 5.5% hydrogen,

23.6% carbon monoxide, and the balance argon. The furnace was turned on and brought to 500°C in about 15 minutes and held there for an additional 15 minutes. The results are tabulated below.

| | | Experimental conditions | | |
| | | Run 1 | Run 2 | Run 3 |
| | | Reducing gas, 7.3% phosphoric acid | Reducing gas, 14.6% phosphoric acid | Inert gas, 14.6% phosphoric acid |
| | | Sulfur analysis (weight percent) | | |
| | | | Char | |
| | Coal | Run 1 | Run 2 | Run 3 |
|---|---|---|---|---|
| Sulfate sulfur | 0.04 | | | |
| Organic sulfur | 1.97 | 1.77 | 1.70 | 1.56 |
| Pyritic sulfur | 1.01 | 0.11 | 0.06 | 0.62 |
| Total sulfur | 3.02 | 1.89 | 1.76 | 2.18 |
| Char Yield percent | | 63.7 | 66.9 | 67.2 |

## Using Molten Cresylic Acid plus Caustic

*L. Robinson and H.F. Bauer; U.S. Patent 3,919,118; November 11, 1975; assigned to Occidental Petroleum Corporation* have discovered that over 80% of the sulfur in coal char can be removed if the char is treated with a hot molten bath comprised of a caustic material and the cresylic acid fraction from the distillation of coal tar, hereinafter referred to as cresylic acid.

In practicing the process, a caustic material is added to cresylic acid until the mixture is basic when tested with litmus paper. The caustic may be, for example, sodium hydroxide, potassium hydroxide, calcium oxide, sodium carbonate, sodium sesquicarbonate, sodium bicarbonate, or any other sodium and potassium acetate alkaline material or combination thereof.

The cresylic acid-caustic mixture produces an alkaline salt of the cresylic acid which is heated to a temperature which will melt the salt and maintain it in a fluid condition. In the preferred embodiment the mixture comprises 1 mol of sodium hydroxide to about 1 mol of cresylic acid and is heated and maintained at a temperature between 500° to 750°F.

The ratio of char to cresylic acid-caustic mixture should be maintained at a level which will effect maximum desulfurization. In the preferred embodiment, this ratio is 1 part of char to at least 2 parts of a mixture comprising 1 part by weight of sodium hydroxide and 3 parts by weight of cresylic acid. In the event the ratio of char to the cresylic acid-caustic mixture is less than 2 parts of mixture to 1 part of char, at least 1 part of a polyphenyl ether or a mixture thereof is preferably added to it for each part of char, to improve the contact between the char and the cresylic acid-caustic mixture. One such polyphenyl ether mixture is sold under the trade name Mobiltherm.

When the char is introduced to the molten bath it is admixed therein by means
of a motor-driven stirrer or other equivalent means which will provide a thorough mixing of the char with the molten cresylic acid-caustic mixture for intimate contact therebetween.  An inert medium, such as nitrogen gas, may be admixed into the mixture to eliminate the presence of oxygen therein and prevent
the possibility of any reactions therewith.  The elimination of oxygen from the
bath, however, is not necessary to the practice of the process.

The period of intimate contact between the char and the mixture of cresylic
acid and caustic is referred to as the residence time and should be about 2 to
60 minutes.  About 30 minutes is preferred for maximum desulfurization.  After the prescribed residence time, the char is allowed to rise to the surface of
the bath where it is decanted or skimmed off by conventional flotation separation means.  The char is thereafter washed with water to remove any inorganic
residues, and then washed with organic solvents, such as acetone, to remove any
organic residues.

The remaining cresylate fraction of the bath may be recovered as cresylic acid
by steam distillation and recovered after the desulfurization is completed and
preferably recycled for further use.  Recycling the cresylic acid is a substantial
improvement over prior methods which consumed these desulfurization reagents
during the process.

The cresylic acid used is derived from coal tars and includes phenols, cresols,
xylenols, and phenolic homologs derived therefrom.  These materials are organic
acids having a thermal stability which is necessary for use in this process, and
form alkaline salts when mixed with the caustic materials.  It is believed that the
alkaline salt of the cresylic acid partially solvates the char thereby rendering the
char in a condition where the sulfur is leached therefrom.  It is believed that the
alkaline salt of any other thermally stable fatty acid, aromatic carboxylic acid or
any other organic acid may be substituted for cresylic acid.

*Example:*  33 g of sodium hydroxide were added to 100 g of cresylic acid.  The
resultant mixture was basic to a test by litmus paper and was in a dark viscous
semisolid condition.  155 g of the semisolid mixture was heated to 750°F under
a nitrogen blanket.  The solid was completely fluid at about 500°F.

A 25 g specimen of 60-mesh char prepared from a high-volatile bituminous coal
was admixed to the fluid mixture and stirred for 30 minutes while the temperature was maintained at 750°F.  The mixture was cooled, the char removed therefrom and washed with water to remove the inorganic residues and washed with
acetone to remove organic residues.  The washed specimen was thereafter dried
at 212°F to remove any acetone or water and then analyzed.

A second run was made on another 25 g specimen of the same char following
the same procedure as set forth above.  Analysis of the char after treatment indicates that the sulfur content of the char was reduced to about 86%.

## With Carbon Dioxide and Water

*L. Robinson and A. Sass; U.S. Patent 4,053,285; October 11, 1977; assigned to
Occidental Research Corporation* have discovered that the increased sulfide sulfur content of char associated with hydrogen treatment at elevated temperatures

may be significantly reduced by treating the char with a gaseous mixture of carbon dioxide and water at temperatures that will not cause any appreciable gasification of the char.  The carbon dioxide-water treatment is also very effective in reducing the sulfide sulfur content of char whether or not the char has been pretreated with hydrogen.

In this process, char is crushed to a particle size suitable for gravity flow through a char feed line and is fed through the line to form a bed of char particles in the reactor.  Hydrogen gas enters the reactor through an inlet pipe, and, depending upon the pressure of the hydrogen gas and the particle size of the char, will cause the reactor to be operated as either a fluidized bed, an entrained bed, or expanded bed system.  The char is held in the reactor for a prescribed residence time sufficient to convert the organic and pyritic sulfur to sulfide sulfur.  Thereafter the reacted char particles flow into the steam-carbon dioxide reactor.

Any excess hydrogen gas and the gases evolved from the hydrogen reaction with the char exit out of the reactor through a cyclone which prevents the char particles from leaving the reactor with the exit gases.  A gaseous mixture of steam and carbon dioxide enters the reactor.  The carbon dioxide and steam are introduced and mixed.

The char is retained in the reactor for a prescribed period and thereafter a slide valve is opened to permit more char to enter the reactor, which will cause the level of the char in the reactor to rise and flow into a standpipe.  A slide valve is opened to permit the char particles to flow by gravity through an outlet line for recovery.  The excess carbon dioxide and steam mixture and the gases evolved from the reaction of char with the mixture exit out of the reactor through a cyclone, which prevents the char particles from leaving the reactor with the exit gases.

In the operation of the hydrogen reactor, hydrogen gas enters through the inlet pipe at pressures up to 500 psia and at flow rates of 5 to 100 cubic feet of hydrogen per pound of char.  The char is held in this reactor for a period of 5 minutes to 2 hours and at a temperature range of 1100° to 1800°F.  In the preferred embodiment the hydrogen treatment is carried on at 1300° to 1600°F, at pressures from 25 to 150 psia, and at a volume of 5 to 100 cubic feet of hydrogen per pound of char.

In the operation of the carbon dioxide-water reactor, the mixture of carbon dioxide to steam is maintained at a temperature of from 500° to 1600°F and at a ratio of 1:5 to 5:1, respectively, but in the preferred embodiment, the ratio is about 1:1.  The flow rate of the carbon dioxide-steam mixture can vary from 5 to 100 cubic feet of the mixture per pound of char, but in the preferred embodiment the flow rate is 25 cubic feet of the mixture per pound of char.  The mixture has a pressure ranging from atmospheric to 500 psia and a temperature ranging from 700° to 1200°F.  In the preferred embodiment the pressure and temperature of the mixture are 50 psia and 1000°F, respectively.  The residence time of the char in the reactor for the carbon dioxide and steam treatment can range from 5 minutes to 2 hours but is preferably 30 minutes.

*Example:*  West Kentucky bituminous coal was carbonized at 1000°F for 2 hours in a nitrogen atmosphere to produce a char.  This char was ground to –200 mesh and a 10 g portion was treated with 0.09 standard cubic feet per minute (SCFM) of hydrogen at 1600°F and at atmospheric pressure. An analysis of the coal prior to conversion into char and an analysis of the char after the treatment with hydrogen was made in terms of sulfur content, moisture content, and ash content and the results noted in the table below.

A 10 g portion of the hydrogen treated char was placed in a 3-foot long stainless steel tube having a ¾-inch internal diameter and was plugged at the top with quartz wool to prevent the char from blowing out. The tube was heated until reaching 1000°F and 0.02 SCFM of nitrogen was purged through the tube from the bottom end to prevent any reaction with atmospheric oxygen. The tube was maintained at 1000°F and a gaseous mixture comprised of 0.10 SCFM of carbon dioxide and 0.10 SCFM of steam (2.28 ml/min of water) were passed into the tube.

The carbon dioxide was metered into the system through a calibrated rotameter and the water was pumped into the system through a calibrated rotameter with a constant displacement pump from a calibrated water reservoir. The carbon dioxide and water were heated to 1000°F in a preheater before being introduced into the stainless steel tube containing the char. The carbon dioxide-steam pressure was maintained at 65 pounds per square inch for 30 minutes. Thereafter the heat and the flow of carbon dioxide and steam were shut off and about 0.02 SCFM of nitrogen was pumped through the system to stop the reaction. The char was removed and analyzed. The analysis is set forth in the table.

| Material | Total Sulfur Wt. % | Sulfide Sulfur Wt. % | Pyritic Sulfur Wt. % | Organic Sulfur Wt. % | Water Content Wt. % | Ash Content Wt. % |
|---|---|---|---|---|---|---|
| 1. Coal prior to conversion to char. | 3.73 | 0.17 | 2.04 | 1.52 | 4.36 | 14.95 |
| 2. Char after hydrogen treatment. | 2.11 | 1.27 | 0.05 | 0.79 | 0.48 | 22.47 |
| 3. Char after hydrogen treatment and $CO_2$-steam treatment. | 1.04 | 0.21 | 0.04 | 0.79 | 0.99 | 22.17 |

## Acid Washing plus Hydrogenation

One desulfurization method used on coal is to treat the coal with hydrogen gas to react with the sulfur in the coal to form hydrogen sulfide gas. This method may work well with coal but not as well with chars made from coal because the sulfur appears to be more resistant to reaction with hydrogen after being converted into char.

A further problem in treating coal or char with hydrogen gas is that trace amounts of hydrogen sulfide in the hydrogen gas will inhibit desulfurization or back-react with the coal or char to increase its sulfur content. Hydrogen gas used in coal or char desulfurization methods is usually recycled and will contain small amounts of hydrogen sulfide and other sulfur-bearing gases produced during desulfurization. Attempts may be made to remove the sulfur-bearing gases from the hydrogen before recycling but such attempts are not too successful unless the gases are subjected to more elaborate and expensive separation techniques. As a result, the hydrogen used for desulfurization purposes may contain small amounts of sulfur-bearing gases and when it does desulfurization is inhibited or a back-reaction may occur to increase the sulfur content of the coal or char. Consequently, in methods that do employ recycled hydrogen gas it is important that all traces of sulfur-bearing gases be removed therefrom, which as a practical matter is difficult to do.

*L. Robinson and A. Sass; U.S. Patent 4,054,421; October 18, 1977; assigned to Occidental Research Corporation* have discovered that greatly improved results in desulfurizing char with hydrogen gas are realized if the char is first washed with an acid before treatment with the hydrogen gas. The acid washing is believed to make the sulfur in the char more reactive with hydrogen gas by reacting with the inorganic compounds therein to form metallic salts which are removed from the char in the acid wash. These inorganic compounds include sulfides, such as iron sulfide and calcium sulfide, and other inorganics which generally comprise the ash content of the char. The organic sulfur in the char is not believed to be affected by the acid wash but the removal of the inorganic compounds by the acid wash appears to make the organic sulfur more susceptible to removal by this process.

It has further been discovered that when char is washed with an acid before treatment with hydrogen gas, the pressure of sulfur-bearing gases, such as hydrogen sulfide, may be tolerated without any substantial inhibition to the desulfurization reaction. It has also been discovered that pretreating char with acid before desulfurization with hydrogen gas will cause the desulfurization to proceed at a greater rate than in the absence of the acid pretreatment.

The process is as follows: Coal is converted into char and thereafter introduced into an acid bath. The char need not be specially treated or prepared but may be introduced into the acid bath in the condition that it exists after it is manufactured. The acid bath is comprised of an aqueous solution of a mineral acid, such as hydrochloric acid, sulfuric acid, nitric acid, phosphoric acid, or any combination thereof.

In the preferred embodiment, the acid concentration is 1 molar, but it may be as low as 0.05 M or lower if the residence time of the char in the acid is increased. The ratio of acid to char should be maintained at a level which will effect maximum reaction between the inorganic compounds present in the char and the acid. It is preferred that this ratio be at least two parts of a 1 molar aqueous solution of acid to one part of char.

When the char is introduced into the acid bath, it is added therein by means of a motor-driven stirrer or other equivalent means which will provide a thorough mixing of the char with the aqueous acid solution for intimate contact therebetween. An inert medium, such as nitrogen gas, may be admixed into the char-acid mixture to eliminate the presence of oxygen therein and to prevent the possibility of any reactions therewith. The elimination of oxygen from the bath, however, is not necessary.

The acid bath may be maintained at room temperature but the rate of reaction with the inorganics will be increased if the bath is maintained at elevated temperatures. The residence time of the char in the acid bath should be at least 5 minutes when using a 1 molar solution of acid at a ratio of 2 parts of acid to 1 part of char. The residence time, however, like the acid-to-char ratio, is a variable depending on the acid concentration, bath temperature, and the acid-to-char ratio.

After the prescribed residence time, the char is removed by filtration from the aqueous acid bath. The char is thereafter washed with water to remove the acid therefrom. The inorganic salts formed by the reaction between the inorganic sulfides in the char and the acid are separated from the acid which is recycled for further use.

After the char has been washed with water, it is introduced into a reactor containing hydrogen gas maintained at a pressure of 15 to 125 psia, at a temperature between 1100° to 1800°F, and is held in the reactor from 1 minute to 2 hours. In the preferred embodiment, the char is treated with hydrogen gas at a pressure of 65 psia and a temperature of 1600°F for 5 minutes.

The hydrogen reacts with the sulfur in the char to form hydrogen sulfide, mercaptans and other sulfur-bearing gases. These sulfur-bearing gases and excess hydrogen are passed out of the reactor and the hydrogen is separated therefrom and recycled for further reaction with the char. Some trace amounts of hydrogen sulfide and other sulfur-bearing gases remain in the recycled hydrogen. After the char is washed with acid the presence of these small amounts of sulfur-containing gases does not cause any reaction with the char to increase its sulfur content, and only slightly reduces the extent of char desulfurization by hydrogen gas. Up to 3 to 5% of sulfur-containing gases in the hydrogen gas may be tolerated, depending upon the particular char, without impairing the commercial acceptability of the process.

*Example:* To demonstrate the effect of acid washing on the desulfurization of char by hydrogen gas, 100 g of Hamilton coal was converted into char by heating at 1600°F for 1 hour. Hamilton coal is a high volatile, bituminous coal from seam No. 11 in Western Kentucky, and was beneficiated by floating the coal in a zinc chloride solution having a specific gravity of 1.55. The beneficiation is not necessary but it allows the pyritic sulfur, which has a specific gravity greater than 1.55, to settle out from the coal floating on the surface and thus reduces the amount of acid needed in the subsequent washing step.

The char was ground and screened to –150 to +400 mesh. Eight 10 g specimens of char were prepared for treatment. Four of the specimens were not washed with any acid, and three of these specimens were treated in a fluidized bed quartz reactor at 1600°F and 65 psia for 4 hours with hydrogen gas or a mixture of hydrogen sulfide-hydrogen gas as described in the table below. The four specimens were analyzed for their sulfur content and iron content and the results made were noted as set forth in the table.

Three of the remaining four specimens were washed with 1 liter of 1 M HCl at 80°C for 5 minutes. The char was separated therefrom and then washed a second time with 1 liter of 1 M HCl for 5 minutes at 80°C. The char was separated after the second wash, washed with water to remove the acid therefrom, and treated in a fluidized bed quartz reactor at 1600°F and 65 psi for 4 hours with hydrogen gas or a mixture of hydrogen gas and hydrogen sulfide as set forth. The specimens were thereafter analyzed for their sulfur content and iron content and the results noted in the table.

| Treatment of Char | Gas Used to Treat Char | Sulfur in Char | Fe in Char | Ash in Char |
|---|---|---|---|---|
| | | . . . . . . . . weight percent . . . . . . . . | | |
| NW* | None | 1.61 | 1.0 | 12.05 |
| NW | $H_2$ | 0.58 | – | – |
| NW | 1.1% $H_2S$ 98.9% $H_2$ | 1.36 | – | – |
| NW | 3.3% $H_2S$ 96.7% $H_2$ | 1.82 | – | – |
| W** | None | 1.60 | 0.73 | 10.68 |
| W | $H_2$ | 0.40 | – | – |
| W | 1.1% $H_2S$ 98.9% $H_2$ | 0.76 | – | – |
| W | 3.3% $H_2S$ 96.7% $H_2$ | 0.95 | – | – |

*NW means not washed with acid.
**W means washed with acid.

## SOLUTION MINING

*M. Perch; U.S. Patent 3,990,513; November 9, 1976; assigned to Koppers Company, Inc.* has developed a method for "solution mining" in which a solvent of coal is pumped through pipes drilled into the earth to a coal seam. The coal is digested or dispersed into the solvent and solvent recycle, and is then pumped as a solution up to the surface and processed to remove the coal from the solvent. The method produces low-ash coal, stripped of all the extraneous ash, leaving the inherent ash, and stripped of all the sulfur except the organic type. The method comprises making a coal solution in situ and bringing it to the earth surface economically, without the hazards and environmental problems of conventional coal mining and without the need of coal washing facilities.

Referring to Figure 7.1, a coal seam **11** is shown at some finite distance beneath the ground level surface **13** from which it is desired to extract coal, without resorting to conventional shaft mining. In accordance with the described process, a first tubular conduit or casing **15** is installed in a bore-hole **17** in the earth and the lower end portion **19** of the conduit is perforated, as at **21**, to allow fluid carried by the conduit to flow therefrom into the coal seam.

The tubular conduit or casing is cemented in place, as at **23**, in the manner of an oil-well casing. This tubular conduit or casing is associated with one, or preferably more, similar tubular conduits or casings **25** spaced some distance away. The tubular conduit or casing **25** is also cemented in a bore-hole **27**, and its lower end portion **31** is also perforated to admit fluid and coal solution to enter it. The conduit or casing **15** is fluidly connected above the ground level surface to the discharge side of a suitable pump **35** which is fluidly connected at its suction side to a conduit **37** which is fluidly connected by means of conduit **39** to a heater **41** of conventional construction that receives fluid and coal solution from the tubular conduit or casing **25**, and to a solvent still **43** by means of conduit **45**.  The solvent still receives both coal in solution and solvent from a centrifuge or filter **47**, which is an optional piece of equipment and from which, when used, mineral cake is removed through a conduit **49**.  Solvent that is separated from the coal in the solvent still **43** flows in a conduit **51** back into the pump **35** and is then forced down the tubular conduit **15** again.

## FIGURE 7.1:  SOLUTION MINING OF COAL

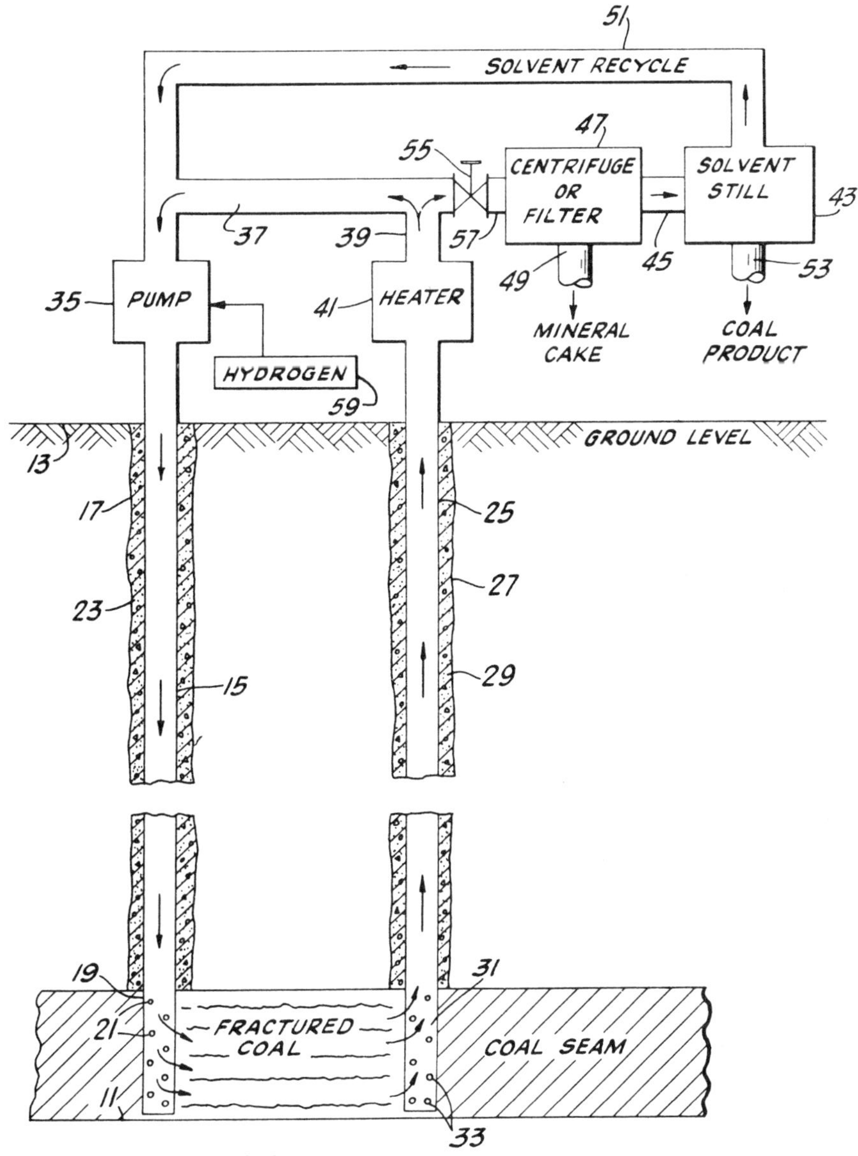


Source:  U.S. Patent 3,990,513

The coal product that is separated from the solvent flows from the solvent still through a conduit 53 into another apparatus (not shown) available for processing the coal.  In order to regulate the flow of solvent and coal in solution into the centrifuge 47, a valve 55 is provided in conduit 57 connecting conduit 39 and the centrifuge.

In some applications it is desirable to add hydrogen from a convenient source of supply 59 to the pump 35, which mixes with the solvent recycle being pumped down the conduit 15.  The hydrogen aids in the digestion or solution of the coal and it also prevents the coal from coking while being processed and handled.

In operation it is desirable and necessary in order to effect any solution of the coal in the seam to bring the coal-solvent interface to a temperature of at least 250°C (480°F) and preferably to a temperature in the range of 300° to 400°C (570° to 750°F).  At the higher end of the temperature scale the rate of dissolving coal is greater than at the lower end of the temperature scale.

Also, it is desirable and necessary in order to contact as much surface of the coal with the solvents as possible to fracture the coal seam in the vicinity of the conduits 15, 25 according to well-known conventional techniques.  Figure 7.1 suggests such a fractured coal seam between the conduits.

Once the method has become operative, a certain amount of solvent and coal are, after heating, returned to the pump through conduit 37, and the balance of the solvent and coal are diverted through the centrifuge 47 and the still 43. Then the recycle solvent is returned to the pump through the conduit 51.

A benefit derived from the method of the process is that heat is brought rapidly to the coal seam face and the pumping action provides agitation that enhances dissolving the coal.  It is contemplated that as much as 85 to 95% or more coal can be dissolved in accordance with this method, depending upon the coal itself and the solvent used.  However, in order to avoid pumping new viscous digestion, it is proposed to operate in the range of from 10 to 20% coal solution.

By using one tubular conduit 15 to inject the hot solvents into the coal seam, and the other conduit 25 to carry the coal solution to the surface and to separating equipment, a great deal of coal surface is exposed to the solvent which results in fast dissolution of the coal.  Further, there is a minimum loss of solvent, and even at the termination of production from a particular coal seam the residual solvent can be recovered by chasing it out with steam or hot water.

It is known from the prior art to hydrogenate coal underground with a hydrogenating agent.  The hydrogenation is carried out at an elevated temperature and pressure which may be controlled from above ground.  The hydrogenation reaction produces a liquid product which is readily recovered as a relatively heavy oil that may be further treated to produce motor fuels and other desired products.

The prior art teaches using free hydrogen or an organic compound capable of liberating hydrogen.  Hydroaromatic oils act as solvents in the liquefaction of coal due primarily to their ability to transfer hydrogen to the coal.  A typical liquid product obtained by underground hydrogenation comprises 60 to 70%

liquid hydrocarbons and 10 to 30% gaseous hydrocarbons. In contrast to the known prior art teachings, the method of the process includes the extraction of the in situ coal by dissolving and not by liquefying. That is to say, the method of this process dissolves and extracts coal in almost its original form whereas the liquefying method known from the prior art extracts a liquid product, not coal, comprising the liquid and gaseous hydrocarbons mentioned previously.

The solvent is an aromatic hydrocarbon or creosote or heavy oil. The molecular configuration of the aromatic solvent is important. Phenanthrene and other 3 and 4 angular member ring compounds, such as fluoranthene, pyrene and chrysene, and N-containing 3-membered ring aromatics, like carbazole and acridine, are suitable as solvents.

These solvents can dissolve 85 to 95% or more coal, while linear aromatics, such as anthracene and fluorene can dissolve less than 25% coal. For example, phenanthrene will dissolve up to 95% of the organic matter in coal, whereas anthracene will dissolve only 24%. For this reason, a middle oil fraction having a boiling point between 325° and 400°C is usually preferred. While tetralin has proven to be a good coal solvent, it acts as a hydrogen donor, reverting partially to naphthalene, and for this reason it is impractical commercially.

The solvent can be any one of the preferred compounds or their mixtures (eutectic composition preferably) and should be heated to above the melting point, or dissolved in a suitable carrier liquid, such as heavy oil, using a minimum of 5% of the preferred solvent in the heavy oil.

Those skilled in the art will also understand that the amount of coal which can be dissolved depends on the rank of the coal as well as the molecular configuration of the solvent. Using the solvents mentioned previously, up to 95% of the organic matter is dissolved, thus permitting the separation of mineral matter and fusain. A substantial portion of sulfur is concomitantly removed.

## DESULFURIZING COAL SLURRIES IN A PIPELINE

This process is one for reducing the pyritic sulfur content of coal while it is being transported in a pipeline in the form of a water slurry. The process preferably is conducted at a temperature of 70° to 100°F at ambient pressure with a concentration of oxidant ranging from 1 to 1.5 times the stoichiometric amount of sulfur estimated present in the coal and with a turbulence ranging from a Reynolds No. of about 2,000 to 3,000.

The process described by *E.L. Cole and M.A. McMahon; U.S. Patent 3,993,456; November 23, 1976; assigned to Texaco Inc.* is conducted in the pipeline normally used to transport the coal. It is particularly applicable to:

(1)    Those coals having a total sulfur content such that when the pyritic sulfur is removed by the subject process, the coal will then meet the EPA limits for $SO_2$ emissions when burned. It is estimated that about 10% of the present coal reserves in the United States have low enough organic and sulfate sulfur content such that with pyrite removal, present emission laws can be met.

(2)    Those coals being pipelined.

The process is further illustrated by Figure 7.2 showing schematically the steps involved. The diagram shows a coal-water slurry preparation plant **10** and a pipeline system **12** together with pumps for moving the slurry. The process involves the injection of the oxidant solution into the pipeline upstream of the dewatering plant **20**. The injection of the oxidant can be made at a single point **22** or at multiple points **24, 26** along the pipeline, the only requirement being that sufficient time be allowed to permit utilization of the oxidant by the pyrite. Advantageously, the point or points of injection are made as close as possible to the dewatering plant due to the corrosive nature of oxidation products. Optionally, the section of the pipeline where oxidation is conducted may be a corrosion-resistant steel or have a corrosion-resistant lining.

## FIGURE 7.2:  DESULFURIZING COAL SLURRIES IN A PIPELINE

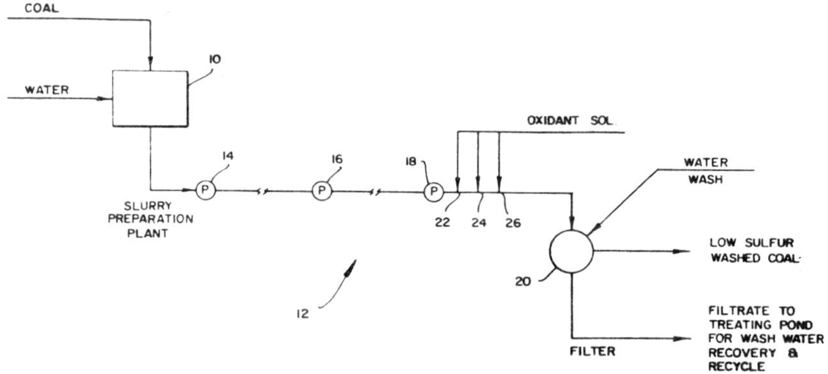


Source:  U.S. Patent 3,993,456

It is important to reduce the total contact time and for this several courses are available. One can raise the temperature, from 70° to 100°F; increase the concentration of oxidant solution, from 1 to 1 stoichiometric to 1.5 times the stoichiometric requirements; increase the mixing by increasing the turbulence in the pipeline, increasing the Reynolds No. from <2,000 to >2,000, particularly above 3,000. A combination of the enumerated variables will be best for particular slurry densities. The process is further illustrated by the following example.

*Example:* A pipeline system as shown in the above figure is being used to transmit 570 tons per hour of coal as a slurry comprising 53% dry coal and 47% water. The size of the coal comprising the slurry is as follows:

| U.S. Standard Sieve | Weight Percent |
|---|---|
| +20 | 12.5 |
| −20+325 | 67.3 |
| −325 | 20.2 |

The analysis of the coal is as follows:  8,020 Btu per pound (as received); 0.70% total sulfur divided as follows:  0.48% organic sulfur, 0.17% pyritic sulfur, 0.05% sulfate sulfur.

The slurry is pumped at a rate of 4 miles per hour for a total distance of 52 miles.  The ambient temperature averages 82°F and the temperature of the slurry is about 80°F.  The slurry is treated with a 20% stoichiometric excess of hydrogen peroxide as a 10% water solution.

The peroxide solution is injected into the pipeline in two equal portions.  The first feed point is 6 miles removed from the dewatering plant, while the second point of injection is 5.5 miles from the dewatering plant.  As the result of adding the oxidant the temperature in the pipeline rose to somewhat over 100°F.  The coal is recovered at the dewatering plant.

Analytical tests on the coal are as follows:  7,950 Btu per pound (as recovered); 0.39% organic sulfur; 0.03% pyrite sulfur and 0.05% sulfate sulfur.


## BENEFICIATION OF BROWN COAL

Numerous materials that can be used as solid fuel contain so many impurities that a beneficiation is required to make them suitable for use.  Examples of such solid fuels are the various kinds of brown coal, such as subbituminous coal, lignite and unconsolidated brown coal, the various kinds of coal, as well as peat, wood, paper, bitumen or asphalt, etc.  All these materials can often only be used as a fuel after a pretreatment.

In some cases the beneficiation aims at greatly reducing the water content of the material (brown coal, peat, wood, paper).  In other cases it is in particular the ash content that is reduced in the beneficiation (brown coal, coal).  Often the sulfur content has to be reduced as well (bitumen, asphalt, brown coal, coal).

It will be clear that beneficiation not only improves combustion properties, but also reduces the cost of subsequent transport of the fuel.  That in some cases this may mean a considerable saving becomes clear when it is considered that, for example, certain kinds of brown coal contain up to 70% by weight of mainly chemically bound water and up to 40% by weight of ash-forming constituents.  In the past it was attempted to find means of removing either the ash, the water or the sulfur compounds.

It has been proposed to remove sulfur compounds from coal by subjecting an aqueous suspension of coal particles at elevated temperature to a heat treatment in the presence of chemicals which cause the water-insoluble sulfur compounds from the coal to be converted into soluble compounds.  However, a considerable drawback of this proposal is that the subsequent separation of the coal particles and the aqueous phase of mechanical means, such as a centrifugal filter, etc., is far from complete.  It yields a moist mass of coal particles containing not less than about 30% by weight of water.  This mass must then be dewatered thermally, the result being that the dissolved sulfur compounds present in this residual water find their way into the coal again.

An aim of the process developed by *E. Verschuur; U.S. Patent 4,080,176; March 21, 1978; assigned to Shell Oil Company* is to obviate this drawback and to make the thermal dewatering step superfluous.  The primary purpose of the process is to conduct the thermal dewatering of brown coal under conditions such that desulfurization is also achieved.

In accordance with the process, a heat treatment at elevated pressure is applied to an aqueous suspension of the solid fuel in a finely divided form in the presence of an additive which, under the prevailing conditions, causes water-insoluble sulfur compounds to be converted into soluble compounds, after which fuel and water are separated.

The separation of fuel and water after the heat treatment is carried out by forming in the suspension, under conditions of turbulence and in the presence of a hydrocarbon-based water-insoluble binder, agglomerates of fuel and binder and by separating the agglomerates formed from the aqueous phase.

Since the agglomerates thus obtained are water-repellent, a complete separation of water and fuel can be effected in a simple manner (for instance, mechanically). An important additional advantage is that in the process the fuel is simultaneously deashed, because the fuel particles are incorporated into the agglomerates, whereas the ash particles are not, so that in the subsequent separation the fine ash particles are removed in the water phase and the agglomerates stay behind.

For the agglomeration of the fuel particles any water-insoluble binder may be used which contains hydrocarbons that wet the fuel particles and which, under turbulent conditions (depending on the conditions), make them ball up to form soft cohesive agglomerates or hard pills. These agglomerates or pills are much easier to separate from the aqueous phase than the individual fuel particles; for instance, by taking them up into a light hydrocarbon fraction and effecting phase separation, or by collecting them on a sieve. Examples of suitable binders are naphtha, fuel oil, bitumen and toluene.

The heat treatment of the aqueous suspension of finely divided solid fuel is usually carried out at temperatures above 150°, and preferably above 250°C, at a pressure that is higher than the vapor pressure of water at that temperature. This renders it unnecessary to provide the heat required for the evaporation of the water.

Preferably, the aqueous suspension comprises substantially particles that are smaller than 2 millimeters and at least 30% by weight of free water. These particles are then small enough and the suspension is sufficiently diluted for a fair degree of beneficiation of the fuel to be effected within an acceptable time.

An eligible additive is, in principle, any substance or combination of substances by means of which free sulfur, organically bound sulfur and/or insoluble inorganic sulfur compounds can be taken into solution. Numerous substances are known which are more or less capable of producing this effect. An example is the combination iron/oxygen compound. Another example is sodium or calcium hydroxide.

This process is particularly suitable for application in the beneficiation of brown coal. Brown coal contains a relatively large percentage of chemically-bound water and often also ash-forming constituents and sulfur compounds. It is known that at a high temperature brown coal not only gives off chemically-bound water, but also undergoes such a change that no complete reabsorption of water will take place. Consequently, in the heat treatment applied, chemically-bound water is irreversibly expelled from the brown coal. This phenomenon is sometimes called carbonization. The process also effects a deashing and a desulfurization.

It is certainly possible to use a permanent binder for agglomerating the solid fuel. This is necessary in particular when it is desired to transport and/or use the agglomerated fuel as such. An advantage here is the simple transport of the agglomerates. A disadvantage may be that the price of the binder is higher than that of the solid fuel itself.

It will, therefore, in some cases be preferred after separation of the aqueous phase and the agglomerates to remove the binder from the agglomerates and to recycle it. This may, for instance, be done by heating the agglomerates in a fluidized bed to a temperature above the boiling point of the binder.

Under certain circumstances it will be preferred to add at least part of the binder to the aqueous suspension before the heat treatment. The contact time between binder and fuel particles will thus be increased, which may accelerate the subsequent agglomeration. In addition, it permits the use of a binder containing sulfur because the sulfur compounds will disappear from the binder during the heat treatment of the aqueous suspension. The use of a binder containing sulfur may mean a considerable saving.

*Example 1:* 500 g of brown coal in lumps was ground with 500 g of water in a colloid mill to a homogeneous suspension. The 500 g brown coal comprised about 300 g water, 180 g coal free of ash and sulfur, 8 g ash, 8 g organic sulfur and 4 g pyritic sulfur. To the suspension obtained was added 66 g heavy fuel oil containing 6 g organic sulfur and 30 g calcium hydroxide.

The mixture obtained was vigorously stirred in a high-pressure autoclave and heated to 300°C in 1 hour, the pressure increasing to about 100 atm. Thereupon the mixture was cooled to 20°C in 2 hours. The gases in the autoclave, which were collected separately and analyzed, were found to consist substantially of $N_2$ and $CO_2$.

The liquid containing the solids was passed through a sieve with an aperture size of 0.5 mm. In the sieve hard globular agglomerates (pills) were left, which after drying at 110°C were found to contain 30 g water. From the dry pills, 62 g liquid hydrocarbon were extracted containing 2 g organic sulfur. It is assumed that this is substantially heavy fuel oil. It contained 170 g coal free of ash and sulfur; 2 g ash, 2 g organic sulfur; and 1 g pyritic sulfur. The calorific value of the ash- and sulfur-free coal was 27.2 MJ/kg before and 31.5 MJ/kg after the treatment described hereinbefore. The percentage of bound and/or unbound oxygen of this coal dropped from 20 to 8% as a result of this treatment.

*Example 2:* To a suspension identical with that of Example 1 were added 70 g of a light, sulfur-free crude oil fraction and 60 g colloidal fine iron oxide powder. This mixture was warmed up to 300°C in 1 hour in a vigorously stirred autoclave and subsequently cooled down to 20°C. The pills formed were also sieved off and, after drying at 110°C (which caused a light hydrocarbon to be released), were found to contain 40 g water.

The dry solids had the following composition: 172 g coal free of ash and sulfur; 4 g ash; 1 g organic sulfur; and 0.2 g pyritic sulfur. The calorific value of the ash- and sulfur-free coal was 27.1 MJ/kg before, and 29.9 MJ/kg after treatment. The oxygen content also dropped from 20 to 8%.

# PHYSICAL DESULFURIZATION
# A CASE STUDY

Material in this chapter was taken from the report written by
Seongwoo Min (IS-ICP-35) for the Energy and Mineral Resources
Research Institute of Iowa State University. The report is covered
in some detail in order to show how the problem of desulfurizing
coal can be studied on a laboratory scale. The results from a study
of this kind can then be applied on a large scale, with appropriate
calculations as to the economics of the various treatments used compared to their effectiveness.


## INTRODUCTION

The development of a practical process for reducing the sulfur content of coal
is of great importance owing to the scarcity of low sulfur coal and the stringent
standards for sulfur dioxide emissions from coal burning power plants. In particular, coal is Iowa's only fossil fuel resource which occurs in large enough reserves to be of much interest as an energy source to meet the growing demand
for energy within the state. Iowa coals, however, are high in sulfur and thus
are not environmentally acceptable in the unbeneficiated state.

Sulfur in coal is chiefly present as organic sulfur and pyritic sulfur. Although
the organic sulfur, being molecularly bound to the coal matrix, cannot be removed by any mechanical cleaning method, it is possible to remove liberated
pyritic sulfur by various mineral beneficiation methods.

The theory and practice of modern methods of coal preparation have been described by Leonard and Mitchell (1). Although there are a large number of
potential methods which can be utilized for removing pyritic sulfur from coal,
the applicability of various methods is heavily dependent on the particle size
of the coal to be treated. The degree of pyritic sulfur reduction depends upon
the amount of free pyrite liberated by comminution prior to cleaning. Most commercial coal cleaning methods rely on the difference in specific gravity between
coal and its impurities, and they are effective only on relatively coarse coal, i.e.,

+35 mesh (2). Many high sulfur coals, however, contain finely disseminated pyrites which can only be liberated by very fine grinding. Several methods, including those based on dense medium cyclones, hydrocyclones and froth flotation are used commercially for washing fine-size coal. Their effectiveness, however, varies widely with the particular coal to be washed. In general, physical cleaning of coal is practically limited to removing approximately one-third of the sulfur and one-half of the mineral matter (3).

Among the commercially applied coal cleaning methods, froth flotation is the only one where the separation is not based upon a difference in specific gravity. In froth flotation, the separation is based on the difference in surface characteristics of coal and inorganic minerals. The underlying principles of froth flotation have been described by Gaudin (4). Although there is a large body of technical literature about froth flotation, the mechanism of coal flotation is not fully understood and the potential for improvement of this method appears to be great.

Another method which utilizes the difference in surface properties between coal and mineral matter to effect a separation is one involving selective oil agglomeration of fine-size coal particles suspended in water (5)(6)(7). The relatively large coal and oil agglomerates can be separated readily from the unagglomerated mineral particles by screening. Although the effectiveness of this method has been demonstrated experimentally, it is not presently used commercially.

In order to effect a better separation of coal and mineral matter, chemical comminution has been proposed as a more effective means of liberating the mineral matter than mechanical crushing (8)(9). This method of comminution involves soaking coal in liquid anhydrous ammonia or other suitable chemical agents so that the coal comes unglued along bedding planes and boundaries between coal and mineral matter and literally falls apart.

The work described in this report was undertaken to compare and evaluate the effectiveness of different coal preparation and cleaning methods including gravity separation, froth flotation, oil agglomeration, chemical comminution and mechanical crushing and grinding to see which ones offer the greatest potential for use in desulfurizing Iowa coal. In addition the oil agglomeration and froth flotation processes were selected for more intensive research and further development.

## MATERIALS AND EQUIPMENT USED

The coal samples used in this work were collected from several Iowa coal mines. The proximate analysis and sulfur distribution of these coal samples are given in Table 8.1. The proximate analysis was determined by ASTM method D271 and the sulfur distribution by ASTM method D2492 (10).

Most of the samples were grab samples which were not necessarily representative of the coal in the mines where they were obtained. In experimental studies of these coals, the practice was to prepare kilogram quantities of the coals with the particle sizes of interest. Representative samples of these kilogram quantities were then taken and analyzed to establish controls for the studies.

## TABLE 8.1: CHEMICAL ANALYSIS OF IOWA COALS

| Coal Mine | ..... Proximate Analysis, %. ..... | | | | .... Sulfur Distribution, % .... | | | |
|---|---|---|---|---|---|---|---|---|
| | Moisture | Ash | Volatile Matter | Fixed Carbon | Pyritic | Organic | Sulfate | Total |
| Star | 0.92 | 10.81 | 43.26 | 45.01 | 3.33 | 2.26 | 0.16 | 5.75 |
| Big Ben | 1.18 | 12.52 | 37.50 | 48.80 | 2.05 | 0.87 | 0.16 | 3.08 |
| ICO* | 4.51 | 8.92 | 43.98 | 42.59 | 2.41 | 0.99 | 0.05 | 3.45 |
| Jude | 8.79 | 14.18 | 39.42 | 37.61 | 2.97 | 3.53 | 0.44 | 6.94 |
| Scott** | 8.26 | 14.31 | 34.52 | 42.91 | 3.64 | 4.00 | 0.59 | 8.23 |

*Channel sample.
**Sample from ISU Demonstration Mine #1.

Throughout this investigation, Iowa State University tap water (11) was used for all experiments except where otherwise noted. For the size reduction and separation experiments, the following equipment was used:

(1) Bench-scale double roll crusher (Smith Engineering Works, Milwaukee, Wisconsin). Fifty pounds of coal per hour could be crushed to ¼ inch top size with this machine.

(2) Mikro-Samplmill (Pulverizing Machinery Division, American-Marietta Co., Summit, New Jersey). Five pounds of coal per hour could be pulverized to –35 mesh size with this device. The size distribution of the pulverized coal is shown in Table 8.2.

(3) Disk Mill (BICO Inc., Burbank, California). Four pounds of coal per hour could be pulverized to –100 mesh size with the disk mill.

(4) Blender (Sears Insta-Blend Model No. 400). One thousand milliliters of slurry could be agitated at a speed of up to 20,000 rpm with the 14 speed blender.

(5) Wemco Flotation Machine (Western Machinery Co.). Two liters of slurry could be agitated at a speed of 1,725 rpm and processed by froth flotation in this unit.

(6) Benco Agitator (Bench Scale Equipment Co., Dayton, Ohio). Up to 6 gallons of slurry could be agitated at different speeds in the range of 0 to 1,100 rpm with this unit.

## TABLE 8.2: SIZE DISTRIBUTION BY WEIGHT OF COALS PULVERIZED BY MIKRO-SAMPLMILL

| Size Tyler Mesh | ............ Distribution, % ............ | | | |
|---|---|---|---|---|
| | Star | Big Ben | Jude | Scott |
| –35+65 | 4.33 | 7.78 | 16.20 | 20.20 |
| –65+100 | 36.90 | 34.50 | 24.39 | 31.73 |
| –100+200 | 30.33 | 35.30 | 30.05 | 19.62 |
| –200 | 28.44 | 22.40 | 29.34 | 28.45 |

## OIL AGGLOMERATION

Oil agglomeration provides a method for collecting coal fines in an aqueous suspension. When a small amount of oil is added to a coal slurry under vigorous agitation, the oil selectively collects hydrophobic coal particles to form agglomerates which can be separated from the aqueous mixture. The water wetting minerals remain suspended in the water and can be rejected. Sirianni et al (5) reported that agglomerates could be produced in a wide variety of equipment such as blenders, paint shakers, and simple agitated tanks. Capes et al (6) found that the reduction of pyritic sulfur accomplished by this process was up to 90% when the coal was pretreated with the iron-oxidizing bacterium, *Thiobacillus ferroxidans*. Investigators at the National Research Council of Canada (7) described a selective oil agglomeration process for cleaning coal fines in two steps, flocculation followed by balling.

In order to evaluate the potential of the oil agglomeration process for cleaning Iowa coal and to further develop and perfect the process, a series of experiments was carried out to demonstrate the process under different conditions. The effectiveness of the process was evaluated in terms of the following quantities which were determined for each run:

$$\text{Yield (\%)} = \frac{\text{Weight of Dry Product}}{\text{Weight of Dry Feed}} \times 100$$

$$\text{Ash Reduction (\%)} = \frac{\text{\% Ash in Feed} - \text{\% Ash in Product}}{\text{\% Ash in Feed}} \times 100$$

$$\text{Sulfur Reduction (\%)} = \frac{\text{\% Sulfur in Feed} - \text{\% Sulfur in Product}}{\text{\% Sulfur in Feed}} \times 100$$

$$\text{Oil Consumption (\%)} = \frac{\text{Weight of Oil Used}}{\text{Weight of Feed}} \times 100$$

### Effect of Oil

Preliminary experiments indicated that the effectiveness of the oil agglomeration method is sensitive to the type and amount of oil used as well as the method of adding the oil to the coal-water slurry.

*Method of Adding Oil:* Tests were made to compare adding oil directly to the aqueous coal slurry with adding it as an oil-water emulsion. Twenty grams of pulverized Scott coal (–35 mesh) were slurried with deionized water in a cylindrical container (2½" diameter x 5" height). Two hundred milliliters of water were used for the straight oil experiments and 100 ml of water for the emulsified oil experiments. After the slurry was agitated for five minutes at a speed of 350 rpm with the Benco agitator using a 2½" diameter pitched blade turbine impeller, a measured amount of kerosene-No. 5 heavy fuel oil mixture (SG = 0.83) was added either directly or as an emulsion of oil in 100 ml of deionized water prepared with an ultrasonic vibrator. The agitation was continued for a specified time and speed. The coal slurry was then poured onto a 60 mesh screen to recover the coal. The collected coal was dried overnight at 100°C, weighed and analyzed for ash and sulfur content.

The results show that when an emulsion of oil and water was used, a high yield of agglomerated coal was obtained over the whole range of agitator speed and/or retention time. On the other hand, when straight oil was used, an agitator speed of 1,150 rpm and retention time of 3 minutes or more were needed to obtain a high yield. Use of emulsified oil did not have much effect on ash and sulfur reduction. It could well be that an oil and water emulsion is a necessary prerequisite for agglomeration of coal because even where straight oil is used the conditions necessary for agglomeration are likely to produce an emulsion.

*Type of Oil:* The effectiveness of eight different petroleum oils (Table 8.3) for agglomerating coal was studied with another series of experiments.

The oil was added as an oil-water emulsion produced from one part of oil and 40 parts of tap water with an ultrasonic vibrator. Kerosene, heater oil and furnace oil were readily emulsified without any emulsifying agent, but heavier oils required a small amount of silicon surfactant (L520, Union Carbide Corp.) as an emulsifying agent. For No. 6 fuel oils, 50 volume percent kerosene was added to reduce the viscosity of the oil.

## TABLE 8.3: SPECIFICATIONS OF PETROLEUM OILS* USED FOR AGGLOMERATING COAL

| Oil | Specific Gravity at 60°F | Viscosity (cs) | Pour Point (°F) |
|---|---|---|---|
| Kerosene | 0.807 | — | — |
| Heater oil | 0.811 | — | — |
| Furnace oil | 0.855 | 2.08 | -20 |
| No. 200 fuel oil, LLS | 0.916 | 13.11 | -20 |
| No. 5 light fuel oil, LLS | 0.934 | 49.70 | -10 |
| No. 5 heavy fuel oil, LLS | 0.946 | 58.89 | 0 |
| No. 6 fuel oil, LLS | 0.959 | 350.43 | +30 |
| No. 6 fuel oil | 0.973 | 297.17 | +30 |

*Source of oils and data was Amoco Oil Co.

Fifteen grams of pulverized coal (-35 mesh) were slurried with 500 ml of tap water in the blender and the pH of the slurry was adjusted to 6 to 7 by adding small amounts of 1 N potassium hydroxide solution. Twenty milliliters of an oil-water emulsion were added to the slurry and the blender was operated for 10 minutes at the lowest speed. If agglomerates formed and floated, the float product was skimmed off onto a 100 mesh sieve to recover the agglomerated coal. If the agglomerates were too small, more of the oil-water emulsion was added until stable agglomerates formed. The recovered coal was dried, weighed and analyzed for ash and sulfur contents.

The experimental results showed that the oil consumption decreased with an increase in the specific gravity but the yield appeared to be constant. Since it appeared that maximum ash and sulfur reductions could be obtained with oil having a specific gravity of 0.82 to 0.84, further tests were made with several oil mixtures having a specific gravity of 0.83.

Jude coal was pulverized to two different top sizes, -35 mesh with the Mikro-Samplmill and -100 mesh with the disk-mill. Twenty grams of pulverized coal

was slurried with 200 ml of tap water in the blender. At the same time, a specified oil mixture was emulsified with 200 ml of tap water in another blender for
5 minutes. The oil-water emulsion was then poured into the coal slurry and the
mixture was agitated for 5 minutes. The slurry was placed in a 1,000 ml separatory funnel and the floated coal was separated by draining off the water containing suspended ash. The recovered coal was filtered, dried and analyzed.

For these tests either kerosene or heater oil was blended with one of the heavier
oils to provide a blend having the desired specific gravity (0.83). Although less
oil was used than during the previous series of tests with –35 mesh Jude coal,
the yield of agglomerated coal (about 80%) was generally as good as with either
lighter or heavier oils. On the other hand, the ash and sulfur reductions were
quite varied, evidently depending upon the particular heavy oil used in the blend.
Thus the use of blends containing either light or heavy No. 6 fuel oils resulted
in poor rejection of sulfur and ash in the case of –35 mesh coal.

The blend of kerosene and No. 5 heavy fuel oil provided the greatest reduction
in sulfur and ash but also gave the lowest yield for –35 mesh coal. The results
with –100 mesh coal were less varied and the yield (about 90%) was uniformly
higher than that obtained with the coarser material.

*Amount of Oil:* A series of tests was carried out to examine the effect of oil
concentration on the agglomeration process. Heater oil and two different coals
were selected for the series. Star coal which had been ground with a disk-mill
and screened through a 100 mesh sieve was one of the coals selected and Scott
coal which had been ground for 16 hours in a 1½ gallon jar mill with 1,900
grams of flint pebbles was the other. An aqueous slurry containing 40% solids
was placed in the ceramic jar mill for the grinding operation; the product was
–400 mesh. For the agglomeration of ground Star coal, 15 grams of coal were
mixed with 500 ml of tap water in the blender and a 1 N solution of potassium
hydroxide was added to raise the pH to about 6.

A measured amount of heater oil which had been emulsified with 200 ml of tap
water by using an ultrasonic vibrator was added to the coal slurry. After blending the mixture for 5 minutes, the agglomerated coal was recovered by pouring
the mixture onto a 100 mesh sieve.

For the agglomeration of ground Scott coal, 500 ml of a 4 weight percent aqueous slurry of the coal was placed in the blender and a measured amount of
heater oil was added directly to the slurry. After 5 minutes agitation, the slurry
was put in a 1,000 ml separatory funnel and the floated coal was recovered.
The coal was then placed back in the blender and slurried again with 200 ml
of fresh tap water. After 2 minutes of agitation, the slurry was again poured
into the separatory funnel and the floated coal was recovered. This washing
procedure was repeated once more and the product was filtered, dried and analyzed.

The experimental results are shown in Table 8.4. The results for –100 mesh
Star coal show that the yield increased with increasing oil concentrations in the
product up to about 14% while the reduction in sulfur and ash contents declined.
On the other hand, for –400 mesh Scott coal the yield appeared to be essentially independent of the amount of oil used over a wide range while the reduction in sulfur and ash contents increased sharply as the amount of oil used in-

creased at first but above 20% oil in the product less sulfur and ash were rejected. However, the maximum reduction in sulfur and ash contents was much larger for Scott coal than for Star coal. This difference in behavior was due more likely to the difference in particle size than to some other difference in the properties of the two coals because the coals were quite similar in other respects.

### TABLE 8.4: OIL AGGLOMERATION TESTS USING DIFFERENT AMOUNTS OF HEATER OIL

| Oil (ml) | . . Star Coal* Product . . . | | | Oil (ml) | . . Scott Coal** Product . . | | |
|---|---|---|---|---|---|---|---|
| | Yield | Ash | Total S | | Yield | Ash | Total S |
| | . . . . . . . . %. . . . . . . . | | | | . . . . . . % . . . . . . . . . . . | | |
| 0.25 | 23.1 | 3.98 | 4.47 | 1 | 77.2 | 11.27 | 6.44 |
| 0.50 | 29.9 | 4.80 | – | 3 | 86.8 | 8.09 | 5.47 |
| 1.00 | 48.4 | 5.46 | 4.96 | 5 | 76.7 | 4.30 | 4.10 |
| 1.25 | 49.2 | 5.39 | – | 10 | 83.6 | 4.65 | 4.27 |
| 1.50 | 69.8 | 5.09 | – | 20 | 83.0 | 4.86 | 4.33 |
| 2.50 | 88.0 | 6.17 | 5.37 | 60 | 88.0 | 6.13 | 5.03 |
| 3.00 | 85.7 | 6.49 | – | – | – | – | – |

*Untreated Star coal contained 10.81% ash and 5.75% total sulfur.
**Untreated Scott coal contained 15.73% ash and 7.58% total sulfur.

## Effect of Particle Size

*Pulverized Coal:* In order to evaluate the effect of coal particle size on the oil agglomeration process, another series of experiments was carried out. First, the coal to be treated was pulverized with the Mikro-Samplmill and separated into four different particle size fractions (see Table 8.2). Fifteen grams of coal of a given size fraction was slurried with 500 ml of tap water in the blender and the pH was adjusted to 6 to 7. Twenty milliliters of an oil-water emulsion, which was prepared from one part of heater oil and 40 parts of water with the ultrasonic vibrator, was added to the coal slurry and the blender was operated for 10 minutes at the lowest speed.

If agglomerates formed and floated, the float product was skimmed off. If the agglomerates were too small to float, more of the oil-water emulsion was added until the agglomerates just floated. The results indicate that the three different coals which were used behaved somewhat differently. Thus better yields were obtained with Star coal than with the other two coals and for both Star coal and Big Ben coal the yield was virtually independent of particle size, whereas for Jude coal the yield changed with particle size. The oil consumption, however, increased as the particle size decreased for all three coals. Furthermore, the maximum percentage reduction in sulfur and ash contents was generally obtained with the smallest particle size fraction. On the other hand, the minimum percentage reduction in sulfur content was obtained in every case with one of the intermediate size fractions (–100+200 mesh).

*Ground Coal:* A study of coal microstructure with a scanning electron microscope by Greer (12) revealed the presence of much finely disseminated pyrite in Iowa coals. Therefore, it appeared that very fine grinding would be required to liberate this pyrite so that it could be separated. Thus, another series of experiments was conducted to see if oil agglomeration of finely ground coal would produce a cleaner product than oil agglomeration of somewhat coarser material.

In addition the effect of grinding finer and finer on the separation obtained with oil agglomeration was examined.

To prepare the coal for the oil agglomeration tests, 200 grams of pulverized (−35 mesh) Jude coal in a 20 weight percent aqueous slurry (or 400 grams of pulverized Scott coal in a 40 weight percent slurry) were ground in a jar mill containing 1,900 grams of flint pebbles. The size distribution of the ground coal was determined with a Coulter Counter. The data show that after 4 hours of grinding, most of the particles were finer than 38 $\mu$ (400 mesh) and as grinding time increased, the particles became smaller and the size distribution became narrower.

Oil agglomeration experiments of Jude coal were conducted in the blender using 300 ml of a 6 weight percent aqueous slurry of ground coal. The coal was agglomerated with an oil-water emulsion containing 1 ml of a mixture of heater oil and No. 5 light fuel oil (SG = 0.83) and 200 ml of tap water. The agglomerated coal was placed in a separatory funnel and recovered. The recovered coal was then put back into the blender and washed twice with 200 ml of fresh tap water.

The experimental results obtained with Jude coal indicate that the cleanest product was obtained with coal which had been ground for 4 hours. This product not only contained the least sulfur and ash, but it also was obtained in greater yield than that provided by the agglomeration of pulverized but unground coal or coal which had been ground for 6 hours. Prolonged grinding for up to 22 hours resulted in larger yields of agglomerated coal at the expense of increased ash and sulfur in the agglomerated product.

Similar experiments were carried out with Scott coal. Five hundred milliliters of a 4 weight percent aqueous slurry of ground Scott coal were placed in the blender and mixed with an emulsion containing 1 ml of blended heater oil and No. 5 light fuel oil (SG = 0.83) and 200 ml of tap water. The agglomerated coal was recovered in a separatory funnel and washed twice with fresh tap water as described above. The results were similar in some respects to those obtained with Jude coal but different in other respects. Thus, a cleaner product in higher yield was obtained in both cases when the coal was ground for several hours before agglomeration than when it was only pulverized in the Mikro-Samplmill. Also prolonged grinding was detrimental to product quality in both cases.

On the other hand, the optimum grinding time appeared to be somewhat longer for Scott coal than for Jude coal. However, this result could have been due to differences in experimental conditions as well as a difference in the two materials. Since twice as much coal was charged in the jar mill when Scott coal was ground as when Jude coal was ground, it may have required longer to reduce Scott coal to the same size as Jude coal. Also the lower slurry concentration used while agglomerating Scott coal may have had some effect on the results.

**Effect of Slurry Concentration**

*Using the Blender:* A series of experiments was carried out to determine the effect of slurry concentration on the oil agglomeration process. A measured amount of pulverized coal (−35 mesh) was slurried with 500 ml of tap water in a blender and agitated for 5 minutes. Then, an emulsion of heater oil and

water, equivalent to 8 weight percent oil based on dry coal, was added to the
slurry and the blender was operated for 5 minutes at the lowest speed. The agglomerated coal was recovered by pouring the slurry onto a 100 mesh sieve.
The product was dried, weighed and analyzed for ash and total sulfur. The
results indicate that the yield increased as the slurry concentration increased
for each of the three coals tested. The reduction in total sulfur content, however, declined as the concentration increased and the reduction in ash content
rose at first and then declined. In other words, the product contained the minimum amount of ash when a slurry concentration of 4 to 5% was employed.

*Using Froth Flotation Cell:* Pulverized coal was agglomerated in a Wemco froth
flotation cell in another series of experiments conducted to investigate the effect of slurry concentration. One thousand milliliters of tap water were placed
in the bowl of the flotation cell and a measured amount of pulverized Scott
coal (–35 mesh) was added. After 5 minutes of agitation, an emulsion of heater
oil and water, equivalent to 8 weight percent oil based on the dry coal, was
added and the slurry was agitated for 10 minutes. The agglomerated coal was
then floated by introducing air bubbles and was recovered by skimming. The
product was poured back into the bowl and washed twice with fresh tap water.

The results were similar to those obtained with the blender in that the yield of
agglomerated coal increased as the slurry concentration increased and also the
reduction in ash content at first increased and then decreased. On the other
hand, the results with the froth flotation cell were different in that the yield
of agglomerated coal was much lower and the reduction in sulfur content much
greater for small slurry concentrations. Also the product contained the least
sulfur and ash when a large slurry concentration (10 weight percent) was used
in the froth flotation cell.

These results seem to indicate that in dilute slurries the distance between particles is too great for interactions which produce agglomerates. Although this
problem may be alleviated by using more concentrated slurries, mechanical entrapment of refuse particles in the agglomerates may then become a problem.
Thus, it may be necessary to balance one factor against the other.

## Effect of pH and Different Chemical Agents

Since the separation of coal and pyrite particles was far from perfect in previous
oil agglomeration experiments, an attempt was made to improve the separation
by changing the pH or adding different chemical agents which might modify the
surface of the pyrite. Thus, if the surface of the pyrite could be made uniformly
hydrophilic through adsorption of some chemical species, there would be less
likelihood of the pyrite particles being agglomerated with the hydrophobic coal
particles.

*Effect of pH:* A number of experiments were carried out to examine the effect
of the initial pH of the water employed for preparing the coal slurries used for
oil agglomeration. For these experiments 500 ml of tap water were placed into
the blender and the pH was adjusted to the desired level by adding a 1 N solution of either hydrochloric acid of potassium hydroxide. Then 15 grams of pulverized coal (–35 mesh) were mixed with the water for 5 minutes at the lowest
speed. Twenty milliliters of an oil-water emulsion, which was made from one
part of heater oil and 40 parts of water with the ultrasonic vibrator, were added

to the slurry and the blender was operated for 10 minutes. If agglomerates formed and floated, the float product was ladled onto a 100 mesh screen. If the agglomerates were too small to float, more of the oil-water emulsion was added and agitation was continued. This operation was repeated until suitable agglomerates were produced. Interestingly enough, the pH of the slurry was always in the range of 7.5 to 8.5 after the agglomerates were produced, regardless of the initial pH of the water. Even though the buffering action of the coal was large enough to control the ultimate pH of the slurry, the initial pH of the water seemed to have an effect on the results. Thus for all three coals tested, the pyritic sulfur left in the product declined as the initial pH of the water was increased. Also for both Star coal and Big Ben coal, the ash content and yield of agglomerated material declined as the pH was raised.

*Different Chemical Agents:* A series of tests was carried out to determine the effects of various chemical agents at different concentrations on the results of oil agglomerating pulverized (–35 mesh) Star coal and Jude coal. For these tests, 15 grams of coal were mixed with 400 ml of tap water in the blender and the desired amount of a 0.2 weight percent solution of the chemical agent was added. After 5 minutes of agitation, the pH of the slurry was adjusted to 6 to 7 by adding a 1 N solution of either hydrochloric acid or potassium hydroxide. The oil agglomeration and recovery procedure were exactly the same as described above in the preceding section. The results were somewhat erratic and rather inconclusive.

*Effect of Chemical Agents at Different pHs:* Since the effect of any given chemical agent might well depend on the pH of its solution, a series of tests was made to determine the effect of various agents at different pH levels on the oil agglomeration process. For these tests 20 grams of pulverized coal were mixed with 400 ml of tap water and 10 ml of a 0.2 weight percent solution of the chemical agent. After 5 minutes of agitation in the blender, the pH was adjusted to the desired level by adding a 1 N solution of either hydrochloric acid or potassium hydroxide. Then an emulsion composed of 2.0 ml of a kerosene-No. 6 heavy fuel oil mixture (SG = 0.83) and 40 ml of tap water was added. The emulsion was prepared with the ultrasonic vibrator. The slurry was agitated for 5 minutes and the coal was recovered by ladling onto a 100 mesh sieve.

The results indicate that more sulfur was rejected from ICO coal by the oil agglomeration process when a small amount of ferric chloride was added to the coal slurry then when either sodium carbonate or sodium silicate were added. Moreover, ferric chloride appeared to have a beneficial effect over a pH range of 5 to 10. However, at higher pH levels the yield of agglomerated coal declined.

Unfortunately, a similar beneficial effect was not found when ferric chloride was used in conjunction with the oil agglomeration of Jude coal. The addition of various chemical agents and the simultaneous control of pH also affected the yield of agglomerated coal. Thus a very high yield of agglomerated ICO coal was obtained when sodium carbonate was added to the system, and relatively high yields of either agglomerated ICO or Jude coal were obtained when sodium silicate was added.

Another series of experiments was carried out to determine the effect of various chemical agents. For this series the chemical agent was ground with Scott coal in the jar mill before oil agglomeration. To prepare the coal, 400 grams of pulverized Scott coal (–35 mesh) and 4 grams of chemical agent were ground for 16 hours in the jar mill with 1,000 ml of tap water and 1,900 grams of flint pebbles. The ground coal slurry was diluted with tap water to a slurry concentration of 4 weight percent. Five hundred milliliters of the diluted slurry were placed in the blender and a small amount of 1 N solution of either sulfuric acid or sodium hydroxide was added. After 5 minutes of agitation the pH of the slurry was measured. The coal was then agglomerated with an emulsion composed of 1 ml of kerosene-No. 5 heavy fuel oil mixture (SG = 0.83) and 200 ml of tap water. The emulsion was prepared with the ultrasonic vibrator.

The agglomerated coal was recovered in a separatory funnel and washed twice with fresh tap water. The results show that a large percentage of sulfur and ash was removed at an optimum pH without any chemical agent at all. None of the chemical agents improved the separation of coal and mineral matter very much. Although coal with the lowest sulfur content was produced when potassium permanganate was added, the yield was also very low.

*Effect of Chemical Pretreatment:*  Another approach which was utilized to modify the surface properties of pyrite in order to improve the separation of pyrite and coal was to oxidize the surface of the pyrite. Presumably an oxide coating would be hydrophilic, whereas the unoxidized pyrite surface tends to be hydrophobic. To oxidize the pyrite surface, coal was pretreated with an aqueous solution containing dissolved oxygen and alkali.

For the pretreatment step, 100 grams of pulverized (–35 mesh) Scott coal was mixed with 500 ml of deionized water in a 1,000 ml reaction flask. A small quantity of an alkali such as sodium carbonate was added to the slurry and the mixture was heated to boiling. Air was then introduced through a sparger at a flow rate of 0.33 ft$^3$/min and the treatment was continued for a specified time at 80°C. After this treatment the slurry was diluted to 1,000 ml by adding more deionized water and cooled to room temperature.

A portion of the coal was then agglomerated by placing 200 ml of the slurry in the blender and mixing it for 5 minutes with an emulsion composed of 2 ml of kerosene-No. 5 heavy fuel oil mixture (SG = 0.83) and 200 ml of deionized water. The slurry was then poured into a 100 mesh sieve to recover the product.

It was found that even with the 5 minute pretreatment, the results were improved significantly. Although the greatest reduction in sulfur content was obtained in a run using 0.5 gram $Na_2CO_3$ + 0.5 gram $Ca(OH)_2$ for 30 minutes at 80°C, the results of a run in which 0.5 gram of NaOH was substituted for the $Ca(OH)_2$, were nearly as good.

*Effect of a Third Component:*  The effect of adding another solid component to a coal slurry undergoing oil agglomeration was investigated. A relatively coarse (+20 mesh) silica sand and a finer loess (–200 mesh) were utilized for this purpose.

To study the effect of sand, Scott coal was first ball milled for 16 hours to reduce its size to –400 mesh. For the agglomeration tests 500 ml of a slurry containing 20 grams of the ground coal were mixed in the blender for 5 minutes with an emulsion composed of 1 ml of a kerosene-No. 5 heavy fuel oil mixture

(SG = 0.83) and 200 ml of tap water.  The emulsion was prepared with the ultrasonic vibrator.  The agglomerated coal was recovered in a separatory funnel and washed twice with fresh tap water as described before.  A blank run was made first in which no sand was added.  In the next run 10 grams of sand were added to the already agglomerated coal.

In the washing stage and in the final run 10 grams of sand were added to the coal slurry before it was agglomerated.  The results show that adding sand during the agglomeration process was beneficial.  The greatest benefit was obtained when the sand was added before agglomeration took place.  Both the ash and sulfur content of the product were reduced by the addition of sand.  It appears that sand particles, which are both hydrophilic and heavy, may adsorb pyrite and gangue mineral slimes and thereby reduce the entrapment of such particles in the coal agglomerates.

To study the effect of adding loess, 20 grams of pulverized (-35 mesh) Scott coal were mixed with 100 ml of tap water and a specified amount of loess.  Mixing and agglomeration were accomplished with the Benco agitator using a speed of 1,050 rpm.  An emulsion composed of 2 ml of a kerosene-No. 5 heavy fuel oil mixture (SG = 0.83) and 100 ml of tap water was prepared with the ultrasonic vibrator and used for agglomeration.

Agglomeration was continued for 5 minutes before the slurry was poured onto a 60 mesh sieve to recover the product.  The results indicate that the addition of loess had a detrimental effect on both product yield and ash content.  As increasing amounts of loess were added, the yield declined and the ash content increased.  The corresponding decrease in the sulfur content of the product could be accounted for by the diluting effect of the increased ash content.

*Production of Larger Agglomerates:*  For ease of drying and material handling it is generally advantageous to produce relatively large agglomerates of coal and oil.  However, the particle diameter of coal agglomerated with a high speed blender is smaller than 1 mm regardless of the amount or type of oil used.  It is apparent that the high speed rotating blade prevents the growth of agglomerates above a certain size.

Some exploratory experiments were conducted to study different techniques for increasing the size of oil-coal agglomerates.  Several different methods of agitation were tested.  It was found that gentle agitation with either a paddle stirrer or a magnetic stirrer could increase the size of coal agglomerates.

Table 8.5 shows some of the experimental results.  This study indicated that heavier or more viscous oil was more effective than lighter oil for increasing the size and strength of agglomerates.  However, the addition of heavy oil to the coal slurry was difficult.  The coal could not be agglomerated by adding No. 6 heavy oil directly because the oil was so viscous that it could not be dispersed by simple agitation.  Thus, the heavy oil had to be broken up into small droplets either by emulsification before it was added or by spraying it into the coal slurry.  In order to facilitate emulsification or spraying, the heavy oil was heated in a boiling water bath.  However, a completely satisfactory method has not been developed yet.

The optimum speed of the cylindrical agitator appeared to be 500 to 1,000 rpm for an emulsified oil and 1,500 to 2,000 rpm for a straight oil depending on the viscosity of the oil.

## TABLE 8.5:  SIZE OF AGGLOMERATES PRODUCED BY VARIOUS TREATMENTS STARTING WITH PULVERIZED (–35 MESH) SCOTT COAL

| Treatment | Average Diameter, mm |
|---|---|
| 50 grams coal was agglomerated with a paddle stirrer using 5 ml of heated No. 6 heavy oil sprayed with an atomizer | 1–2 |
| 50 grams coal was agglomerated in a blender with 5 ml of heated and emulsified No. 6 heavy oil and then was agitated with a paddle stirrer | 5 |
| 50 grams coal was agglomerated in a blender with 2 ml of emulsified kerosene-No. 5 heavy oil mixture and balled with cylindrical agitator using 4 ml of kerosene-No. 5 heavy oil mixture | 5 |
| 20 grams coal was agglomerated with cylindrical agitator using 6 ml of emulsified kerosene-No. 5 heavy oil mixture | 2–3 |
| 20 grams coal was agglomerated with cylindrical agitator using 6 ml of emulsified No. 1-No. 6 heavy oil mixture | 5 |
| 100 grams coal was floated with 0.4 ml MIBC and agglomerated with 10 ml of emulsified heated No. 6 heavy oil in a froth flotation cell and then agitated with a magnetic stirrer | 2–3 |
| 100 grams coal was floated with 0.4 ml MIBC and agglomerated with 10 ml of No. 1 fuel oil in a froth flotation cell.  The agglomerated coal was divided into four parts and balled with cylindrical agitator using different amounts of No. 1-No. 6 heavy oil mixture: | |
|     1 ml | 1–2 |
|     2 ml | 3–4 |
|     4 ml | 4–5 |
|     6 ml | 5–10 |

## FROTH FLOTATION

In a modest way froth flotation has long been used as a beneficiation method for fine-size coal.  To apply this method, finely divided coal is mixed with water and a frothing agent and sometimes other chemical reagents.  Air bubbles are generated in the mixture and the hydrophobic coal particles become attached to the air bubbles and are buoyed to the surface where they are recovered in a froth.  Since the ash forming mineral particles tend to be hydrophilic, they remain in the aqueous suspension.

The effectiveness of froth flotation varies widely with coal characteristics, particle size, pulp density, pH value and flotation agents (13).  In particular, pyrite is frequently troublesome in the flotation of coal, owing to its tendency to float with coal.  Yancy and Taylor (14) showed that the soluble oxidation products of pyrite were powerful depressants for pyrite.  Baker and Miller (15) reported that the colloidal hydroxides precipitated from ferric chloride, aluminum chloride, and cupric sulfate solutions by sodium hydroxide would act as pyrite depressants under controlled conditions of pH and concentration.  One of the most promising developments is a two-stage process which has been demonstrated by U.S. Bureau of Mines personnel (16). In the first stage, coal is floated while coarse, free pyrite and other minerals are rejected. In the second stage, coal is depressed while fine-size pyrite is floated.

In order to evaluate the desulfurization potential of the froth flotation method, a series of bench scale experiments was carried out with several Iowa coals. Some important results are presented below.

### Evaluation of Two-Stage Froth Flotation Process

The Bureau of Mines two-stage froth flotation process was tested to evaluate its effectiveness for cleaning Iowa coal. Experiments were carried out with a Wemco laboratory froth flotation cell using the following steps, except where otherwise stated:

(1) 200 grams of coal was placed in the bowl and 1,300 ml of tap water was added.

(2) The coal slurry was agitated for 15 minutes and 2.5 ml of amyl alcohol was added.

(3) After 30 seconds' agitation, air was turned on at a flow rate of 0.33 cubic feet per minute and 1,000 ml of tap water was added.

(4) Froth was collected until it looked clean (about 5 minutes).

(6) The first froth was poured back into the bowl and 1,000 ml of tap water was added.

(7) Coal depressant (Aero depressant 633) and pyrite collector (potassium amyl xanthate) were added and the coal slurry was agitated for 10 minutes.

(8) 1 ml of amyl alcohol was added and the slurry was agitated for 30 seconds before air was turned on (0.33 ft$^3$/min).

(9) The second froth was collected in 2 minutes.

(10) The tails were filtered to recover clean coal.

The results are summarized in Table 8.6 and show that the two-stage froth flotation process reduced the ash content 15 to 50% and the pyritic sulfur content 15 to 55% depending on the coal source. However, it was not as effective for Iowa coals as shown by Miller (16) for Pennsylvania coals. In particular, the yield was generally low and relatively large amounts of frothing and depressing agents were required for Iowa coals.

### TABLE 8.6: TWO-STAGE FROTH FLOTATION TEST OF PULVERIZED IOWA COALS (–35 MESH)

| Coal | Product | Weight | Ash | Pyritic S | . . Treatment, wt %. . . | |
|------|---------|--------|-----|-----------|--------------------------|---|
|      |         | . . . . . . . . . .%. . . . . . . . . . | | | | |
| Big Ben | Clean coal | 70.38 | 7.39 | 0.97 | 1st stage | |
|      | Reject 2 | 19.36 | 9.29 | 3.91 | amyl alcohol | 1 |
|      | Reject 1 | 10.26 | 64.52 | 8.14 | 2nd stage | |
|      | Feed | 100.00 | 13.62 | 2.27 | Aero 633 | 0.1 |
|      |  |  |  |  | xanthate | 0.05 |
|      |  |  |  |  | amyl alcohol | 0.4 |

(continued)

**TABLE 8.6: (continued)**

| Coal | Product | Weight | Ash | Pyritic S | Treatment, wt %. | |
|------|---------|--------|-----|-----------|-------------------|---|
| Big Ben | Clean coal | 73.87 | 6.45 | 0.90 | 1st stage | |
| | Reject 2 | 15.72 | 11.98 | 4.60 | amyl alcohol | 1 |
| | Reject 1 | 10.41 | 67.29 | — | 2nd stage | |
| | Feed | 100.00 | 11.79 | 2.05 | Aero 633 | 0.15 |
| | | | | | xanthate | 0.03 |
| | | | | | amyl alcohol | 0.04 |
| Jude | Clean coal | 78.68 | 11.75 | 2.34 | 1st stage | |
| | Reject 2 | 9.80 | 12.80 | 4.24 | amyl alcohol | 1 |
| | Reject 1 | 11.52 | 74.27 | — | 2nd stage | |
| | Feed | 100.00 | 19.00 | 3.12 | Aero 633 | 0.15 |
| | | | | | xanthate | 0.03 |
| | | | | | amyl alcohol | 0.04 |
| Jude | Clean coal | 79.06 | 12.17 | 2.18 | 1st stage | |
| | Reject 2 | 14.63 | 12.49 | 3.77 | amyl alcohol | 1 |
| | Reject 1 | 6.31 | 78.98 | — | 2nd stage | |
| | Feed | 100.00 | 16.43 | 3.12 | Aero 633 | 0.25 |
| | | | | | xanthate | 0.03 |
| | | | | | amyl alcohol | 0.04 |
| Star | Clean coal | 40.08 | 6.89 | 1.63 | 1st stage | |
| | Reject 2 | 14.37 | 11.85 | 5.94 | amyl alcohol | 0.32 |
| | Reject 1 | 45.55 | 18.33 | 3.68 | 2nd stage | |
| | Feed | 100.00 | 12.74 | 3.18 | Aero 633 | 0.05 |
| | | | | | xanthate | 0.05 |
| | | | | | amyl alcohol | 0.16 |
| Star | Clean coal | 57.47 | 6.02 | 2.14 | 1st stage | |
| | Reject 2 | 20.41 | 5.10 | 2.13 | amyl alcohol | 1 |
| | Reject 1 | 22.12 | 23.28 | 6.28 | 2nd stage | |
| | Feed | 100.00 | 11.96 | 3.05 | Aero 633 | 0.25 |
| | | | | | xanthate | 0.03 |
| | | | | | amyl alcohol | 0.04 |

*Effect of Different Frothing Agents:* Several different agents were tested for flotation of –35 mesh ICO coal with a Wemco laboratory froth flotation cell. The experimental procedure used is as follows:

(1) 100 grams of pulverized coal was placed in the bowl and 1,000 ml of tap water was added.

(2) The coal slurry was agitated for 5 minutes and 1 ml of frothing agent was added.

(3) After 30 seconds of agitation, air was turned on at a flow rate of 0.33 ft$^3$/min and 1,000 ml of tap water was added.

(4) Froth was collected until the water phase looked clean (about 2 to 3 minutes) and filtered.

(5) The resultant filter cake was dried, weighed and analyzed.

Although there seemed to be no major differences in the results produced by different frothing agents, methyl isobutyl carbinol (MIBC) gave the best ash reduction, while Dowfroth No. 1012 gave the greatest sulfur reduction.

*Effect of pH:* A series of tests was carried out to determine the effect of pH on froth flotation of Iowa coals. First, 100 grams of pulverized (–35 mesh) coal and 1,000 ml of tap water were placed in the bowl of the Wemco flotation cell, and the slurry was agitated for 5 minutes. The pH of the slurry was adjusted to the desired level and 0.4 ml of kerosene was added. After 5 minutes of agitation, the coal was floated with 0.05 ml of MIBC using an air flow rate of 0.33 ft³/min. The froth was collected until the water phase appeared free of coal. Similar tests were conducted subsequently with deionized water to compare the effect of water quality.

The data indicate that the two coals which were used responded differently. The yield of Scott coal was larger than the yield of ICO coal over the entire pH range. Also the sulfur and ash contents of ICO coal were reduced more than the sulfur and ash contents of Scott coal. In addition the greatest reduction in the ash content of ICO coal was obtained at a pH of 3 to 4 whereas the greatest reduction in the ash content of Scott coal was obtained at a pH of 7 to 8. Furthermore, the reduction in sulfur content of ICO coal was relatively constant over the pH range from 1 to 10 while the reduction in sulfur content of Scott coal rose gradually over this pH range.

The results obtained when deionized water was used for the froth flotation of Scott coal were superior to those obtained when tap water was used. Thus the yield was larger over a pH range of 3 to 10 and the product had less sulfur and ash.

*Wettability of Coal and Pyrite:* A series of qualitative tests was carried out to determine the wettability of coal and pyrite particles over a pH range from 2 to 12 in solutions of various chemical reagents. Scott coal (−20+40 mesh) was cleaned by floating it in a liquid having a specific gravity of 1.3 and pulverized to provide a −200 mesh sample. Pyrite was carefully hand-picked from another portion of Scott coal and ground to −200 mesh. The wettability of coal and pyrite particles was determined by the following procedures:

(1) 5 grams of coal or pyrite was slurried with 500 ml of 0.01 mol/l chemical solution for 10 minutes in a blender. Deionized water was used to make the chemical solution.

(2) 10 ml of the slurry was placed in a glass test tube and the pH adjusted to the desired value by adding 1 N hydrochloric acid or sodium hydroxide solution.

(3) 5 ml of No. 1 fuel oil was added to the test tube and the test tube was shaken vigorously.

(4) The test tube was then allowed to stand for a while and the distribution of coal or pyrite between oil and water phases was observed.

It was found that the coal stayed almost completely at the oil-water interface over the entire pH range regardless of the chemical solutions used. Thus, it can be concluded that the wettability of Scott coal is constant and unaffected by the pH or chemical additives tested. On the other hand, the pyrite behaved differently depending upon the pH and chemical environment.

The wettability tests indicated conditions where coal and pyrite have different wettability, and, thus, where they might be separated effectively. These conditions were chosen and applied to the single stage froth flotation of pulverized Scott coal. The results, however, were disappointing, since the addition of various chemicals did not seem to make any significant improvement in sulfur reduction. The results obtained by the addition of ferric chloride, for example, were very poor, and seemed to contradict what would have been expected from the wettability test. The poor separation may have been due to the following reasons. Firstly, the pyrite in the pulverized Scott coal may not have been liberated sufficiently. Thus, a large portion of the pyrite may have been attached to coal particles which were carried into the froth product. Secondly, the soluble salts contained in the run of mine coal may have changed the chemical environment for the froth flotation tests from that of the wettability tests. Thirdly, the inner pore surface of the coal may have adsorbed the chemical reagents so that they were unavailable for depressing the pyrite.

*Effect of Chemical Pretreatment:* Pulverized (–35 mesh) Scott coal was subjected to a mild chemical pretreatment which was designed to oxidize the surface of the pyrite particles and thereby render the surface hydrophilic. After the pretreatment the coal was separated by the single stage froth flotation method.

The results showed that in most cases chemical pretreatment improved the sulfur reduction obtained by froth flotation over that obtained in the first run where the coal was not pretreated. On the other hand, pretreatment also appeared to reduce the yield.


## CHEMICAL COMMINUTION

Chemical comminution involves fracturing or breaking coal by application of specific chemical agents such as anhydrous liquid ammonia rather than by application of mechanically applied forces. Chemical comminution seems to induce cleavage at coal-mineral interfaces and, thus, appears to be more selective in its action than mechanical breakage. Thus, it may be possible to liberate mineral particles through chemical comminution without resorting to the same degree of size reduction required with mechanical crushing.

To test the effectiveness of chemical comminution for liberating the minerals in Iowa coal, several experiments were conducted where coal was soaked in a known comminuting agent, dried, and then passed through a double roll crusher. After this treatment, the material was separated by a float-sink procedure and the results were compared with those obtained by mechanical crushing and float-sink separation.

Two comminuting agents were utilized, anhydrous liquid ammonia and a 1 M solution of sodium hydroxide in methanol. For applying the treatment, 500 grams of lump coal (1½" x 0) was placed in a 2-liter Erlenmeyer flask and the liquid comminuting agent was added until the lumps were completely covered. When liquid ammonia was used, the flask was kept in a cold bath (–70°C) consisting of dry ice and methanol.

After the coal had soaked in the ammonia for 1 hour, the flask was removed from the cold bath and placed in a well ventilated hood where the ammonia evaporated. It was left in the hood until the odor of ammonia could no longer be detected. When the methanol solution of caustic was used as the comminuting agent, the coal was allowed to soak in the solution at room temperature for 2 hours. The solution was then decanted off and the coal was washed with warm water until the caustic was completely removed. The coal was subsequently dried.

The treated coal was passed through a double roll crusher to reduce it to 1¼"
top size and next was mixed with 2 liters of tetrachloroethylene liquid (SG =
1.613). After standing for 30 minutes to allow the material to separate, the
float fraction was skimmed off into a 100 mesh sieve and the remaining suspension was filtered to recover the sink fraction. The two fractions were heated
in a drying oven at 100°C for 4 hours to evaporate the organic liquid and then
were weighed and analyzed. Untreated coal was also reduced to 1¼" top size
by the double roll crusher and then separated in the same manner.

In the case of ICO coal it appears that chemical comminution resulted in somewhat lower yields of float coal and only slightly greater sulfur reductions than
those obtained by straight mechanical crushing. But in the case of Jude coal,
chemical comminution seemed to provide somewhat higher yields of float coal
with a lower ash content than was provided by mechanical crushing. However,
there was not much difference in the sulfur content. Since these results were
rather inconclusive, it appears that either a different type or a more sensitive
test is needed to show whatever advantages chemical comminution may have
over mechanical crushing.


## COMPARISON OF DIFFERENT BENEFICIATION METHODS

In order to compare the relative effectiveness of various coal beneficiation
methods and different combinations of these methods, a series of sixteen different coal benefication treatments was applied to two Iowa coals. These treatments involved different combinations of size reduction methods (crushing,
pulverizing, grinding, and chemical comminution) and of physical separation
methods (float-sink, froth flotation and oil agglomeration) previously described.
The results are compared on the basis of product yield and percentage reduction
in sulfur and ash brought about by the treatment.

### Materials

Coal from two Iowa strip mines was used for comparing different treatments.
A channel sample from the ICO mine and a run of mine sample from the Jude
mine were the source of the materials used. The proximate analysis and sulfur
distribution of each of these samples are shown in Table 8.1. Although the sulfur and ash contents of these samples were widely different, the samples represented coal of the same rank (high volatile C). Investigation of the coal microstructure with a seaming electron microscope revealed substantial amounts of
finely disseminated microcrystals of iron pyrites (12).

Each coal sample was crushed to 1½" top size and then divided into three size
fractions (1½" x ⅜", ⅜" x 48 M, and 48 M x 0). Each size fraction was then
float-sink tested at various specific gravities using organic liquids of known specific gravity. The standard Bureau of Mines procedure was employed for this
test (17). The data for the different size fractions were combined to provide
a composite washability analysis for 1½" x 0 coal.

### Experimental Procedure

The sequence of steps involved in each of the sixteen treatments which were
applied to each of the two coals is shown in Figure 8.1. Thus the first treat-

FIGURE 8.1:  FLOW DIAGRAM WITH DIFFERENT TREATMENTS

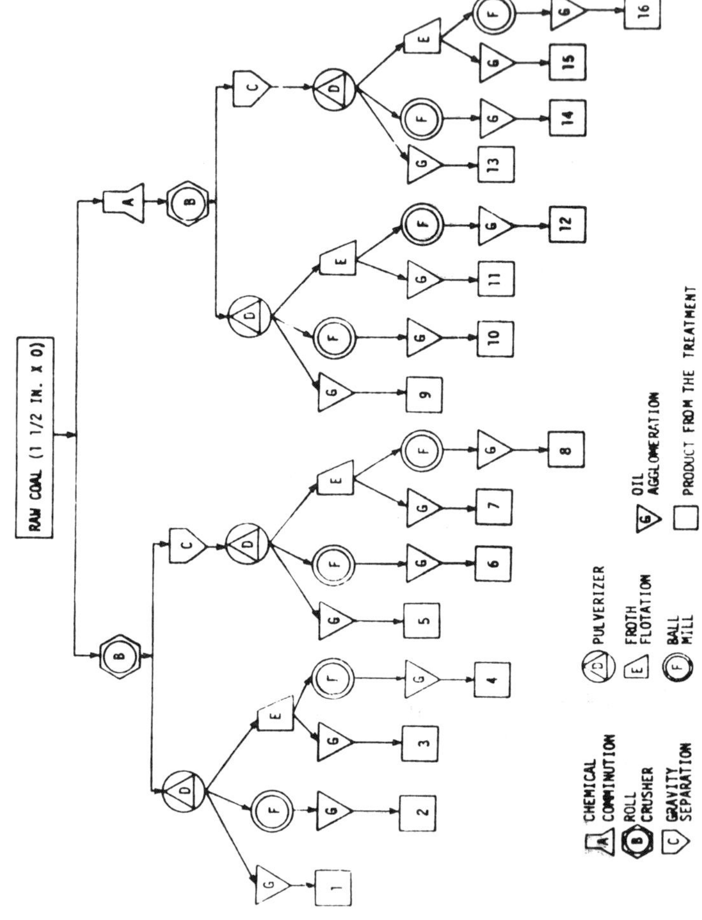


Source:  IS-ICP-35

ment was the simplest and involved crushing with the roll crusher, pulverizing with the Mikro-Samplmill, and oil agglomeration. The second treatment included a ball milling step in addition to the other steps. The third and fourth treatments included a froth flotation step. In the fifth through eighth treatments the crushed coal was subjected to gravity separation before being pulverized and otherwise treated as in the first four treatments.

In the last eight treatments the coal was chemically comminuted before being conducted through the roll crusher. Following the chemical comminution step the pattern of treatments was the same as for the first eight treatments. The final step of each treatment was an oil agglomeration step.

After each separation step within any given treatment, the weight of coal recovered was measured after drying overnight in an oven at 80° to 100°C and a small sample was taken for analysis. The samples were subsequently analyzed for ash, pyritic sulfur and total sulfur.

**Conclusions**

The laboratory application of sixteen different treatments involving size reduction and physical separation to high sulfur coal containing substantial amounts of finely disseminated microcrystals of iron pyrites provided several interesting and important results. Comparison of these results with a standard washability analysis showed that most of the treatments produced a cleaner product for a given yield than could be obtained by gravity separation alone of 1½" x 0 size coal. In this regard the treatments which failed to produce a product with a lower sulfur content were generally those which involved only size reduction and and oil agglomeration.

A comparison of the results of individual steps of different experimental treatments showed that gravity separation at 1.613 of crushed (¼" x 0) coal was more effective than oil agglomeration of pulverized (-35 mesh) coal for reducing the ash and sulfur content and it was more effective than froth flotation applied to one of the pulverized coal samples (Jude) but not the other (ICO) where it was about equal.

The results of individual froth flotation and oil agglomeration steps applied to pulverized coal showed that froth flotation produced a higher yield of product with a lower sulfur content than oil agglomeration but it did not always produce a lower ash content whereas with one of the coal samples (ICO) froth flotation produced the lower ash content, with the other (Jude) oil agglomeration did. Moreover, oil agglomeration proved more effective when it was preceded by fine grinding.

Thus, a combination of fine grinding and oil agglomeration gave results which were more like those obtained with froth flotation. When chemical comminution was included with the other size reduction methods, the subsequent separation of coal and mineral matter by any of the separation methods was facilitated.

The use of two and even three different separation methods in series proved more effective than any single method by itself. When the float product recovered by gravity separation at 1.613 was pulverized and subjected to either froth flotation or oil agglomeration, the ash and sulfur content of the material was reduced noticeably by the second separation step and the recovery of product was generally high. Although froth flotation provided a greater marginal reduction in pyritic sulfur content than oil agglomeration, the latter generally resulted in a larger marginal reduction in ash content.

Also if the float product was finely ground before applying oil agglomeration, the marginal reduction in both ash and sulfur content given by oil agglomeration was generally greater than that given by froth flotation of pulverized but not ground material.

When oil agglomeration was used to treat material which had been cleaned first by froth flotation or by a combination of gravity separation and froth flotation, a further reduction in the ash and sulfur content was produced by the oil agglomeration step. When the oil agglomeration step was applied without further size reduction, the decrease in ash and sulfur content which it provided was small but when it was preceded by fine grinding the reduction in ash and sulfur content was large. Consequently the treatments which included all three separation steps in series with size reduction between each separation step generally produced the cleanest coal.

These results indicate not only that froth flotation and oil agglomeration compliment gravity separation as a means for beneficiating coal but that oil agglomeration also compliments froth flotation especially if the oil agglomeration step is preceded by fine grinding. Moreover, oil agglomeration can produce spherical agglomerates which are dust-free and large enough for easy dewatering and handling.

Of course, these results should be considered preliminary in nature and not necessarily indicative of what might result from applying the same treatments to high sulfur coal from other regions. Obviously similar experiments should be carried out on many different samples of coal from a number of regions. Furthermore, the oil agglomeration and froth flotation methods themselves should be optimized because it was unlikely that they were applied in an optimum manner in this investigation. Finally, an experimental investigation such as this needs to be supported by a careful economic evaluation to see if the improvement resulting from additional size reduction and separation steps can be justified economically.

## REFERENCES


(1) Leonard, J.W., and Mitchell, D.R., (editors), *Coal Preparation,* Am. Inst. of Mining, Metallurgical, and Petroleum Engineers, New York, 1968.
(2) Leonard, J.W., and Cockrell, C.F., "Basic Methods of Removing Sulfur from Coal," *Mining Congress Journal,* 56 (No. 12), 65 (1970).
(3) Gorin, E. and Lebowitz, H.E., "Removing Sulfur and Mineral Matter from Coal," in *Coal Processing Technology,* pp. 64-68, Am. Inst. Chem. Engr., New York, 1974.
(4) Gaudin, A.M., *Flotation,* 2nd ed. McGraw-Hill, New York, 1957.
(5) Sirianni, A.F., Capes, C.E., and Puddington, I.E., "Recent Experience with the Spherical Agglomeration Process," *Can. J. Chem. Eng.,* 47, 166 (1969).
(6) Capes, C.E., McIlhinney, A.E., Sirianni, A.F., and Puddington, I.E., "Bacterial Oxidation in Upgrading Pyritic Coals," *Canadian Mining and Metallurgical (CIM) Bulletin,* 66, 88 (1973).

(7) Capes, C.E., Smith, A.E., and Puddington, I.E., "Economic Assessment of the Application of Oil Agglomeration to Coal Preparation," *CIM Bulletin,* 67, 115 (1974).

(8) Aldrich, R.G., Keller, D.V., and Sawyer, R.G., "Chemical Comminution and Mining of Coal," U.S. Patent 3,815,826 (June 11, 1974).

(9) Howard, P., Hanchett, A., and Aldrich, R.G., "Chemical Comminution for Cleaning Bituminous Coal," Symposium II, Clean Fuels from Coal, Institute of Gas Technology, Chicago, June 23-27, 1975.

(10) *1974 Annual Book of ASTM Standards,* Part 26, "Gaseous Fuels; Coal and Coke; Atmospheric Analysis," American Society for Testing and Materials, Philadelphia, 1974.

(11) Malik, A.M., "Air and Water Backwashing of Granular Filters," M.S. thesis, Iowa State University, Ames, Iowa (1972).

(12) Greer, R.T., "Nature and Distribution of Pyrite in Iowa Coal," Joint Meeting of the Electron Microscopy Society of America and Microbeam Analysis Society, Miami, Florida, August 9-13, 1976.

(13) Zimmerman, R.E., "Froth Flotation in Modern Coal Preparation Plants," *Mining Congress Journal,* 50 (No. 5), 26 (1964).

(14) Yancey, H.F., and Taylor, J.A., *Froth Flotation of Coal: Sulphur and Ash Reduction,* Report of Investigations 3263, U.S. Bureau of Mines, 1935.

(15) Baker, A.F., and Miller, K.J., *Hydrolyzed Metal Ions as Pyrite Depressants in Coal Flotation: A Laboratory Study,* Report of Investigations 7518, U.S. Bureau of Mines, 1971.

(16) Miller, K.J., *Flotation of Pyrite from Coal: Pilot Plant Study,* Report of Investigations 7822, U.S. Bureau of Mines, 1973.

(17) *Methods of Analyzing and Testing Coal and Coke,* Bulletin No. 638, U.S. Bureau of Mines, 1967.

# COAL CLEANING WITH SCRUBBING
# FOR SULFUR CONTROL

The material in this chapter is from a report by E.C. Holt, Jr.
of Hoffman-Muntner Corporation and A.W. Deurbrouck of the
Bureau of Mines, U.S. Department of Interior. It was prepared
for the U.S. Environmental Protection Agency (EPA-600/9-77-017;
for details see bibliography on page 301).


## INTRODUCTION

In this country, the continued generation of power in a reliable, cost-effective,
and environmentally acceptable fashion is critical to economic and social stability.
Much of this power is produced by coal-fired plants. During 1975, fifty-five
percent of the utility industry's fossil fuel requirements were met with coal. In
view of the rapidly diminishing reserves and rising prices of alternate fossil fuels,
the percentage provided by coal will increase. If these plants are going to adhere
to the environmental standards set forth by EPA and others relative to $SO_x$ emissions, it will often be necessary to consume coals of less than 1% sulfur and/or
invest in costly flue gas desulfurization systems. Therefore, in the interest of
overall national economics and preserving our limited Eastern supply of low
sulfur coal, it is important that economic means be found and quickly imple-
mented to encourage the utility industry to make use of our vast resources of
higher sulfur-content coal.

According to data published by the Bureau of Mines, the United States had
minable underground and surface coal reserves of nearly 437 billion tons on
1 January 1974 (see Figure 9.1). Of this number, approximately 386 billion tons
had been categorized according to sulfur content; 52% had 1% or less, 24% had
between 1.0 and 3.0%, and 24% had greater than 3.0%. Although over half of
these reserves fall into the low sulfur content range, 88% of the low sulfur coal
is found west of the Mississippi (concentrated in such states as Montana and
Wyoming), far removed from the electric power demand centers of the east.
In the east only 24% of the coal is low sulfur. Therefore, in developing a reliable source of coal (long term supply not too far from point of consumption)
to meet the demands of the eastern utilities, the medium and high sulfur content
coals must be considered. It was the purpose of this study to identify an economic means of keeping such higher sulfur content coals in the energy market.

Data generated by the Bureau of Mines under the Federal Interagency Energy/
Environment Research and Development Program indicate that some Eastern
coals will show a significant reduction in ash and sulfur contents when physically
cleaned to a 90% weight yield. With current technology, this level of physical
cleaning can often be accomplished at an attractive cost. These coals with reduced ash and sulfur levels are often not too far removed from the sulfur content required to meet environmental standards in the geographic areas historically
served by these coals.

When coal can be physically cleaned to a sulfur content close to that required
to meet governing emission standards, a flue gas desulfurization (FGD) system
treating only a portion of the flue gas would permit the coal burning facility
to comply with such emission regulations. In many cases, the net cost of physical cleaning followed by FGD is substantially less than that associated with
meeting standards exclusively with a larger capacity FGD system. This results
from the fact that the savings associated with being able to use a smaller FGD
system more than cover the net cost of physically cleaning the coal. This study
quantified this relationship under assumed conditions based upon current environmental standards being met by utility power plants using specific coals which are
economically obtainable in the particular geographic areas considered.

The overall study findings covering new utility plants using physically cleaned
coal followed by FGD indicated a savings of 2 to 112% as compared to meeting
standards by FGD alone. The results were even more impressive for existing
plants, where study assessments indicated a 13 to 140% savings for physical
cleaning followed by FGD as compared to FGD alone.

## FIGURE 9.1: LOW SULFUR COAL RESERVES VERSUS STEAM ELECTRIC POWER GENERATION

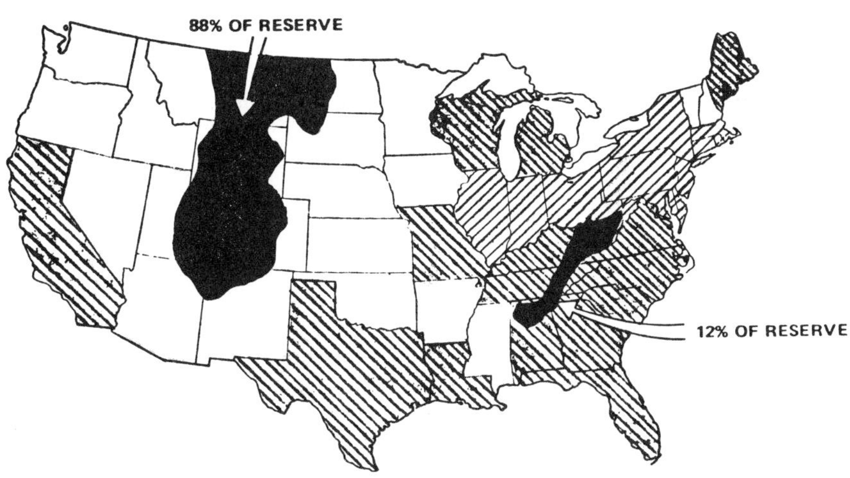


Hawaii:  Negligible reserves and consumption

Alaska:  Potentially large reserves, minimal
         consumption

>1.5 million kWh per square mile
0.5-1.5 million kWh per square mile
0-0.5 million kWh per square mile

## BACKGROUND

Coal is a heterogeneous material containing organic combustible matter and mineral matter. The mineral matter (i.e., impurities) may be broadly divided into two categories, those forming ash and those that contribute sulfur. Such ash-forming and sulfur-containing impurities may be further divided into two groups, (1) those that are chemically a part of the coal and cannot be removed by mechanical processes, and (2) those that are not chemically bound to the coal and can be removed to varying degrees by mechanical means. It is toward this latter category that physical coal cleaning (coal beneficiation, preparation, or washing) is directed.

### Physical Cleaning

Physical cleaning is accomplished by first crushing the coal to liberate those impurities that are not chemically bound with the coal. These impurities are set free upon crushing since the seam where they are stuck to the coal is generally weaker than either the coal or the impurity. After the raw material (as mined coal) is crushed, the impurities can be separated out by mechanical means.

The most common processes for achieving this separation are based upon the difference between the specific gravities of the impurities and that of the coal. Simply stated, coal has a specific gravity of around 1.3 which is less than that of its impurities. Thus, when the crushed particles (coal and impurities) are distended in water or other media a separation occurs as the heavier undesirable particles settle at a faster rate than the coal. In commercial scale cleaning operations, it is not uncommon to process 500 to 1,000 tons of coal per hour using a variety of equipment designed and arranged around the makeup of the particular raw coal and the desired end product.

Physical cleaning of coal has been used for many years. In the past its principal purpose was to reduce the ash-forming impurities. However, today cleaning is of significant value in reducing the sulfur content of certain coals. Its applicability in this regard is not universal due to the various forms of sulfur occurring in coal. Sulfur exists in coal in two principal forms, organic and inorganic.

Organic sulfur is one of those impurities referred to earlier which is chemically bound to the coal and cannot be removed by physical means. On the other hand, inorganic sulfur (pyritic sulfur) is not bound chemically and may be physically removed to varying degrees from the coal. The extent to which pyritic sulfur can be removed economically is a function of pyrite size and distribution. Once this information is obtained through careful laboratory analysis, then economic considerations will influence the level to which the coal is crushed and subjected to the cleaning operation.

Physical cleaning processes, while removing impurities from the coal, also reduce the total Btu available (i.e., a portion of the heat content of the raw coal is lost with the impurities). However the Btu content per unit weight of the cleaned coal increased due to the removal of the lower heat value impurities. In practice, an economic balance must be achieved between the Btu loss and the improvement in coal quality for various coals and applications. Certainly, this balance is further influenced by market and environmental considerations.

## Problems

With regard to the environmental aspects, the oxides of sulfur ($SO_x$) created when coals are burned have long been recognized as a real threat to both the ecosystem and human health. Federal Standards, designed to protect against serious harm, have been established for new electric power plants which burn coal. To meet these standards, several methods are available to control sulfur oxide emissions from coal-fired combustion sources. These control methods include:

    (1)   The use of low sulfur coal, either naturally occurring or physically cleaned;

    (2)   Chemical treatment to extract sulfur from coal;

    (3)   Removal of sulfur oxides from the combustion flue gas (stack gas scrubbing);

    (4)   Conversion of coal to clean fuel by such processes as gasification and liquefaction.

Of these methods, for certain coals, physical cleaning to reduce sulfur content is the lowest cost and has the most developed technology. However, as noted earlier, physical cleaning can remove significant amounts of inorganic sulfur (pyritic) from certain coals, but has no influence upon the organic sulfur content. Therefore, when trying to come up with an environmentally acceptable product, physical cleaning may only be a partial step. This leads to another approach to controlling $SO_x$ emissions, which could be the combined use of physical cleaning followed by stack gas scrubbing (i.e., flue gas desulfurization).

The purpose of a flue gas desulfurization (FGD) system is to limit the sulfur compounds escaping into the atmosphere. One of the most common FGD systems accomplishes its mission by forcing the $SO_x$ in the flue gas to react with a limestone slurry. The cost of these systems goes up substantially with the amount of flue gas treated, which, in turn, is a function of the sulfur concentration of the flue gas stream. Therefore, significant savings can be realized if the sulfur concentration of the stream can be limited to the point where less than half of the total flue gas needs to be treated by the FGD system to meet standards. This becomes the basic economic concept of combining the two sulfur reduction technologies since such a combined approach could mean the ability to use higher sulfur content coals and minimal to moderate stack gas scrubbing.

## Physical Cleaning with Scrubbing

This concept of physical coal cleaning combined with flue gas desulfurization is not new. For some time there have been discussions, speculations, and some very preliminary assessments addressing the possible benefits of physical coal desulfurization followed by FGD (see Figure 9.2). Past opinions based on a general appreciation of some of the cost and benefit factors associated with such an approach have led to the expressed belief that economic advantage in many instances could be attained. However, the associated specific economics had not previously been fully addressed. Therefore, to more completely define the potential economics associated with such a combined approach as a means of increasing the usefulness of some of our higher sulfur content coals, this study was initiated. It was conducted by the Bureau of Mines under the auspices of the EPA-coordinated Federal Interagency Energy/Environment R&D Program.

## FIGURE 9.2: PHYSICAL COAL CLEANING WITH FLUE GAS DESULFUR-IZATION VERSUS FLUE GAS DESULFURIZATION ALONE

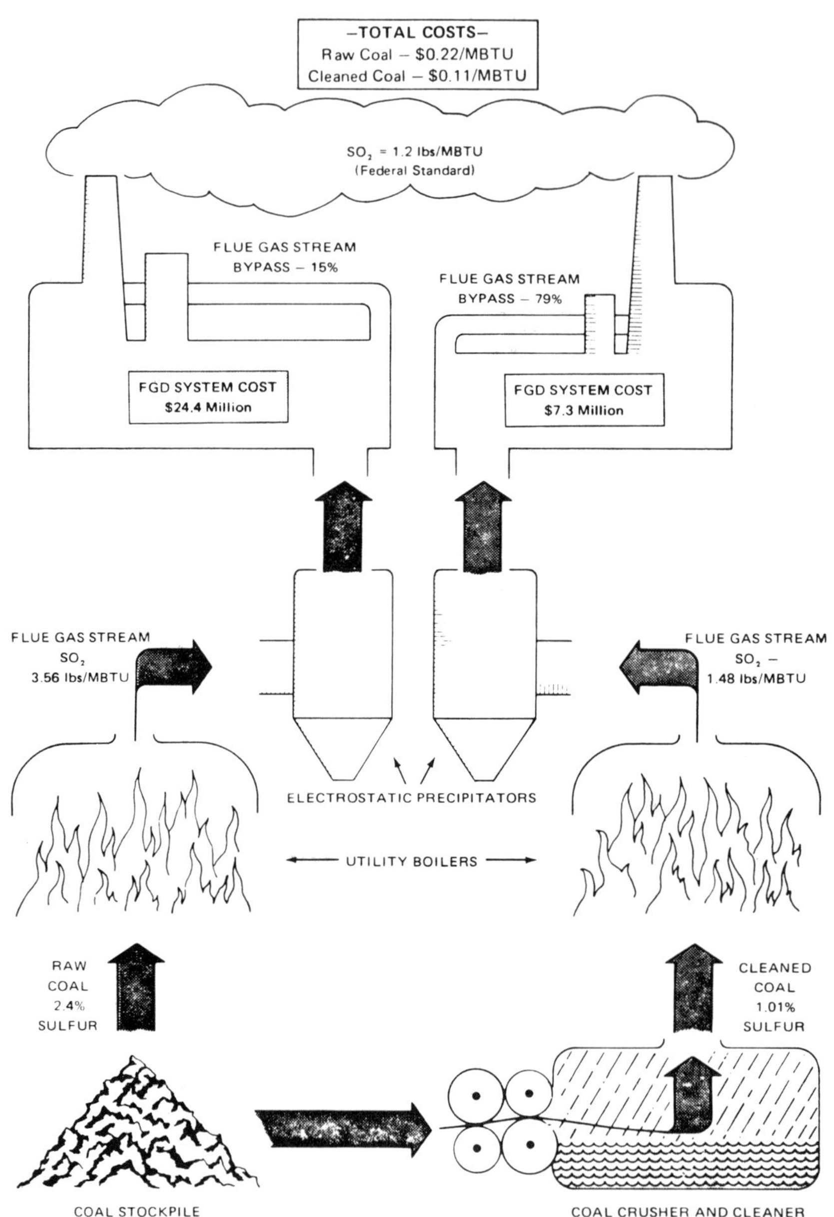

For purposes of this study, higher sulfur coals from the Northern Appalachian and Eastern Interior Regions were selected since they have been shown to have reasonable physical cleaning potential (i.e., economic ash and pyritic sulfur reduction). Then, possible users of these coals in the electric power generating industry were identified along with the current environmental constraints in their respective localities. This realism provided the framework within which to study and compare the economics associated with meeting sulfur emission standards in two alternative ways. The study considered both new and existing plants using either physically cleaned coal followed by stack gas scrubbing, or sulfur clean-up exclusively by stack gas scrubbing.

## ECONOMIC ANALYSES

In order to determine the relative economics of meeting environmental standards via two different approaches, the factors influencing the costs and benefits of each element were identified. These are summarized as Table 9.1.

For physical coal cleaning, the costs are principally dependent upon such factors as coal composition and plant size/cost. Offsetting these costs somewhat are the benefits of using cleaned coal. The magnitude of these benefits is influenced most by the increased heat content of the processed product brought on by the removal of some of the low heat content impurities. In the case of flue gas desulfurization (FGD), costs are sensitive to the sulfur composition of the flue gas stream.

## TABLE 9.1: KEY COST/BENEFIT FACTORS

### COST OF CLEANING COAL

—Amortization of Cleaning Plant Capital Cost
—Operation and Maintenance Cost of Cleaning Plant
—Cost of Raw Coal Lost in Cleaning Process
—Taxes and Insurance Costs of Cleaning Plant

### COSTS OF FLUE GAS DESULFURIZATION

—Amortization of Flue Gas Desulfurization (FGD) System Capital Cost
—Fuel and Electricity Cost Associated with FGD System
—Operating and Maintenance Costs of FGD System, Including:

Limestone • Fixation Chemicals • Operating Labor
Maintenance (Labor and Materials) • Supplies • Overhead

### BENEFITS OF USING CLEANED COAL

—Increased Heat Content of Cleaned Coal
   (Greater BTU Content Per Unit Weight)
—Transportation Savings (Less Weight to Ship for Same BTU Content)
—Ash Disposal Cost Saving (Clean Coal Leaves Less Ash)
—Pulverizing Cost Savings
   (Less Cleaned Coal Needs to be Pulverized for Same BTU Content Required to Meet Output)
—Boiler and Related Equipment Maintenance Savings
   (Clean Coal Is Less Corrosive)

Having identified the key cost/benefit factors associated with the two approaches, case analyses were performed based, in part, upon the major study factors appearing in Table 9.2. As appropriate, these factors were based upon actual industry performance and practice.

## TABLE 9.2: MAJOR STUDY FACTORS

The coals considered were those for which the Bureau of Mines has performed float-sink analyses. Bureau of Mines estimates of ash and sulfur levels versus top size and yield values were used.

The top size coal considered was 3/8 inch.

Economic assessments were based on a range of raw coal selling prices and a range of mining costs for a ton of coal.

A coal cleaning plant cost of $18,000 per ton hour input capacity for plants of 500 tons per hour or greater.

A coal cleaning plant financed by a 15 year equal payment self-liquidating loan.

A coal cleaning plant use factor of 38.5% (i.e., plant operates 260 days per year at 13 hours per day).

A coal cleaning plant property tax and insurance level equal to two percent of the initial investment.

A power plant ash disposal cost of $4 per ton.

A power plant coal pulverizing cost of $0.50 per ton.

Annual stack gas scrubber capital charges that are based on the original investment and the remaining boiler life.

Stack gas scrubber size/cost values based on recent EPA funded studies and a size/cost exponential relationship of 0.8.

Stack gas scrubber operating labor cost that does not vary with plant size (i.e., there is a minimum practical labor level).

General power plant locations that are consistent with actual conditions.

Coal producing areas serving assumed user locations consistent with past patterns.

Coal transportation parameters consistent with historical patterns and economical coal delivery (to insure conservative economic relationships).

An assumed power plant boiler size of 500 Mw with an annual utilization of 7,000 hours for a new plant and 5,000 hours for a 10 year old plant.

Stack gas scrubber systems of the minimum "practical" sizes necessary to meet standards.

A sample case study is presented as Table 9.3. The key element in showing economic advantage in favor of the combined approach is in the significant difference in FGD costs. This significant difference results from only 21% of the flue gas being processed by the FGD system as opposed to 85% when cleaned coal is not used. A graphic representation of these approaches is presented in Figure 9.2.

## TABLE 9.3:  CASE STUDY

**Problem:**
Determine most cost-effective approach for new coal burning utility plant to meet emission standard of 1.2 lbs $SO_2$ per million Btu (MBTU)

**Case Conditions:**
Coal Use Area:        Tonawanda (Buffalo), New York
Coal Source Area:   Cambria County, Pennsylvania        Coalbed:  Lower Freeport
Raw Coal Characteristics:        11.4% Ash,    2.4% Sulfur
Cleaned Coal Characteristics:        6.7% Ash,    1.01% Sulfur

**Costs of Alternate Approaches to Meeting Standard:**

| | Physical Cleaning<br>Followed by FGD | FGD Alone |
|---|---|---|
| **Coal Cleaning Cost** | | |
| Amortization of Cleaning Plant Capital Cost | $        0.68/Ton | $ -0- |
| O & M Cost of Cleaning Plant | 0.75/Ton | -0- |
| Cost of Coal Lost During Cleaning | 1.56 To 2.00/Ton | -0- |
| Taxes & Insurance of Cleaning Plant | 0.12/Ton | -0- |
| Total Cleaning Cost | $  3.11 To 3.55/Ton | $ -0- |
| **Cost of Flue Gas Desulfurization (FGD)** | | |
| Amortization of FGD Capital Cost | $        0.92/Ton | $2.92/Ton |
| Fuel & Electricity of FGD System | 0.17/Ton | 0.66/Ton |
| O & M Cost of FGD System | 0.72/Ton | 2.24/Ton |
| Total FGD Cost | $        1.81/Ton* | $5.82/Ton* |
| **Benefits of Using Cleaned Coal** | | |
| Increased Heat Content | $  1.21 To 1.40/Ton | $ -0- |
| Transportation Savings | 0.30/Ton | -0- |
| Ash Disposal Savings | 0.21/Ton | -0- |
| Pulverizing Savings | 0.02/Ton | -0- |
| Maintenance & Other Savings | 0.23/Ton | -0- |
| Total Benefit of Cleaning | $  1.97 To 2.16/Ton | $ -0- |
| **Net Cost (Costs Less Benefits)** | $  2.76 To 3.39/Ton | $5.82/Ton |
| **Converted to per MBTU** | $  0.10 To 0.12 | $0.22 |

---

**SOLUTION:**

**The combined approach is the most cost-effective, compared to a 100% additional cost for  meeting the standard by FGD alone.**

---

*Significant difference in cost of FGD results from 21% of the flue gas being processed versus 85% in the more expensive approach to meet the required emission standard.

## Additional Benefits

In addition to the benefits covered by this study, there are other positive features associated with combining physical desulfurization and FGD to meet emission standards.  Such a benefit accruing from the combination of these two technologies is a decrease in the time required for FGD installation.  This occurs for two reasons. First, the total sludge disposal at the plant site can be reduced to about one-quarter of what it would be in the absence of coal cleaning. This will, in some cases, reduce the time required for, and the difficulty with, obtaining legal permits and site acquisition for ponding and/or landfill. Second, as covered by this study, the cleaned coal reduces the volume of flue gas which must be scrubbed to meet emission standards, thereby requiring smaller, or perhaps fewer, FGD units with an associated reduction in construction requirements and time.

Another benefit is the more intangible aspect of physical cleaning in the event of FGD shut-down. Although it cannot be readily quantified, there is a definite environmental benefit associated with the use of physically cleaned coal in the event of a failure in the $SO_2$ scrubbing system. This benefit is in the form of lower toxic emissions, particularly $SO_2$, than would have been the case if some sulfur and other contaminants had not been previously removed through physical cleaning.

*Continued Prospects:* The Federal government as well as the coal industry has established a production goal of more than one billion tons of coal by 1985. Although this means an increase of roughly 400 million tons over current production, the majority of this additional coal will be consumed by the approximately 250 new coal-fired facilities scheduled to be on line by that time.

This substantial projected increase in coal utilization has precipitated much discussion concerning the possibility of revising EPA's new source performance standard for coal-fired plants. Although the extent of such a change is unknown at this time, one figure under consideration is to go from the current 1.2 pounds of $SO_2$ per million Btu to 0.8 pounds. Whereas now a coal having a heat content of approximately 25 million Btu's per ton and a sulfur content of 0.8% can meet the standard, only those coals of 0.5% or less sulfur content could meet the 0.8 pound of $SO_2$ per million Btu standard. If the standard is revised in this manner, there will be a significant reduction in the amount of coal which can be consumed without the use of $SO_2$ emission control devices. This would include both the coals naturally occurring with a low enough sulfur content and those capable of meeting the current standard through physical cleaning alone.

However, should the standard be amended, it will not negate the benefits associated with coal cleaning as set forth in this report (Table 9.1). What it will mean is a reevaluation of emission control strategy on the part of potential coal users.

Since coal having a sulfur content of 0.5% or less (either occurring naturally or after cleaning) is in limited supply and high in price due to pressure from the metallurgical market, it is impractical to consider the widespread burning of coal in new facilities without any $SO_2$ emission control devices under a 0.8 pound of $SO_2$ per million Btu standard. This being the case, in searching for the most cost-effective approach to meeting any revised standard, the impact of coal preparation on reducing emission control costs as outlined by this report should be carefully considered.

After such appraisal, users may find that, under their particular circumstances, the use of physically cleaned coal in combination with stack gas scrubbing will show a definite economic advantage even under more restrictive emission standards.

## Summary of Case Results

A total of forty-eight case analyses were performed.  Each case examines one of twelve selected coals and possible use areas from the standpoint of both a new and an existing utility plant using either a combination of physically cleaned coal followed by stack gas scrubbing, or sulfur cleanup exclusively by stack gas scrubbing (i.e., flue gas desulfurization).

The cases are grouped according to coal and use area.  This permits ease of comparison between like plants using the same coal in the same area but utilizing two different approaches to meeting existing or projected environmental standards.

In most cases, the emission standards used were applicable as of January 1976. However, where knowledge of projected changes was present, those standards were used and identified accordingly.

In Table 9.4 on the following pages are summaries of each of the forty-eight cases analyzed.  As stated above, these summaries are grouped in sets for comparison of the costs associated with meeting emission standards by the two alternate approaches.

For example, Case Numbers 1A, 1B, 1C and 1D are based upon an actual coal coming from Sullivan County, Indiana which is assumed to be used in a utility plant in the Knoxville, Tennessee area.

Cases 1A and 1B approach the analysis on the basis of an assumed new facility, whereas Cases 1C and 1D assume an existing facility.  However, Case Numbers 1A and 1C address the analysis using a combination of physically cleaned coal followed by stack gas scrubbing, whereas Case Numbers 1B and 1D approach the situation from the standpoint of using stack gas scrubbing alone.

As can be readily seen from the summarization of these cases, the cost to meet the applicable sulfur emission standard is less in Case Numbers 1A and 1C which are the plans using physically cleaned coal followed by flue gas desulfurization (FGD) in new and existing facilities, respectively.

In each set of cases, the relative economic advantage (or disadvantage) is expressed as a percent for comparative purposes.  The manner of stating economic advantage (less costly) or disadvantage (more costly) is dependent upon what if any action the coal-using plant has taken to meet environmental standards.

For example if, as in the case of 1C, the utility is using physically cleaned coal followed by FGD, then economic advantage could be stated as their cost is 25% less than by FGD alone.

If, as in case 1D, the utility is using FGD alone, then economic disadvantage could be stated as their cost is 33% more than by the combined FGD/coal cleaning approach.

## TABLE 9.4: CASE RESULTS

**Case Numbers:**　1A, 1B, 1C, & 1D

**Case Conditions**

Coal Use Area: Knoxville (Clinton), Tennessee

Coal Source Area: Sullivan County, Indiana　　　Coalbed: Number VII

Raw Coal Characteristics: 10.5 % Ash　1.87 % Sulfur

Clean Coal Characteristics: 7.3 % Ash　1.11 % Sulfur

**Comparison of Costs**

| CASE NO. | TYPE OF PLANT & APPROACH | EMISSION STANDARD | COST TO MEET EMISSION STD. | ECONOMIC ADVANTAGE |
|---|---|---|---|---|
| 1A — | New Plant PC Followed by FGD | 1.2 lbs $SO_2$ per MBTU | $0.15-0.17 per MBTU | 16% less than by FGD alone |
| 1B — | New Plant FGD Alone | 1.2 lbs $SO_2$ per MBTU | $0.19 per MBTU | 19% more than by PC & FGD |
| 1C — | Existing Plant PC Followed by FGD | 1.2 lbs $SO_2$ per MBTU | $0.23-0.25 per MBTU | 25% less than FGD alone |
| 1D — | Existing Plant FGD Alone | 1.2 lbs $SO_2$ per MBTU | $0.32 per MBTU | 33% more than by PC & FGD |

**Case Numbers:**　2A, 2B, 2C, & 2D

**Case Conditions**

Coal Use Area: Tonawanda (Buffalo), New York

Coal Source Area: Cambria County, Pennsylvania　　　Coalbed: Lower Freeport

Raw Coal Characteristics: 11.4 % Ash　2.4 % Sulfur

Clean Coal Characteristics: 6.7 % Ash　1.01 % Sulfur

**Comparison of Costs**

| CASE NO. | TYPE OF PLANT & APPROACH | EMISSION STANDARD | COST TO MEET EMISSION STD. | ECONOMIC ADVANTAGE |
|---|---|---|---|---|
| 2A — | New Plant PC Followed by FGD | 1.2 lbs $SO_2$ per MBTU | $0.10-0.12 per MBTU | 50% less than by FGD alone |
| 2B — | New Plant FGD Alone | 1.2 lbs $SO_2$ per MBTU | $0.22 per MBTU | 100% more than by PC & FGD |
| 2C — | Existing Plant PC Followed by FGD | 1.4 lbs $SO_2$ per MBTU | $0.06-0.09 per MBTU | 50% less than by FGD alone |
| 2D — | Existing Plant FGD Alone | 1.4 lbs $SO_2$ per MBTU | $0.15 per MBTU | 100% more than by PC & FGD |

**Case Numbers:**　3A, 3B, 3C & 3D

**Case Conditions**

Coal Use Area: Essexville (Saginaw), Michigan

Coal Source Area: Harrison County, Ohio　　　Coalbed: Lower Freeport

Raw Coal Characteristics: 10.4 % Ash　2.30 % Sulfur

Clean Coal Characteristics: 4.8 % Ash　1.26 % Sulfur

**Comparison of Costs**

| CASE NO. | TYPE OF PLANT & APPROACH | EMISSION STANDARD | COST TO MEET EMISSION STD. | ECONOMIC ADVANTAGE |
|---|---|---|---|---|
| 3A — | New Plant PC Followed by FGD | 1.2 lbs $SO_2$ per MBTU | $0.12-0.14 per MBTU | 38% less than by FGD alone |
| 3B — | New Plant FGD Alone | 1.2 lbs $SO_2$ per MBTU | $0.21 per MBTU | 62% more than by PC & FGD |
| 3C — | Existing Plant PC Followed by FGD | 1.6 lbs $SO_2$ per MBTU | $0.11-0.14 per MBTU | 58% less than FGD alone |
| 3D — | Existing Plant FGD Alone | 1.6 lbs $SO_2$ per MBTU | $0.30 per MBTU | 140% more than by PC & FGD |

(continued)

## TABLE 9.4:  (continued)

**Case Numbers:**  4A, 4B, 4C, & 4D
**Case Conditions**
Coal Use Area: Boston, Massachusetts
Coal Source Area: Clearfield County, Pennsylvania    Coalbed: Upper Kittanning
Raw Coal Characteristics:   9.3 % Ash    0.85 % Sulfur
Clean Coal Characteristics:   7.0 % Ash    0.45 % Sulfur

**Comparison of Costs**

| CASE NO. | TYPE OF PLANT & APPROACH | EMISSION STANDARD | COST TO MEET EMISSION STD. | ECONOMIC ADVANTAGE |
|---|---|---|---|---|
| 4A — | New Plant PC Followed by FGD | 0.28 lbs sul. per MBTU | $0.12-0.14 per MBTU | 24% less than by FGD alone |
| 4B — | New Plant FGD Alone | 0.28 lbs sul. per MBTU | $0.17 per MBTU | 31% more than by PC & FGD |
| 4C — | Existing Plant PC Followed by FGD | 0.28 lbs sul. per MBTU | $0.17-0.19 per MBTU | 38% less than by FGD alone |
| 4D — | Existing Plant FGD Alone | 0.28 lbs sul. per MBTU | $0.29 per MBTU | 61% more than by PC & FGD |

**Case Numbers:**  5A, 5B, 5C, & 5D
**Case Conditions**
Coal Use Area: Grand Rapids, Michigan
Coal Source Area: Preston County, W. Virginia    Coalbed: Upper Freeport
Raw Coal Characteristics:   18.5 % Ash    2.24 % Sulfur
Clean Coal Characteristics:   11.9 % Ash    1.25 % Sulfur

**Comparison of Costs**

| CASE NO. | TYPE OF PLANT & APPROACH | EMISSION STANDARD | COST TO MEET EMISSION STD. | ECONOMIC ADVANTAGE |
|---|---|---|---|---|
| 5A — | New Plant PC Followed by FGD | 1.6 lbs $SO_2$ per MBTU | $0.07-0.10 per MBTU | 53% less than by FGD alone |
| 5B — | New Plant FGD Alone | 1.6 lbs $SO_2$ per MBTU | $0.18 per MBTU | 112% more than by PC & FGD |
| 5C — | Existing Plant PC Followed by FGD | 1.6 lbs $SO_2$ per MBTU | $0.12-0.14 per MBTU | 57% less than FGD alone |
| 5D — | Existing Plant FGD Alone | 1.6 lbs $SO_2$ per MBTU | $0.30 per MBTU | 131% more than by PC & FGD |

**Case Numbers:**  6A, 6B, 6C, & 6D
**Case Conditions**
Coal Use Area: Springfield, Massachusetts
Coal Source Area: Armstrong County, Penn.    Coalbed: Upper Freeport
Raw Coal Characteristics:   13.0 % Ash    2.53 % Sulfur
Clean Coal Characteristics:   7.2 % Ash    1.09 % Sulfur

**Comparison of Costs**

| CASE NO. | TYPE OF PLANT & APPROACH | EMISSION STANDARD | COST TO MEET EMISSION STD. | ECONOMIC ADVANTAGE |
|---|---|---|---|---|
| 6A — | New Plant PC Followed by FGD | 0.55 lbs sulfur per MBTU | $0.11-0.13 per MBTU | 48% less than by FGD alone |
| 6B — | New Plant FGD Alone | 0.55 lbs sulfur per MBTU | $0.23 per MBTU | 92% more than by PC & FGD |
| 6C — | Existing Plant PC Followed by FGD | 0.55 lbs sulfur per MBTU | $0.18-0.21 per MBTU | 49% less than by FGD alone |
| 6D — | Existing Plant FGD Alone | 0.55 lbs sulfur per MBTU | $0.38 per MBTU | 95% more than by PC & FGD |

(continued)

## TABLE 9.4: (continued)

**Case Numbers:** 7A, 7B, 7C, & 7D

**Case Conditions**

Coal Use Area: _Lansing, Michigan_

Coal Source Area: _Jefferson County, Ohio_     Coalbed: _Pittsburgh_

Raw Coal Characteristics: _9.8_ % Ash     _2.82_ % Sulfur

Clean Coal Characteristics: _6.0_ % Ash     _2.03_ % Sulfur

**Comparison of Costs**

| CASE NO. | TYPE OF PLANT & APPROACH | EMISSION STANDARD | COST TO MEET EMISSION STD. | ECONOMIC ADVANTAGE |
|---|---|---|---|---|
| 7A — | New Plant PC Followed by FGD | 1.6 lbs SO$_2$ per MBTU | $0.19-0.22 per MBTU | 6.8% less than by FGD alone |
| 7B — | New Plant FGD Alone | 1.6 lbs SO$_2$ per MBTU | $0.22 per MBTU | 7.3% more than by PC & FGD |
| 7C — | Existing Plant PC Followed by FGD | 1.6 lbs SO$_2$ per MBTU | $0.30-0.32 per MBTU | 11% less than FGD alone |
| 7D — | Existing Plant FGD Alone | 1.6 lbs SO$_2$ per MBTU | $0.35 per MBTU | 13% more than by PC & FGD |

**Case Numbers:** 8A, 8B, 8C, & 8D

**Case Conditions**

Coal Use Area: _Nashville (Gallatin), Tennessee_

Coal Source Area: _Vigo County, Indiana_     Coalbed: _Number VII_

Raw Coal Characteristics: _12.0_ % Ash     _1.54_ % Sulfur

Clean Coal Characteristics: _7.7_ % Ash     _0.90_ % Sulfur

**Comparison of Costs**

| CASE NO. | TYPE OF PLANT & APPROACH | EMISSION STANDARD | COST TO MEET EMISSION STD. | ECONOMIC ADVANTAGE |
|---|---|---|---|---|
| 8A — | New Plant PC Followed by FGD | 1.2 lbs SO$_2$ per MBTU | $0.11-0.13 per MBTU | 25% less than by FGD alone |
| 8B — | New Plant FGD Alone | 1.2 lbs SO$_2$ per MBTU | $0.16 per MBTU | 33% more than by PC & FGD |
| 8C — | Existing Plant PC Followed by FGD | 1.2 lbs SO$_2$ per MBTU | $0.16-0.18 per MBTU | 39% less than by FGD alone |
| 8D — | Existing Plant FGD Alone | 1.2 lbs SO$_2$ per MBTU | $0.28 per MBTU | 65% more than by PC & FGD |

**Case Numbers:** 9A, 9B, 9C, & 9D

**Case Conditions**

Coal Use Area: _Burlington, New Jersey_

Coal Source Area: _Garrett County, Maryland_     Coalbed: _Upper Freeport_

Raw Coal Characteristics: _13.8_ % Ash     _2.37_ % Sulfur

Clean Coal Characteristics: _8.8_ % Ash     _1.6_ % Sulfur

**Comparison of Costs**

| CASE NO. | TYPE OF PLANT & APPROACH | EMISSION STANDARD | COST TO MEET EMISSION STD. | ECONOMIC ADVANTAGE |
|---|---|---|---|---|
| 9A — | New Plant PC Followed by FGD | 0.30 lbs SO$_2$ per MBTU | This coal will not meet emission standards for new plants in the State of New Jersey with either combined physical cleaning and FGD or FGD alone. | |
| 9B — | New Plant FGD Alone | 0.30 lbs SO$_2$ per MBTU | | |
| 9C — | Existing Plant PC Followed by FGD | 1% sulfur by weight per MBTU | $0.23-0.25 per MBTU | 25% less than FGD alone |
| 9D — | Existing Plant FGD Alone | 1% sulfur by weight per MBTU | $0.32 per MBTU | 33% more than by PC & FGD |

(continued)

## TABLE 9.4:  (continued)

**Case Numbers:**　10A, 10B, 10C, & 10D
**Case Conditions**

Coal Use Area: Milwaukee, Wisconsin
Coal Source Area: Franklin County, Illinois　　Coalbed: Number 6
Raw Coal Characteristics: 14.8 % Ash　1.12 % Sulfur
Clean Coal Characteristics: 7.1 % Ash　0.95 % Sulfur

**Comparison of Costs**

| CASE NO. | TYPE OF PLANT & APPROACH | EMISSION STANDARD | COST TO MEET EMISSION STD. | ECONOMIC ADVANTAGE |
|---|---|---|---|---|
| 10A — | New Plant PC Followed by FGD | 1.2 lbs SO$_2$ per MBTU | $0.07-0.10 per MBTU | 29% less than by FGD alone |
| 10B — | New Plant FGD Alone | 1.2 lbs SO$_2$ per MBTU | $0.12 per MBTU | 41% more than by PC & FGD |
| 10C — | Existing Plant PC Followed by FGD | 1.2 lbs SO$_2$ per MBTU | $0.12-0.15 per MBTU | 33% less than by FGD alone |
| 10D — | Existing Plant FGD Alone | 1.2 lbs SO$_2$ per MBTU | $0.20 per MBTU | 48% more than by PC & FGD |

**Case Numbers:**　11A, 11B, 11C, & 11D
**Case Conditions**

Coal Use Area: Concord, New Hampshire
Coal Source Area: Greene County, Pennsylvania　　Coalbed: Sewickley
Raw Coal Characteristics: 11.4 % Ash　3.45 % Sulfur
Clean Coal Characteristics: 8.1 % Ash　2.20 % Sulfur

**Comparison of Costs**

| CASE NO. | TYPE OF PLANT & APPROACH | EMISSION STANDARD | COST TO MEET EMISSION STD. | ECONOMIC ADVANTAGE |
|---|---|---|---|---|
| 11A — | New Plant PC Followed by FGD | 1.2 lbs SO$_2$ per MBTU (projected standard) | $0.25-0.28 per MBTU | 2% less than by FGD alone |
| 11B — | New Plant FGD Alone | 1.2 lbs SO$_2$ per MBTU (projected standard) | $0.27 per MBTU | 2% more than by PC & FGD |
| 11C — | Existing Plant PC Followed by FGD | 1.5 lbs sulfur per MBTU | $0.12-0.14 per MBTU | 54% less than FGD alone |
| 11D — | Existing Plant FGD Alone | 1.5 lbs sulfur per MBTU | $0.28 per MBTU | 115% more than by PC & FGD |

**Case Numbers:**　12A, 12B, 12C, & 12D
**Case Conditions**

Coal Use Area: Dickerson (Montgomery County), Maryland
Coal Source Area: Marion County, W. Virginia　　Coalbed: Pittsburgh
Raw Coal Characteristics: 11.0 % Ash　3.80 % Sulfur
Clean Coal Characteristics: 5.9 % Ash　2.16 % Sulfur

**Comparison of Costs**

| CASE NO. | TYPE OF PLANT & APPROACH | EMISSION STANDARD | COST TO MEET EMISSION STD. | ECONOMIC ADVANTAGE |
|---|---|---|---|---|
| 12A — | New Plant PC Followed by FGD | 1% sulfur by weight | $0.20-0.23 per MBTU | 20% less than by FGD alone |
| 12B — | New Plant FGD Alone | 1% sulfur by weight | $0.27 per MBTU | 26% more than by PC & FGD |
| 12C — | Existing Plant PC Followed by FGD | 1% sulfur by weight | $0.32-0.34 per MBTU | 23% less than by FGD alone |
| 12D — | Existing Plant FGD Alone | 1% sulfur by weight | $0.43 per MBTU | 30% more than by PC & FGD |

A report (PB-277 273) entitled *Desulfurization of Various Midwestern Coals by Flotation,* by K.J. Miller of the Coal Preparation and Analysis Laboratory of the Bureau of Mines in Pittsburgh, Pennsylvania gives the results of laboratory tests made on nine coals from western Kentucky and Illinois. The tests were carried out for the purpose of evaluating the Bureau of Mines coal-pyrite flotation process.

The results showed that up to 70% pyritic sulfur reduction could be achieved by flotation. However, it was found that many of the coals retained far too much sulfur even after second-stage flotation treatment, leading the author to the conclusion that these midwestern coals are more difficult to desulfurize by flotation than the Appalachian coals which had been tested previously.

# ADDENDUM

On June 5, 1978, a symposium entitled "Advances in Desulfurization of Coal" was scheduled as part of the 85th meeting of the American Institute of Chemical Engineers in Philadelphia, with J.M. Holmes of Oak Ridge National Laboratory as chairman and D.W. Furstenau of the University of California, Berkeley as cochairman of the symposium. The symposium was planned to examine the technology for the removal of sulfur from coal prior to combustion in coal-fired facilities.

Papers scheduled are as follows:


**Advances in Desulfurization of Coal**                      Paper No. 2a
    S.P.N. Singh, Oak Ridge National Laboratory

Brief reviews of the currently used and potential coal beneficiation processes used to lower the sulfur and ash levels of coal prior to combustion. Included is a comparison of the relative costs of beneficiating the coal by several of the processes reviewed.


**Coal Flotation**                                           Paper No. 2b
    F.F. Aplan, Pennsylvania State University

Coal flotation is a rapidly developing method for the removal of pyrite from coal. The influence of the kind and amount of reagents, flotation cell operating conditions, and the use of kinetics to enhance separation are discussed.


**Relative Flotability of Coal and Pyrites**                 Paper No. 2c
    H.V. Le, T.K. Ho, and T.D. Wheelock, Iowa State University

In order to enhance the separation of coal and pyrite by froth flotation, various surface treatments were applied to these materials, and the effectiveness of the treatments was determined by measuring the relative flotability of the treated products. Among various surface treatments, an oxidative chemical pretreatment proved to be the most effective. This procedure apparently selectively oxidizes the surface of the pyrite, and thereby reduces the flotability of the pyrite particles while not affecting the flotability of the coal.


**Coal Beneficiation with Magnetic Fluids**                        Paper No. 2d
   T.A. Sladek and C.H. Cox, Colorado School of Mines Research Institute

Magnetic fluids are described, and their potential application to coal desulfurization is discussed. A program is described in which magnetic fluids were used to enhance magnetic susceptibility of pyrite and other inorganics in coal. Experimental results are discussed, and preliminary economics are presented for a commercial scale coal cleaning plant.


**Is Coal Preparation Cost Effective for Sulphur Removal
in Electric Power Generation?**                        Paper No. 2e
   P.P. DeRienzo, R. Destefanis and C.M. Trapani, Jr.,
   Gibbs and Hill, Inc., New York, NY

The comparative bus bar generation costs of ROM and washed steam coals in achieving varying sulfur dioxide emissions are investigated. A high sulfur midwestern coal and a medium sulfur eastern coal are examined. Costs are estimated for coal preparation plants for varying levels of beneficiation. Flue gas treatment system cost estimates include capital costs, operating and maintenance expenses and, where applicable, credits for the use of washed coal. The cost of sulfur removal is computed for varying sulfur reduction levels. The sensitivity of bus bar generation costs to the cost of ROM coal, capacity factor and the cost of replacement electricity is also presented.

# BIBLIOGRAPHY

The following reports used in the preparation of this book are available from:

National Technical Information Service
U.S. Department of Commerce
5285 Port Royal Road
Springfield, Virginia 22151


EPA-600/9-77-017    *Coal Cleaning with Scrubbing for Sulfur Control,* by E.C. Holt, Jr. of Hoffman-Muntner Corporation and A.W. Deurbrock of the Bureau of Mines, for Environmental Protection Agency Office of Research and Development, August 1977.

IS-ICP-35    *Physical Desulfurization of Iowa Coal,* by S. Min, Energy and Mineral Resources Research Institute of Iowa State University, June 1976.

LBL-5216    *Low Temperature Processes for Coal Desulfurization,* by E. Mendizabal, Lawrence Berkeley Laboratory of the University of California, for the U.S. Energy Research and Development Administration, August 1976.

ORNL-tr-4330    *Reduction of the Sulfur Content in Coals,* by V.S. Kaminskii and T.K. Yagodkina translated by L. Kobylenski, 1975.

PB-211 505    *Survey of Coal Availabilities by Sulfur Content,* by L. Hoffman, F.J. Lysy, J.P. Morris and K.E. Yeager, all of The Mitre Corporation, McLean, Virginia, for the Environmental Protection Administration Office of Research and Monitoring, May 1972.

PB-232 011    *An Interpretative Compilation of EPA Studies Related to Coal Quality and Cleanability,* by L. Hoffman, J.B. Truett and S.J. Aresco, all of The Mitre Corporation, McLean, Virginia, for the Environmental Protection Agency Office of Research and Development, May 1974.

PB-252 965    *Sulfur Reduction Potential of the Coals of the United States,* by J.A. Cavallaro, M.T. Johnston and A.W. Deurbrouck of the Bureau of Mines, U.S. Department of the Interior, for Environmental Protection Agency, Office of Air Programs, April 1976.

PB-262 716    *Coal Preparation Environmental Engineering Manual,* by D.C. Nunenkamp of J.J. Davis Associates, McLean, Virginia, for Environmental Protection Agency Office of Research and Development, May 1976.

# COMPANY INDEX

The company names listed below are given exactly as they appear in the patents, despite name changes, mergers and acquisitions which have, at times, resulted in the revision of a company name.

# INVENTOR INDEX

# U.S. PATENT NUMBER INDEX

# NOTICE

Nothing contained in this Review shall be construed to constitute a permission or recommendation to practice any invention covered by any patent without a license from the patent owners. Further, neither the author nor the publisher assumes any liability with respect to the use of, or for damages resulting from the use of, any information, apparatus, method or process described in this Review.

# COAL RESOURCES, CHARACTERISTICS AND OWNERSHIP IN THE U.S.A. 1978

### Edited by Robert Noyes

Coal currently occupies less than eighteen percent of the nation's energy market. This is despite the fact that coal reserves are adequate for at least a century at present consumption rates, while the reserves of the alternative fossil fuels, petroleum and natural gas, are probably measurable only in decades. The probable course of the petroleum supply picture is further beclouded by economic and political factors, making any kind of probability forecast next to impossible.

In the late 1960s, the abrupt rise of environmentalism and the ensuing passage of legislation, setting standards of air quality, imposed direct restrictions on the sulfur content of coal used for the generation of heat. The scarcity of coals with low sulfur content in the Eastern U.S. and the lack of an effective and economic method of sulfur removal have helped to limit the redevelopment of the nation's most abundant and available energy source. The variations in properties of coal have given rise to geographical patterns of coal mining operations for specific industrial purposes.

This book presents an accurate picture of U.S. coal reserves, the nature and composition of the coal and of the land and minerals ownership. It is based on federally funded studies and is thus a valuable tool in the all-out attack on the energy crisis. A partial and condensed table of contents follows here.

**1. U.S. COAL DEVELOPMENT**
Problems and Promises
The U.S. Energy Problem
Is Coal the Answer?
How Much Coal is Needed?
Other Fuels from Coal
Coal Resources & Reserves
The Coal Industry
Productivity & Capacity

**2. GEOLOGY & GEOGRAPHY OF AMERICAN COAL**
Coal & Its Origin
Geology & Age
Coal as a Rock
The Minerals
The Macerals (Plant Debris)
Characteristics of Coal Seams
Geological Distribution
Extractability & Usability

**3. THE NATURE OF NORTH AMERICAN COAL**
Current Knowledge of
 Coal Chemistry
Petrography
Structural Differences
 Between Macerals
Physical Properties
Organic Structure of Coal
Other Minerals & Elements
 in Coal

**4. ENVIRONMENTAL STUDY OF THE MAJOR U.S. COAL REGIONS**
Physical Environment
Climate & Meteorology
Availability of Water
Geology & Soils
Biological and
 Sociological Environment
Land Use
Water Use
Demography
Economic and
 Social Conditions of the
 Appalachian Region
 Eastern Interior Region
 Fort Union Region
 Powder River Region
 Four Corners Region
 Oil Shale Region

**5. OWNERSHIP AND LAND USE CONSTRAINTS ON COAL MINING**
Coal Production History
Coal Reserves
Measurement & Evaluation
Land Use Patterns & Controls
Land & Minerals Ownership
Tenures Affecting Coal Recovery
Land Ownership Records
Surface Ownership Patterns
Coal Ownership

**6. OWNERSHIP IN THE WEST**
14 Maps and 19 Tables
Showing the Ownership
 of Coal Lands & Oil Shale
 Regions in the States of
 North & South Dakota, Idaho,
 Wyoming, Colorado, New
 Mexico and Arizona.

ISBN 0-8155-0698-8

346 pages

# UNDERGROUND COAL GASIFICATION 1977

## by George H. Lamb

*Energy Technology Review No. 14*

This book presents an overall view of underground coal gasification starting with its history, thereby explaining the theoretical basis for the different methods, the typical processes, including U.S. program studies, the problems encountered, special instrumentation and equipment required, foreign advances, economical aspects and future outlook.

Although *in situ* gasification of coal will not eliminate all energy problems, it appears to have great promise in alleviating some of the problems associated with the production, transportation, and burning of coal.

While not expected to replace coal mining, underground gasification offers an alternative method of bringing the energy value of the coal to the surface. If economically feasible, this clean form of energy could supplement the natural gas supply, particularly in the industrial sector where it could be used also to produce electricity.

Most of the information presented in this book is based on federally funded studies. Here are condensed vital data that are scattered and difficult to pull together. Experimental equipment and set-ups are reviewed and detailed by actual case histories. Each chapter is followed by its own list of references to the most recent literature. A partial and condensed list of contents follows here.

## 1. BACKGROUND & HISTORY

## 2. THEORETICAL ASPECTS
Composition & Reactions
Fundamental Processes

## 3. GENERAL METHODS
Shaft Methods
Chamber Method
Borehole Method
Stream Method
Shaftless Methods
Percolation Methods
Variations of Above
Combination Methods
Underground Gallery and
   Surface Drill Holes

## 4. U.S. PROGRAMS
LERC (Laramie Energy
   Research Center)
LLL (Lawrence Livermore Lab.)

MERC (Morgantown Energy
   Research Center)
Hanna Coal Project
LVW (Linked Vertical Wall)
TPB (Thick Packed Bed)
LWG (Long Wall Generator)

## 5. ALTERNATIVE METHODS
Blind Borehole Fill System
Multiple Borehole
Continuous Filling Process
Flame Front Channeling
Liquid Plugging
Underground Movable Furnace Wall
Addition of $CO_2$ and Steam
Regulation of Fracture Network

## 6. TECHNIQUES & DESIGNS
Site Selection
High Pressure Techniques
Material Survival Stress
Combustion Efficiency
Use of Lasers in
   Drilling Links

## 7. PROBLEMS & LIMITATIONS
Combustion Control
Roof Collapse Control
Flame Front Coning

## 8. MEASUREMENTS AND INSTRUMENTATION SYSTEMS
*In Situ* Information Needs
Monitoring Coal Consumption
SANDIA Labs. Instrumentation

## 9. FOREIGN EFFORTS
Western Europe
Russia
Czechoslovakia

## 10. ENVIRONMENTAL IMPACTS
Hoe Creek Experiments
Control and Disposal
   of Pollutants
Liquid Reinjection
Management of Emergencies

## 11. ECONOMICS
General Benefits
Cost Estimates by
   Lawrence Livermore Labs.

## 12. FUTURE TRENDS
Conjectures of Feasible
   Developments in the U.S.
Advance Indications
   and the Energy Crisis

ISBN 0-8155-0670-8

255 pages